Lecture Notes in Computer Sc

T0250633

Edited by G. Goos and J. Hartmanis

Advisory Board: W. Brauer   D. Gries   J. Stoer

553

Lecture Notes in Computer Science

Edited by G. Goos and J. Hartmanis

Advisory Board: W. Brauer D. Gries J. Stoer

H. Bieri H. Noltemeier (Eds.)

Computational Geometry– Methods, Algorithms and Applications

International Workshop on
Computational Geometry CG '91
Bern, Switzerland, March 21-22, 1991
Proceedings

Springer-Verlag

Berlin Heidelberg New York
London Paris Tokyo
Hong Kong Barcelona
Budapest

Series Editors

Gerhard Goos
Universität Karlsruhe
Postfach 69 80
Vincenz-Priessnitz-Straße 1
W-7500 Karlsruhe, FRG

Juris Hartmanis
Department of Computer Science
Cornell University
Upson Hall
Ithaca, NY 14853, USA

Volume Editors

Hanspeter Bieri
Institut für Informatik und angewandte Mathematik, Universität Bern
Länggassstraße 51, CH-3012 Bern, Switzerland

Hartmut Noltemeier
Institut für Informatik I, Universität Würzburg
Am Hubland, W-8700 Würzburg, FRG

CR Subject Classification (1991): E.1, F.2.1-3, G.2, H.3.3, I.1.2, I.2, I.3.5, J.6

ISBN 3-540-54891-2 Springer-Verlag Berlin Heidelberg New York
ISBN 0-387-54891-2 Springer-Verlag New York Berlin Heidelberg

Typesetting: Camera ready by author
Printing and binding: Druckhaus Beltz, Hemsbach/Bergstr.
45/3140-543210 - Printed on acid-free paper

Preface

The 7th International Workshop on Computational Geometry (CG '91) was held at the University of Berne, Switzerland, March 21/22, 1991.

Previous workshops in this series took place at various universities in Germany and Switzerland:

1983	Zürich	J. Nievergelt
1984	Bern	W. Nef
1985	Karlsruhe	A. Schmitt and H. Müller
1988	Würzburg	H. Noltemeier
1989	Freiburg	T. Ottmann
1990	Siegen	K. Hinrichs

These workshops, or rather mixtures between conference and workshop, try to combine the internationality imperative for high-level research with a local, personal atmosphere. In Berne about sixty scientists participated, most of them from Germany and Switzerland, but also some from Austria, the Netherlands and the United States.

Computational geometry is not a precisely defined field. Often, it is understood as a nearly mathematical discipline, dealing mainly with complexity questions concerning geometrical problems and algorithms. But often too, and perhaps increasingly, questions of more practical relevance are central, such as applicability, numerical behavior and performance for all kinds of input size. CG '91 considered most aspects of computational geometry, and the following list is not exhaustive:

- Generalizations and applications of the Voronoi diagram
- Problems with rectangular objects
- Path determination
- Moving objects
- Visibility questions
- Layout problems
- Representation of spatial objects and spatial queries
- Problems in higher dimensions
- Implementation questions
- Relations to artificial intelligence

Of the 25 contributions presented at the workshop, 21 are contained in this volume, most of them with some modifications. The papers by J. Nievergelt et al. and H. Noltemeier are additional. The following colleagues helped us review the contributions: H. Alt, J. Blömer, M. Eichenberger, H. Hagen, K. Hinrichs, G. Jäger, D. Lasser, P. Mani, H. Müller, W. Nef, J. Nievergelt, B. Nilsson, T. Ottmann, A. Schmitt, S. Schuierer and O. Schwarzkopf.

We are indebted to all participants in the workshop, especially to the contributors to this volume, and to all referees.

Several members of the Institut für Informatik und angewandte Mathematik helped organize the workshop and prepare this volume, primarily Igor Metz. The Max & Elsa Beer-Brawand-Fonds of the University of Berne granted financial support. We thank all of them for their help in making CG '91 a success.

H. Noltemeier edited a predecessor of this volume, namely "Computational Geometry and its Applications", Lecture Notes in Computer Science, Volume 333, which contains the proceedings of the 1988 workshop in Würzburg. Again, the editors thank Springer Verlag – and in particular Alfred Hofmann and Hans Wössner – as well as the series editors of the Lecture Notes in Computer Science for their valuable support.

Bern and Würzburg
September 1991

Hanspeter Bieri
Hartmut Noltemeier

Contents

The Post Office Problem
for Fuzzy Point Sets*

Franz Aurenhammer

Institut für Informatik, Fachbereich Mathematik, Freie Universität Berlin,
Arnimallee 2–6, D-1000 Berlin 33
Germany

Gerd Stöckl

Institute für Informationsverarbeitung, Technische Universität Graz,
Schiesstattgasse 4a, A-8010 Graz
Austria

Emo Welzl

Institut für Informatik, Fachbereich Mathematik, Freie Universität Berlin,
Arnimallee 2–6, D-1000 Berlin 33
Germany

Abstract

The post-office problem for n point sites in the plane (determine which site is closest to a later specified query point) is generalized to the situation when the residence of each site is uncertain and it is described via uniform distribution within a disk. Two probabilistic concepts of neighborhood – expected closest site and probably closest site – are discussed and the resulting Voronoi diagrams are investigated from a combinatorial and computational point of view.

1   Introduction

The so-called *post-office problem* is one of the earliest and most popular problems in computational geometry. Given n points $x_1, \ldots, x_n$ in the Euclidean plane E^2 (called post offices or *sites*), it asks for a site x_i that is closest to a later specified point q (called the *query point*). The post-office problem has been first posed by Knuth [9] who observed that its solution is no longer trivial when queries occur frequently on a fixed set of sites. In order to reduce the *query time* (the time required to report a site x_i closest to q), several attempts have been made to organize the sites into an appropriate data structure. An efficient and widely used approach is called the *locus approach* and is due to Shamos [11]: Each site x_i is associated with the locus of the plane closer to x_i than to any other site x_j, that is, with the region

$$reg(x_i) = \left\{ x \in E^2 \mid \delta(x, x_i) < \delta(x, x_j), j \neq i \right\}$$

*The first author's work was supported by the ESPRIT II Basic Research Action of the EC under contract no. 3075 (project ALCOM)

where δ denotes the Euclidean distance function. This partitions the plane into regions of equal answer with respect to the post-office problem. $q \in reg(x_i)$ if and only if x_i is a site closest to q. The resulting partition is called the *Voronoi diagram* of $x_1, \ldots, x_n$. After having constructed the Voronoi diagram it suffices to locate the region that contains q. Several data structures supporting such *point locations* have been developed; see [7,4,5]. They imply a solution of the post-office problem in $O(n \log n)$ preprocessing time, $O(n)$ storage, and $O(\log n)$ query time which is asymptotically optimal.

The locus approach applies to various more general versions of the post-office problem. Voronoi diagrams for sites more general than points and for distance functions more general than the Euclidean have been considered. The interested reader is referred to the survey paper [1].

In this paper we generalize the post-office problem to the situation when the position of each point site is uncertain and it is described by some density function on E^2. More specifically, the residence of each site x_i is given via uniform distribution within a disk s_i whose radius is particular to that site. For example, sets of sites with numerical errors or sets of moving sites might lead to this type of scenario. Two probabilistic concepts of closeness are discussed for this model, leading to the following types of queries:

1. Find the site x_i whose probability of being closest to the query point q is maximal.

2. Find the site x_j whose expected distance from the query point q is minimal.

Partitioning the plane into regions of equal answer with respect to these queries leads to two interesting types of Voronoi diagrams. Section 2 and 3 are devoted to their investigation. We show that the regions are star-shaped, and that the edges separating two adjacent regions are one-dimensional curves in both cases, provided the disks $s_1, \ldots, s_n$ are pairwise disjoint. Hence these diagrams can be viewed as planar graphs with $O(n)$ edges, which implies their usefulness as supporting data structures for the corresponding post-office problems. We further show that the diagram of type (2) is an abstract Voronoi diagram in the sense of [8], suitable for randomized incremental construction in time $O(n \log n)$; see [10]. The structure of the diagram of type (1) appears to be more complex. Its edges are not defined locally (by the two adjacent disks) but depend on all disks $s_1, \ldots, s_n$ simultaneously. This unusual property outrules common construction methods like incremental insertion or divide-and-conquer.

2 Probably-closest disks

Let $S = \{s_1, \ldots, s_n\}$ be a set of n pairwise disjoint disks in E^2, not neccessarily of equal radius. We conceptually assume that each disk s_i contains one site x_i whose position is described by uniform distribution within s_i. Hence the position of x_i can be expressed by the following density function on E^2:

$$f_i(x) = \begin{cases} \frac{1}{\mu(s_i)} & \text{for } x \in s_i, \text{ and} \\ 0 & \text{otherwise,} \end{cases}$$

where $\mu(s_i)$ denotes the area of s_i. Note that

$$\int_{x \in E^2} f_i(x)\, dx = 1.$$

Since a site is specified completely by its disk (and the corresponding density function) we will talk of disks rather than of sites in the following discussion.

Let now $q \in E^2$ be an arbitrary point. We are interested in the probability that a particular disk $s_i \in S$ is closest to q among all disks in S. Let $P_i(q)$ denote this probability. The *p-region* of s_i, *p-reg*(s_i), is defined to contain all points q such that $P_i(q) > P_j(q)$ for all $j \neq i$. We shall first derive a formula for $P_i(q)$ and then study some properties of the induced p-regions.

What is the probability that some point $x \in E^2$ is closer to q than a fixed disk s_k? Let $D(q, x)$ be the disk with center q and radius $\delta(q, x)$. Since we deal with uniform distributions, this probability is given by

$$A_k(q, x) = \frac{\mu(s_k \setminus D(q, x))}{\mu(s_k)}, \tag{1}$$

i.e., the relative portion of s_k that is further from q than x is. This immediately gives:

$$P_i(q) = \mathbf{Prob}\{s_i \text{ is closer to } q \text{ than all } s_k,\ k \neq i\}$$

$$= \frac{1}{\mu(s_i)} \int_{x \in s_i} \prod_{k \neq i} A_k(q, x)\, dx \,.$$

The formal proof of $\sum_{i=1}^{n} P_i(q) = 1$ is left to the interested reader. Clearly $P_i(q)$ depends on the whole set S of disks. This is important to note because $P_i(q)$ plays the role of some 'distance' between s_i and q in the definition of the regions of a Voronoi diagram; distances usually are defined locally between the two corresponding objects. For example, this property of $P_i(q)$ gives rise to the somewhat surprising situation below.

Consider the three disks s_1, s_2, and s_3 in Figure 1, the first one having radius zero (s_1 is a point) and the others being equal-sized. Let q be the center of a circle that passes through s_1 and that simultaneously cuts s_2 and s_3 into halves. (Note that such a circle always exists.) If we take $S = \{s_1, s_i\}$ for $i = 2$ or $i = 3$ then

$$P_1(q) = A_i(q, s_1) = \frac{1}{2}$$

by construction. On the other hand, if we take $S = \{s_1, s_2, s_3\}$ then

$$P_1(q) = A_2(q, s_1) A_3(q, s_1) = \frac{1}{4}$$

and consequently

$$P_2(q) = P_3(q) = \frac{3}{8}$$

since the three probabilities add up to 1. Hence, if s_1 is slightly moved towards q then s_1 is closest to q with the highest probability among s_1 and s_2 (or among s_1 and s_3), while s_1 is closest to q with the lowest probability in the situation where all three disks are considered simultaneously.

Our next aim is to provide a technical lemma which will allow us to shed light into the behaviour of the p-regions. Note that P_i and P_j in the statement refer to probabilities considering all the disks in S.

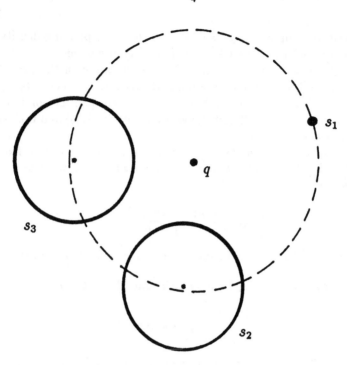

Figure 1:

Lemma 1 *For two disks $s_i, s_j \in S$ let q be a point with $P_i(q) \geq P_j(q) > 0$, and consider some point q' interior to the line segment joining q and m_i, the center of s_i. Then we have $P_i(q') > P_j(q')$.*

Before we prove the lemma, we discuss some of its consequences. First it implies that, for each point $x \in p\text{-}reg(s_i)$, the line segment joining x and m_i is entirely contained in that region. In other words, $p\text{-}reg(s_i)$ is star-shaped as seen from m_i which particularly implies its simple connectedness. Note at this place that p-regions cannot be empty: By the pairwise disjointness of the disks we have $A_k(m_i, x) = 1$ for all $x \in s_i$ and $k \neq i$, and hence $P_i(m_i) = 1$ so that $m_i \in p\text{-}reg(s_i)$.

Lemma 1 also implies that the p-regions partition E^2 up to a set of area zero. Let the *p-separator* of two disks be defined as

$$p\text{-}sep(s_i, s_j) = \left\{ q \in E^2 \mid P_i(q) = P_j(q) \right\}.$$

If $p\text{-}reg(s_i)$ and $p\text{-}reg(s_j)$ are two adjacent regions then the part of $p\text{-}sep(s_i, s_j)$ which separates them cannot be two-dimensional by Lemma 1. Observe that p-separators need not be curves in general; we may have $P_i(q) = P_j(q) = 0$ for all points q of a two-dimensional set.

In conclusion, the Voronoi diagram formed by the p-regions of n pairwise disjoint disks has a combinatorial complexity of $O(n)$; it can be viewed as a planar graph with as many edges. This shows its suitability for postprocessing with respect to point location, and hence its usefulness for reporting quickly a probably-closest disk.

Note, however, that each p-separator simultaneously depends on all n disks, an undesirable (and rather unusual) property which outrules common Voronoi diagram construction methods like incremental insertion or divide-and-conquer. An attempt to make $p\text{-}sep(s_i, s_j)$ independent of $S \setminus \{s_i, s_j\}$ results in the phenomenon below. If three disks s_i, s_j, and s_k are chosen suitably then there exists a point q such that

$$P_i(q) > P_j(q) \text{ when } S = \{s_i, s_j\},$$

$$P_j(q) > P_k(q) \text{ when } S = \{s_j, s_k\},$$

$$P_k(q) > P_i(q) \text{ when } S = \{s_k, s_i\}.$$

Consequently, the three p-separators defined by s_i, s_j, and s_k in this restricted manner do not intersect in some common point, and there occurs a 'no-mans land' belonging to neither region. (It is hard to illustrate this phenomenon with a figure because such separators are almost straight lines that almost concur in a common point. We have found an example using numerical methods.)

The remainder of this section contains a

Proof of Lemma 1. We have assumed that the point q satisfies $P_i(q) \geq P_j(q)$, i.e.,

$$\frac{1}{\mu(s_i)} \int_{x \in s_i} \prod_{k \neq i} A_k(q, x) \, dx \geq \frac{1}{\mu(s_j)} \int_{y \in s_j} \prod_{k \neq j} A_k(q, y) \, dy. \qquad (2)$$

By splitting up the domains of integration, and by extracting the expressions A_j and A_i, respectively, we get that (2) is equivalent to

$$\frac{1}{\mu(s_i)} \int_{r \geq 0} \int_{x \in b_i(r)} A_j(q, x) \prod_{k \neq i, j} A_k(q, x) \, dx \, dr$$

$$\geq \frac{1}{\mu(s_j)} \int_{r \geq 0} \int_{y \in b_j(r)} A_i(q, y) \prod_{k \neq i, j} A_k(q, y) \, dy \, dr \qquad (3)$$

where $b_i(r) = s_i \cap C(q, r)$, the circle around q with radius r. Now concentrate on the point q' that is defined to lie between q and m_i. We make use of the following properties of the relative portions A_k.

Claim 1 For all $x \in b_i(r)$, $y \in b_j(r)$, and $k \neq i, j$ we have

$$A_k(q, x) = A_k(q, y) \text{ and } A_k(q', x) \geq A_k(q', y).$$

The first assertion is trivial. The second assertion holds because the disks are pairwise disjoint: For each $x \in b_i(r)$, the part of the circle $C(q', \delta(q', x))$ that does not lie within the circle $C(q, r)$ is entirely contained in s_i. So $b_j(r)$ lies entirely outside of $C(q', \delta(q', x))$, which implies $\delta(q', x) < \delta(q', y)$ for all $x \in b_i(r)$ and all $y \in b_j(r)$. By (1) this gives $A_k(q', x) > A_k(q', y)$ unless none of the two circles above intersects s_k in which case we may have $A_k(q', x) = A_k(q', y) = 0$ or $A_k(q', x) = A_k(q', y) = 1$. See also Figure 2.

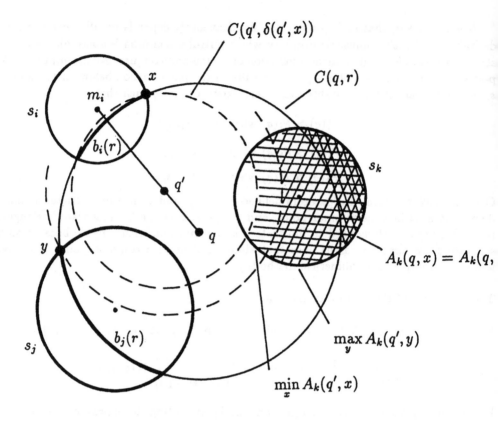

$C(q', \delta(q', x))$

x

m_i

s_i

$b_i(r)$

$C(q, r)$

q'

q

s_k

y

$A_k(q, x) = A_k(q,$

s_j

$b_j(r)$

$\max_y A_k(q', y)$

$\min_x A_k(q', x)$

Figure 2:

By observing that the integrals above increase monotonically with A_k we get that by Claim 1, (3) implies:

$$\frac{1}{\mu(s_i)} \int_{r \geq 0} \int_{x \in b_i(r)} A_j(q, x) \prod_{k \neq i, j} A_k(q', x) \, dx \, dr$$

$$\geq \frac{1}{\mu(s_j)} \int_{r \geq 0} \int_{y \in b_j(r)} A_i(q, y) \prod_{k \neq i, j} A_k(q', y) \, dy \, dr$$

Claim 2 *For all $x \in b_i(r)$, we have*

$$A_j(q, x) \leq A_j(q', x).$$

Furthermore, for all $y \in b_j(r)$, we have

$$A_i(q, y) \geq A_i(q', y).$$

Arguments similar to those used for Claim 1 show the validity of Claim 2 which now implies

$$\frac{1}{\mu(s_i)} \int_{r \geq 0} \int_{x \in b_i(r)} A_j(q', x) \prod_{k \neq i, j} A_k(q', x) \, dx \, dr$$

$$\geq \frac{1}{\mu(s_j)} \int_{r \geq 0} \int_{y \in b_j(r)} A_i(q', y) \prod_{k \neq i,j} A_k(q', y) \, dy \, dr \, ,$$

that is, $P_i(q') \geq P_j(q')$. In fact, strict inequality holds: Recall that we have required $P_i(q) \geq P_j(q) > 0$. Hence there must be some non-trivial interval I such that, for all radii $r \in I$, the circle $C(q,r)$ intersects both of s_i and s_j. Consequently strict inequality holds for all $r \in I$ in Claim 2.

This completes the proof of Lemma 1. We summarize the implications of Lemma 1 as discussed above.

Theorem 1 *Let S be a set of n pairwise disjoint disks in the plane with associated uniform distributions. The Voronoi diagram of S with respect to probably-closest disks consists of n star-shaped and non-empty regions that partition the plane up to a set of area zero.*

3 Expected-closest disks

Let S be a set of disks (with associated density functions) as defined in the preceding section. Given some point $q \in E^2$, we are now interested in the disk $s_i \in S$ that minimizes the expected distance to q.

The expected distance of s_i to q is given by

$$e_i(q) = \frac{1}{\mu(s_i)} \int_{x \in s_i} \delta(x, q) \, dx \, .$$

The *e-region* of a disk s_i, $e\text{-}reg(s_i)$, is defined as the locus of all points q such that $e_i(q) < e_j(q)$ for all $j \neq i$. This concept of a region is obviously different from the previous one. $e_i(q)$ is determined by s_i and q and does not depend on $S \setminus \{s_i\}$ as it was the case for $P_i(q)$. This already reveals that the new concept is less complex.

As a simple observation, the center, m_i, of s_i is always contained in $e\text{-}reg(s_i)$: Since s_i is assumed to be disjoint from all other disks s_j, we have $\delta(x, m_i) < \delta(y, m_i)$ for all $x \in s_i$, $y \in s_j$ and hence $e_i(m_i) < e_j(m_i)$. Actually, $e_i(m_i) = \frac{2}{3} r_i$, where r_i is the radius of s_i. Also not very surprisingly, the following analog of Lemma 1 holds for the expected distance.

Lemma 2 *Let q be a point with $e_i(q) \leq e_j(q)$. For each point q' interior to the line segment joining q and m_i, we have $e_i(q') < e_j(q')$.*

Lemma 2 can be proved by splitting up the domains of integration into the circular arcs $b_i(r)$ and $b_j(r)$ as in the proof of Lemma 1, and by observing that $\delta(x, q) = \delta(y, q) = r$ but $\delta(x, q') < \delta(y, q')$ for all $x \in b_i(r)$, $y \in b_j(r)$. So we again have the nice property that e-regions are star-shaped as seen from the disk centers. Furthermore, the *e-separator* of two disks, defined as

$$e\text{-}sep(s_i, s_j) = \left\{ q \in E^2 \mid e_i(q) = e_j(q) \right\},$$

is a one-dimensional curve that extends to infinity. It follows that the Voronoi diagram induced by S under the expected distance contains $O(n)$ edges.

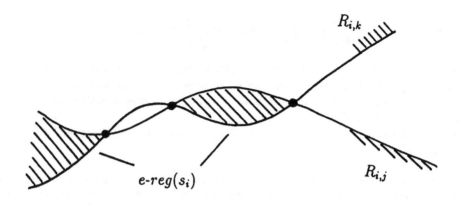

Figure 3:

From the computational viewpoint it is also important that e-separators do not intersect too often. Let us prove that, for three disks s_i, s_j, and s_k, $e\text{-}sep(s_i, s_j)$ and $e\text{-}sep(s_i, s_k)$ can cross at most twice. Denote by $R_{i,j}$ the set of all points q with $e_i(q) < e_j(q)$. Then the region of s_i in the Voronoi diagram of $\{s_i, s_j, s_k\}$ is given by

$$e\text{-}reg(s_i) = R_{i,j} \cap R_{i,k}.$$

Assuming that the separators above cross more than twice implies that $e\text{-}reg(s_i)$ is disconnected; see Figure 3. This is a contradiction.

In conclusion, the diagram fulfils all the properties that allow for a randomized incremental construction in time $O(n \log n)$ and space $O(n)$, see [10]. Clearly these bounds are valid only provided all the elementary construction steps – such as intersecting two e-separators or determining the side of an e-separator on which a given point lies – are considered constant-time operations.

Theorem 2 *Let S be a set of n pairwise disjoint disks in the plane with associated uniform distributions. The Voronoi diagram of S with respect to expected-closest disks consists of n star-shaped and non-empty regions that partition the plane up to a set of area zero. The diagram is computable in time $O(n \log n)$ and space $O(n)$ using randomized incremental insertion.*

Let us finally address a slightly different concept of expected-closeness which leads to a well-known type of Voronoi diagram. Consider the *expected squared distance* of a disk s_i to a point q,

$$\varepsilon_i(q) = \frac{1}{\mu(s_i)} \int_{x \in s_i} \delta^2(x, q) \, dx \,.$$

Simple calculations give

$$\varepsilon_i(q) = \delta^2(m_i, q) + \frac{1}{2} r_i^2 \,,$$

which is the power function with respect to the point m_i with weight $\frac{1}{2} r_i^2$; see e.g. [2]. Hence the resulting diagram is the power diagram of n weighted points in the plane.

Power diagrams behave like classical Voronoi diagrams in many respects. In particular, their regions are convex polygons, and they can be constructed in $O(n \log n)$ time and $O(n)$ space.

4 Discussion

We consider the contributions of this paper as first steps in the investigation of post-office problems where closeness is defined in a probabilistic manner. Basic combinatorial and algorithmic properties of Voronoi diagrams induced by disks under uniform distribution have been proved. The obvious open problem in this respect is an efficient construction of the Voronoi diagram formed by p-regions.

Various modifications of the presented concepts are possible. Let us here address briefly the situation where disjoint line segments under uniform distribution are taken to define a Voronoi diagram. A simple example shows that p-regions as well as e-regions may be disconnected. Consider the three segments s_1, s_2, and s_3 in Figure 4, the latter two having length zero. For the midpoint m of $s_1 = \overline{ab}$ we have

$$P_1(m) = \frac{2\delta(m, s_2)}{\delta(a, b)} < \frac{1}{3} \text{ and } P_2(m) = P_3(m) > \frac{1}{3}$$

but for the endpoints $x = a, b$ of s_1 we have

$$P_1(x) = \frac{\delta(x, s_2)}{\delta(a, b)} > \frac{1}{2}.$$

It is now easy to see that p-reg(s_1) splits into two parts containing a and b, respectively. Concerning the expected distance, observe that

$$e_1(m) = \frac{1}{2}\delta(m, a) > \delta(m, s_2) = e_2(m) = e_3(m)$$

and on the other hand

$$e_1(x) = \delta(x, m) < \delta(x, s_2) = e_2(x) = e_3(x).$$

So e-reg(s_1) behaves similarly as the region above. We leave it as an open problem whether these types of diagrams have linear size.

In order to get an alternative search structure for determining expected-closest objects, the property below can be exploited.

Claim 3 *Let s_i be a line segment or a disk with midpoint m_i. For all $q \in E^2$, $e_i(q) \geq \delta(m_i, q)$.*

The easy proof is left to the reader. Claim 3 has the following consequence.

Lemma 3 *Let $C(q, r)$ be the smallest circle centered at q that entirely encloses s_i. Only objects s_j having their midpoint m_j in $C(q, r)$ can be expected-closer to q than s_i.*

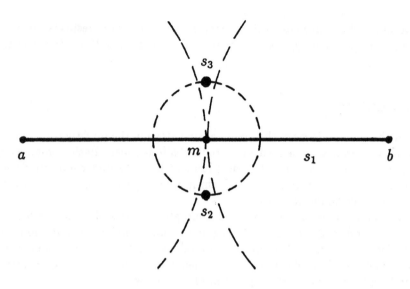

Figure 4:

This can be seen by assuming that m_j is not enclosed by $C(q, r)$ which implies $e_i(q) < r < \delta(m_j, q) \leq e_j(q)$.

Lemma 3 enables us to determine an expected-closest line segment (or disk) as below. For the query point q, the circle $C(q, r)$ having minimal radius r and entirely enclosing some segment is determined. All segments having their midpoint in $C(q, r)$ are reported. Among these, the expected-closest segment to q is calculated directly. Finding all midpoints enclosed by $C(q, r)$ is a circular range search problem which can be solved in $O(\log n + k)$ time with $O(n \log^2 n)$ space, where k is the size of the output; see [3]. Finding $C(q, r)$ can be reduced to $O(\log n)$-time point-location in a Voronoi diagram with respect to the following distance.

Define the *Hausdorff distance* between a point q and an object s_i as:

$$h(s_i, q) = \max \{\delta(x, q) \mid x \in s_i\}$$

Clearly, q falls into the region of s_i in this Voronoi diagram if and only if s_i is the first object that gets fully enclosed by a circle growing around q.

If the objects are disjoint line segments then we can show that the regions are simply connected. Moreover, the diagram is constructable using randomized incremental insertion [10] in $O(n \log n)$ time and $O(n)$ space.

If the objects are disks then $h(s_i, q) = \delta(m_i, q) + r_i$, for m_i and r_i being the center and the radius of s_i, respectively. Hence h is the additively weighted Euclidean distance to point-sites. Its Voronoi diagram is well investigated; in particular, the regions are star-shaped, and they can be computed in $O(n \log n)$ time and $O(n)$ space using the plane-sweep technique [6].

The advantage of the just described data structure for finding expected-closest objects is its ability of handling objects more general than disks. A disadvantage is clearly the query time of $O(\log n + k)$. It is easy to construct a set of segments or of disks such that $k = \Theta(n)$ for a bad choice of q.

References

[1] Aurenhammer, F. Voronoi diagrams – a survey of a fundamental geometric data structure. ACM Computing Surveys, to appear. Also available: Rep. 90-09, Institut für Informatik, FU Berlin, Germany.

[2] Aurenhammer, F. Power diagrams: properties, algorithms, and applications. SIAM J. Computing 16 (1987), 78-96.

[3] Chazelle, B., Cole R., Preparata, F.P., Yap, C.K. New upper bounds for neighbor searching. Information and Control 68 (1986), 105-124.

[4] Edelsbrunner, H., Guibas, L.J., Stolfi, J. Optimal point location in a monotone subdivision. SIAM J. Computing 15 (1986), 317-340.

[5] Edelsbrunner H., Maurer, M.A. Finding extreme points in three dimensions and solving the post-office problem in the plane. Inf. Process. Lett. 21 (1985), 39-47.

[6] Fortune, S. A sweepline algorithm for Voronoi diagrams. Algorithmica 2 (1987), 153-174.

[7] Kirkpatrick, D. Optimal search in planar subdivisions. SIAM J. Computing 12 (1983), 28-35.

[8] Klein, R. Concrete and Abstract Voronoi Diagrams. Springer LNCS 400, 1989.

[9] Knuth, D.E. The Art of Computer Programming, Vol III: Searching and Sorting. Addison-Wesley, 1973.

[10] Mehlhorn, K., Meiser, S., O'Dunlaing, C. On the construction of abstract Voronoi diagrams. Discrete Comput. Geometry, to appear.

[11] Shamos, M.I. Geometric complexity. Proc. 7th Ann. ACM Symp. STOC (1987), 224-233.

References

[1] ...

[2] Aurenhammer, F., Power diagrams: properties, algorithms and applications. SIAM J. Computing 16 (1987) 78–96.

[3] ...

[4] ...

[5] ...

[6] ...

[7] ...

[8] ...

[9] ...

[10] ...

[11] ...

An Optimal Algorithm for Approximating a Set of Rectangles by Two Minimum Area Rectangles*

Bruno Becker[†]   Paolo Giulio Franciosa[‡]   Stephan Gschwind[†]
Thomas Ohler[†]   Gerald Thiemt[†]   Peter Widmayer[†]

Abstract

In this paper we face the problem of computing a conservative approximation of a set of isothetic rectangles in the plane by means of a pair of enclosing isothetic rectangles. We propose an $O(n \log n)$ time algorithm for finding, given a set M of n isothetic rectangles, a pair of isothetic rectangles (s, t) such that s and t enclose all rectangles of M and $area(s) + area(t)$ is minimal. Moreover we prove an $O(n \log n)$ lower bound for the one-dimensional version of the problem.

1 Introduction

Computing approximated, concise representations of complex shapes is a standard problem in computer graphics, pattern recognition and robotics. [1] shows how a convex polygon can be approximated by means of circles, rectangles or k-gons. An algorithm for computing a pair of parallel circumscribed and inscribed rectangles for a convex polygon with n vertices is described in [6]; its complexity is $O(\log^3 n)$ if vertices are given in a sorted array. In [3] a convex polygon with n vertices is approximated by means of a minimum area enclosing triangle in time $O(n)$, starting from the sorted list of the vertices; a very simple $O(n \log n)$ time algorithm for computing the minimum area enclosing circle is given in [7].

In this paper we face the particular approximation problem in which the object to be approximated is a set of isothetic rectangles (which could be the bounding boxes of parts that form an object), and we want to approximate the object by means of a pair of isothetic rectangles enclosing it. Our goal is to find, among all the pairs (s, t) of enclosing rectangles, one for which the sum of the areas of s and t is minimal.

We use the following notations in this paper: A *closed one-dimensional interval* $i = [i.l, i.r] \subseteq \mathcal{R}$, $i.l \in \mathcal{R}$, $i.r \in \mathcal{R}$, is given by its left and right boundaries. In the rest of this paper, *interval* always denotes a *closed interval*. The *length* of a one-dimensional interval i is defined as $length(i) = i.r - i.l$. An *isothetic rectangle* in the plane, i.e. a two-dimensional interval, is defined as the

---

*Work partially supported by grants no. Wi810/2–5 from the Deutsche Forschungsgemeinschaft, by the ESPRIT II Basic Research Actions Program of the European Community, Working Group "Basic GOODS", and by the Italian PFI National Project "Obiettivo Multidata".

[†] Institut für Informatik, Universität Freiburg, Rheinstraße 10–12, D–7800 Freiburg, Germany.

[‡] Istituto di Analisi dei Sistemi ed Informatica, viale Manzoni 30, I–00185 Roma, Italy.

cartesian product of two one-dimensional intervals. The *area* of an isothetic rectangle is given by the product of the lengths of its one-dimensional intervals. In the following, we use the term *rectangle* instead of *isothetic rectangle*. We define the union (intersection) of two intervals as the closure of the union (intersection) of their interiors; these are the regularized operators of [8].

Now we define our approximation problem more formally as follows: Given a set M of rectangles in $\mathcal{R}^2$, find a pair of rectangles s and t such that

$$s \cup t \supseteq \bigcup_{r \in M} r,$$

minimizing $area(s) + area(t)$.

We provide an $O(n \log n)$ worst case time and $O(n \log \log n)$ space algorithm for solving this problem, and we prove an $O(n \log n)$ lower time bound by reduction to the *maximum gap* problem [5].

The algorithm distinguishes three types of enclosing rectangle pairs, on the basis of the disjointness of s and t and of their relative position. The optimal solution is computed as the best one among the minimum area solutions for each type. The best solutions for two of the types directly can be found in $O(n \log n)$ time and $O(n)$ space, while the best solution for the third type is found by applying the compact interval tree datastructure described in [2], resulting in an $O(n \log n)$ worst case time and $O(n \log \log n)$ space performance.

2 A Lower Time Bound

We show that in the algebraic decision tree model [5] $\Omega(n \log n)$ worst case time is needed to find the minimum area solution for our problem. We prove this lower bound in the one-dimensional case; it trivially extends to the two-dimensional problem.

Lemma 2.1 *Given a set M of n one-dimensional intervals $i_1, i_2, \ldots, i_n$, $\Omega(n \log n)$ worst case time is needed for finding two intervals s and t such that*

$$s \cup t \supseteq \bigcup_{j=1}^{n} i_j,$$

minimizing $length(s) + length(t)$.

Proof. We exploit an $\Omega(n \log n)$ lower bound for the *maximum gap* problem [5], by showing that the maximum gap problem can be reduced to the one-dimensional interval approximation problem of Lemma 2.1.

The maximum gap problem is defined in [5] as follows:
Given a set $A = \{a_1, a_2, \ldots, a_n\}$ of n reals, find the maximum γ such that there exist i and j for which $a_j - a_i = \gamma$ and $\forall k \ a_k \leq a_i$ or $a_j \leq a_k$.
Let $max = \max_{i=1}^{n}(a_i)$ and $min = \min_{i=1}^{n}(a_i)$.
A simple argument shows that for $\epsilon = \frac{max - min}{n}$, we certainly have $\gamma > \epsilon$. Let us now define

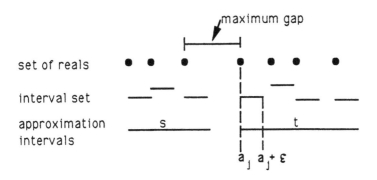

Figure 1: Reduction from *maximum gap* to interval approximation.

$M = \{[a_i, a_i + \epsilon] \mid i = 1, \ldots, n\}$. Note that there is a gap of size at least $\delta = max - min - x + 2\epsilon$ in A, $\delta > \epsilon$, if and only if the pair (s, t) is a feasible solution for M and $length(s) + length(t) = x$ (see Figure 1). Hence maximizing δ corresponds to finding the minimum of $length(s) + length(t)$.

□

3 An Optimal Algorithm

We separately look for solutions of three different types, finding the minimum area solution for each type, and then we select the best one.

For the given set M of rectangles let $m = box(M)$ denote the bounding box of M, i.e., the minimum rectangle containing M. Since $s \cup t$ must enclose all rectangles in M and $area(s) + area(t)$ has to be minimal, for each horizontal (vertical) side of m there must exist a horizontal (vertical) side of s or t having the same coordinate value.

We disregard reflections, rotations and the exchange of s with t. W.l.o.g. let s have at least as many sides in common with m as t; of course, s must have at least two such sides.

We can therefore distinguish different types of solutions, depending on the sides of s that touch sides of m.

We partition the solutions into the following three types (see Figure 2 for an illustration):

1. s has at least two adjacent sides on the boundary of m, and s and t have no common interior points;

2. s has at least two adjacent sides on the boundary of m, and s and t have some common interior point;

3. s has no two adjacent sides on the boundary of m.

We describe in the following how the best solution for each type is computed.

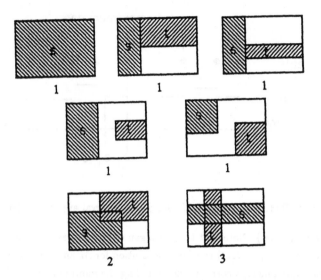

Figure 2: Different types of solutions.

3.1   Type 1 Solutions

For Type 1, w.l.o.g. s and t can be separated by a vertical line, and s lies to the left of t. We show how the minimum area solution (s, t) can be computed for this type of solution.

Given a fixed vertical line $x = v$, let $l(M, v)$ denote the set of rectangles that intersect the left halfplane $H_l(v) = \{(x, y) | x \leq v\}$, and let $r(M, v)$ denote the set of rectangles that intersect the right halfplane $H_r(v) = \{(x, y) | x \geq v\}$.

Each value v defines a feasible solution (s, t), with $s = box(l(M, v)) \cap H_l(v)$ and $t = box(r(M, v)) \cap H_r(v)$.

Lemma 3.1 *If (s, t) is a feasible solution defined by a value v that is not the abscissa of the side of a rectangle, then there exists an abscissa v' of a rectangle side defining a solution (s', t') such that*

$$area(s') + area(t') \leq area(s) + area(t) .$$

Proof. Let a be the largest abscissa of a rectangle side to the left of v and b the smallest abscissa of a rectangle side to the right of v (see Figure 3).

We will distinguish two cases:

- no rectangle intersects the line $x = v$: in this case the abscissa a (or b) defines the same solution as v;

- the line $x = v$ intersects some rectangles: let h_s and h_t be the height of respectively s and

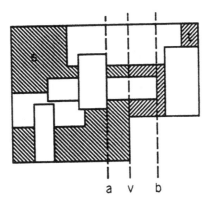

Figure 3: Proof of Lemma 3.1.

t. If $h_s \geq h_t$, abscissa a defines a solution whose area is

$$area(s) + area(t) - (h_s - h_t) \cdot (v - a) \leq area(s) + area(t).$$

Similarly, if $h_s < h_t$, then abscissa b defines a solution with no larger area.

□

Lemma 3.1 allows us to only evaluate solutions defined by vertical lines with the same abscissa as the right side or the left side of some rectangle in M, since other solutions cannot have a smaller area.

The *sweep line* algorithm we propose for this task starts at the left end of M. s (resp. t) is the bounding box of all parts of rectangles to the left (resp. right) of the sweep line. s is initialized to the empty rectangle, while $t = m = box(M)$. The line sweep scans each left and right side of all rectangles in ascending order of abscissa values. The left side of s is fixed at the left side of m and the sweep line defines the right side of s. The coordinate of the upper side of s is the highest coordinate of all upper sides of the set of rectangles whose left sides are to the left of the sweep line. Because rectangles are only added to this set while moving the sweep line from left to right, the upper side of s moves monotonously upwards. A similar argument holds for the lower side of s, which moves monotonously downwards.

If the sweep line intersects some rectangles, the left side of t is defined by the sweep line. Otherwise the left side of t is defined by the lowest coordinate of all left sides of the set of rectangles, whose left sides are not left of the sweep line. The right side of t is fixed at the right side of m, the upper side moves downwards and the lower side upwards, and the left side moves from left to right as the sweep line proceeds.

Because all sides of s and t are either fixed or move monotonously in one direction (of course not all in the same direction), we maintain rectangles and coordinates in the following doubly linked lists:

- One (unsorted) list of all rectangles

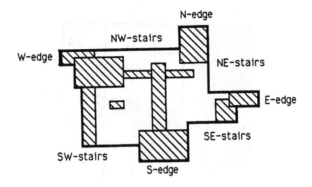

Figure 4: NE(NW/SW/SE)-staircase of a set of rectangles.

- Three sorted lists of left, top and bottom coordinates of sides of the rectangles, the *t-lists*

- One sorted list of left and right rectangle coordinates, the sweep line halting points.

Each element of a sorted list is linked with its rectangle element of the rectangle list by pointers in both directions.

The worst case time needed for building all lists and linking them equals the cost of sorting, i.e. $O(n \log n)$. One step in the line sweep can be reflected in the lists by a change of constant cost. The left, bottom and top coordinates of t can always be found in the head elements of *t-lists*. Therefore the total time needed after preprocessing is $O(n)$. Hence the following theorem holds:

Theorem 3.2 *The overall worst case time complexity to compute an optimal solution of Type 1 is $O(n \log n)$.*

3.2 Type 2 Solutions

A polygon is *orthoconvex* if its intersection with any horizontal or vertical line is either a line segment or is empty [4]. The *orthoconvex hull* of a set of rectangles is the smallest orthoconvex polygon which covers all the rectangles. The algorithms for computing solutions of Types 2 and 3 basically use the boundary of the orthoconvex hull of a given set M of rectangles, *boch(M)* for short. Let N, S, E, W be the four compass directions. We call the northernmost edge of *boch(M) N-edge*. E-edge, S-edge and W-edge are defined accordingly. If there exist edges between the N-edge and the W-edge, we call this part of *boch(M) NW-staircase(M)* or *NW-stairs(M)* for short. NE-, SE- and SW-staircase(M) are defined correspondingly (see Figure 4).

The NE(SE/SW/NW)-staircase of a set of rectangles can be determined in $O(n \log n)$ worst case time, as described in [5].

For Type 2 w.l.o.g. we will only consider solutions in which the lower-left vertex of s coincides with the lower-left vertex of *box(M)* and the upper-right vertex of t coincides with the upper-right vertex of *box(M)*.

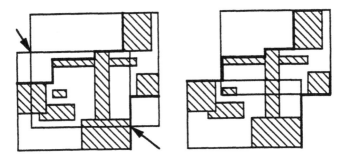

Figure 5: Proof of Lemma 3.3.

We first prove that the minimum area solutions for Type 2 can be found by exploring two opposite staircases of M.

Lemma 3.3 *If (s,t) is a minimum area solution of Type 2 for M, then the boundaries of s and t intersect NW-stairs(M) in the same concave vertex; the same is true for SE-stairs(M).*

Proof. If (s,t) is a solution of Type 2, then the upper side of s and the left side of t both intersect NW-stairs(M).

Suppose now that the upper side of s and the left side of t do not intersect in a concave vertex in NW-stairs(M): in this case the upper side of s can be shifted downwards and/or the left side of t can be moved rightwards, obtaining a better solution; hence (s,t) is not a minimum area solution.

The same argument applies to the right side of s and the lower side of t, with respect to SE-stairs(M) (see Figure 5).

It is simple to see that the only case in which the left side of t and the right side of s cannot be shifted to a concave vertex occurs when the solution becomes of Type 1. The same argument holds for the top side of s and the bottom side of t. □

Each solution of interest is therefore identified by a pair of concave vertices (p,q), with $p \in$ NW-stairs(M) and $q \in SE$-stairs(M).

Before stating the algorithm for finding the minimum area solution for Type 2 we need to further reduce the set of vertices explored in SE-stairs(M) for each concave vertex $p \in NW$-stairs(M). Given $p = (p_x, p_y) \in NW$-stairs(M), p defines a set of feasible vertices in SE-stairs(M), that will be named $active(M,p)$. This is the set of concave vertices $q = (q_x, q_y)$ with $q_x > p_x$ and $q_y < p_y$ (see Figure 6).

The *convex hull* of a set of points P is the smallest convex polygon covering all the points. The *upper boundary* is the upper part of the boundary of the convex hull between an extreme left and an extreme right point on the boundary. By $ubvert(P)$ we denote the vertices of the upper boundary of P.

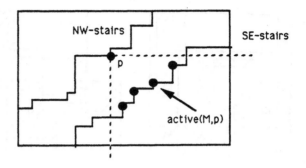

Figure 6: An active interval for p in *SE-stairs*.

We will show that only vertices of $ubvert(active(M,p))$ must be taken into account for the computation of an optimal solution of Type 2 (see Figure 7) .

Lemma 3.4 *If (s,t) is a minimum area solution of Type 2 identified by vertices (p,q), with $p \in NW\text{-}stairs(M)$, then $q \in ubvert(active(M,p))$.*

Proof. Let w and h be respectively the width and the height of $box(M)$, and let $p = (p_x, p_y)$ and a vertex $q' \in active(M,p)$, $q' = (q'_x, q'_y)$; (p,q') identifies a solution with area (see Figure 7)

$$A_{p,q'} = q'_x \cdot p_y + (w - p_x) \cdot (h - q'_y) .$$

The set of points $v = (v_x, v_y)$ such that $A_{p,v} = A_{p,q'}$ is a straight line l whose equation is

$$v_y - q'_y = \frac{p_y}{(w - p_x)} \cdot (v_x - q'_x) .$$

The minimum area solution (with fixed p) is given by a vertex q in $active(M,p)$ through which a line parallel to l is closest to p.

The claim is now proved, since this can be seen as a two-dimensional linear programming problem whose admissible region is the convex hull of $active(M,p)$; and it is well known that the optimal solution can always be chosen on a vertex of the admissible region. □

A set of Type 2 solutions which contains the optimal one can be found by maintaining $ubvert(active(M,p))$ under insertions and deletions of points which enter and leave the set of active vertices while p moves along concave vertices in $NW\text{-}stairs(M)$ and searching an optimal vertex $q \in ubvert(active(M,p))$ for each fixed p.

We just sketch an algorithm for finding this set of solutions:

- set p to the leftmost concave vertex in $NW\text{-}stairs(M)$;

- while p is in $NW\text{-}stairs(M)$ do

    - set $u = ubvert(active(M,p))$;

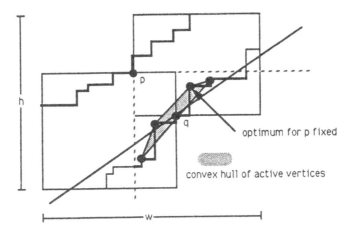

Figure 7: The point with minimum area is on the convex hull.

- find the optimal vertex q on u;
- evaluate the solution given by (p, q);
- set p to the next concave vertex in NW-$stairs(M)$.

While p moves rightwards on NW-$stairs(M)$, both extremes of $active(M, p)$ can only move rightwards. Hence the total number of deletions and insertions while p scans the whole NW-$stairs(M)$ is $O(n)$.

So all we have to do is to maintain the *ubvert* under the following operations:

- insertion of vertices to the right of the current set;

- deletion of the leftmost vertex in the set.

Because SW-$stairs$ is a simple path, it is possible to apply the concept of *Compact Interval Trees*, proposed in [2]. With this data structure, which needs $O(n \log \log n)$ space, it is possible to rebuild the convex hull of the active vertices after each insertion or deletion in time $O(\log n)$. A compact interval tree allows binary search in such a way, that an optimal vertex q for a given vertex p can be found in time $O(\log n)$ after the convex hull is computed.

Theorem 3.5 *The overall worst case time complexity to compute an optimal solution of Type 2 is $O(n \log n)$.*

3.3 Type 3 Solutions

We first prove that, as at Type 2 solutions, the minimum area solutions of this type can be found by exploring two opposite staircases of M.

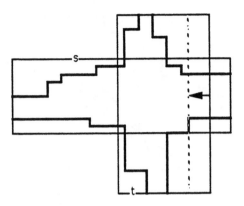

Figure 8: Proof of Lemma 3.6.

Lemma 3.6 *If (s,t) is a minimum area solution of Type 3 for M, with s having the width of box(M) and t having the height of box(M), then*

- *the top side of s and the left side of t intersect in a point of NW-stairs(M), and the bottom side of s and the right side of t intersect in a point of SE-stairs(M); or*

- *the top side of s and the right side of t intersect in a point of NE-stairs(M), and the bottom side of s and the left side of t intersect in a point of SW-stairs(M).*

Proof. It is immediate to see that if (s,t) is a solution of Type 3 then the two vertical (horizontal) sides of s (t) intersect the four staircases of M (see Figure 8).

We can now prove the claim by reduction to the absurd. We choose a side of one of the two solution rectangles, w.l.o.g. the right side of t, and we show that at least one of the intersections of this side with the top and the bottom side of s must lie on a staircase of M. Otherwise, (s,t) could not be a minimum area solution, since the right side of t could be shifted to the left, obtaining a better solution. The claim directly follows. □

Moreover, it is simple to see that points identifying minimum area solutions must be points of the grid defined by the extremal points of rectangles in M, and there are at most $O(n)$ of such points on each of the staircases.

So the minimum area solution is identified by two *grid* points on opposite staircases. Our algorithm will, w.l.o.g., sequentially scan these points in NW-stairs(M), and for each of them will directly find the point in SE-stairs(M) giving the best possible solution.

For each grid point p in NW-stairs(M) all the feasible solutions are identified by one of the *active* grid points q in SE-stairs(M), i.e. the points contained in the interval defined by "reflecting" on NE-stairs(M) and on SW-stairs(M) the horizontal and vertical rays from p (see Figure 9).

The algorithm for finding a minimum area solution of Type 3 is sketched below.

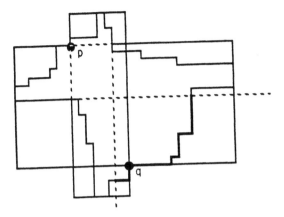

Figure 9: Area of solutions of Type 3.

- set p to the leftmost point in NW-$stairs(M)$;

- build the set of points in the active interval corresponding to p;

- while p is in NW-$stairs(M)$ do:

  - find point q defining the minimum area solution for p;

  - evaluate the solution defined by (p, q);

  - move p to the next point on the grid in NW-$stairs(M)$ and update the active interval.

Let us now evaluate the area $A(p, q)$ of the solution identified by a given pair of points (p, q), $p = (p_x, p_y)$, $q = (q_x, q_y)$, respectively on the NW-$stairs(M)$ and SE-$stairs(M)$. Let w and h be respectively the width and the height of $box(M)$.

The following equality holds:

$$\begin{aligned} A(p, q) &= (p_y - q_y)w + (q_x - p_x)h \\ &= p_y w - p_x h + q_x h - q_y w \end{aligned}$$

Hence minimizing $A(p, q)$ for a fixed p corresponds to minimizing $q_x h - q_y w$; this value does not depend upon p.

It is important for the efficiency of our algorithm that while point p in NW-$stairs(M)$ moves to the right, the corresponding interval of active points on SE-$stairs(M)$ only can move to the left. Hence each point in SE-$stairs(M)$ can be inserted into and deleted from the set of active points just once. Also, the reflection points for p_x on SW-$stairs(M)$ and for p_y on NE-$stairs(M)$ move along in one direction respectively, namely from top to bottom and from right to left.

The set of active points can be maintained in a partially ordered structure, say a heap, according to the values $q_x h - q_y w$. This allows us to perform detection of minimum in constant time, and insertion and deletion of points in $O(\log n)$ worst case time (deletion will be performed by using a pointer to the node representing a point in the heap).

Hence we can state the following:

Theorem 3.7 *The overall worst case time complexity for computing an optimal solution of Type 3 is $O(n \log n)$.*

4   Conclusion and Further Work

We presented an algorithm for finding, given a set of n isothetic rectangles in the plane, a pair of isothetic rectangles whose union encloses the given rectangles, such that the sum of the areas of the two rectangles is as small as possible.

The pair of minimum area enclosing rectangles can be seen as a conservative approximation for the given set of rectangles, which, depending on the shape of the set of rectangles, can be much closer to these than the simple bounding box, but still requires only constant size storage. Hence this approach can be useful in cases where we are looking for a concise representation of complex objects, as in the design of spatial data structures.

The solutions are partitioned into three types, and the algorithm works by finding the minimum area solution for each type. Its worst case running time is $O(n \log n)$, and it needs $O(n \log \log n)$ space.

Currently we are investigating the possibility of extending the same approach to the approximation of objects other than rectangles, the extension to objects in higher dimension, and the approximation by a higher, but fixed, number of objects.

References

[1] H. Alt, J. Blomer, M. Godau, and H. Wagener. Approximation of convex polygons. In *Automata, Languages and Programming (Proc. of the 17th ICALP, Univ. of Warwick, England, July 1990), Lecture Notes in Computer Science 443*, pages 703–716, 1990.

[2] L. Guibas, J. Hershberger, and J. Snoeyink. Compact interval trees: A data structure for convex hulls. *International Journal of Computational Geometry & Applications*, 1:1–22, 1991.

[3] J. O'Rourke, A. Aggarwal, S. Maddila, and M. Baldwin. An optimal algorithm for finding minimal enclosing triangles. *Journal of the Algorithms*, 7:258–269, 1986.

[4] Th. Ottmann, E. Soisalon-Soininen, and D. Wood. On the definition and computation of rectilinear convex hulls. *Information Sciences*, 33:157–171, 1984.

[5] F. P. Preparata and M. I. Shamos. *Computational Geometry, an Introduction*. Springer-Verlag, New York, 1985.

[6] O. Schwarzkopf, U. Fuchs, G. Rote, and E. Welzl. Approximation of convex figures by pairs of rectangles. Report B 89-15, Freie Universitaet Berlin, Department of Mathematics, 1989.

[7] S. Skyum. A simple algorithm for computing the smallest enclosing circle. Report DAIMI 314, Aarhus University, Computer Science Department, 1990.

[8] R. B. Tilove. Set membership classification: A unified approach to geometric intersection problems. *IEEE Trans. on Computers*, pages 874–883, 1980.

An On-Line Algorithm for Constructing Sweep Planes in Regular Position

Hanspeter Bieri* and Peter-Michael Schmidt†

Abstract

An algorithm SWEEPPLANE is presented to be used as an auxiliary algorithm by space-sweep algorithms requiring a sweep plane in regular position, i.e., a sweep plane which never meets ≥ 2 event points at the same time. SWEEPPLANE is an on-line algorithm which ensures that the regular position of the sweep plane is never destroyed by a new event point created during the sweep. If necessary, it dynamically replaces the actual sweep plane by a more suitable one, but always in such a way that the induced order of the event points already swept is maintained.

1   General remarks on the space–sweep method

The *space–sweep method* (in the 2–dimensional case *plane–sweep method* or *scan–line method*) has proven very useful for a large number of problems in Computational and Combinatorial Geometry [LePr84, Mehl84, PrSh85] as well as in Computer Graphics [BeMe86, Müll88]. [Mehl84, Wood84] consider the method important enough to call it *sweep paradigm* and *sweeping search paradigm*, respectively. For some problems which have been solved before by other methods, the sweep approach has lead to simpler and more robust implementations without sacrificing the optimal asymptotic time complexity [Fort87, HiNS88, HiNS90]. One of the most elegant and beautiful examples of the space–sweep method is given in [BeOt79] which has probably been the most cited paper in Computational Geometry so far.

There exists no general definition of the space–sweep technique but it always works basically as follows: A *sweep plane* (in the 2–dimensional case *sweep line* or *scan line*) sweeps across the space $I\!R^d$ which contains a set of geometric objects (mostly points but also line segments, polygons, etc.). This *sweep* is always *greedy*, i.e. without any backtracking. At certain problem-dependent positions, often called *event points*, the sweep is interrupted and a part of the given problem is locally solved. At the end of the sweep the overall problem is solved using the results obtained at all or some of the event points.

[Hadw55] is the earliest paper we know which uses the sweep method. Almost all papers applying the space–sweep method use a *translational sweep*. One early exception is [Hadw68] where a *rotational sweep* is used. We will only consider translational sweeps and take as our basis the following conditions and notations:

---

*Universität Bern, Institut für Informatik und angewandte Mathematik, Länggassstrasse 51, CH–3012 Bern, Switzerland

†Friedrich-Schiller-Universität, Mathematische Fakultät, Universitätshochhaus, 17. OG, D-O–6900 Jena, Germany

We start from the space $I\!R^d$ with its natural basis and call any nonzero linear function $g : I\!R^d \to I\!R$ a *sweep function*. g shall be defined by real numbers $\lambda_1, \ldots, \lambda_n$ (not all $= 0$):

$$g(x) := \sum_{i=1}^{d} \lambda_i \xi_i \qquad \text{for all } x = (\xi_1, \ldots, \xi_d) \in I\!R^d. \qquad (1)$$

For every $t \in I\!R$, $G(t) := g^{-1}(t)$ is a hyperplane in $I\!R^d$. We consider t as a variable and call $G(t)$ the *sweep plane* corresponding to g regardless whether t has been given a specific value or not. t may be interpreted as representing time.

We will only consider situations where the set of geometric objects to be swept is — or can be represented by — a finite number of points: $S = \{p_1, \ldots, p_n\} \subset I\!R^d$. Furthermore, we will confine ourselves to applications for which the following fairly general *sweep scheme* may be applied:

(a) Choose a suitable sweep function g.

(b) Sort the initially known event points — i.e. all or some of the points S — according to increasing g–values and store them in a suitable *data structure Q*.

(c) `while` $Q \neq \emptyset$ `do`

(d) $p := \text{DELETEMIN}(Q)$ $\qquad\qquad\qquad\qquad\qquad\qquad\qquad\qquad\qquad\qquad\qquad (2)$

(e) Process the information locally available at $G(g(p))$. Every new event point p' found in this step must satisfy $g(p') > g(p)$.

(f) Update Q: If new event points have been found in step (e), insert them into Q using a procedure INSERT.

(g) `endwhile`

(h) Determine the final result.

DELETEMIN and INSERT are operations whose implementations depend on the data structure Q. For many applications [BeOt79, HiNS90] Q is a *priority queue* [PrSh85], which guarantees that one execution of DELETEMIN or INSERT, respectively, only costs $O(\log \text{card } Q)$ time. With many other applications no new points can be found in step (e), so Q may be an ordinary 1–dimensional array.

A sweep always induces an *order* on the considered set of geometric objects. Therefore, geometric *counting problems* are good candidates for applying the space–sweep method [KeWe78]. There are certain problems which can be solved by the space–sweep method but which actually only need the induced order, not the sweep plane $G(t)$ itself. Trivial examples are all 1–dimensional variants of higher–dimensional problems solved by the sweep technique. But many sweep algorithms require the existence of a sweep plane at each event point and not just an order of the event points, and some even require that the normal vector of $G(t)$ never changes its direction during the sweep [BiNe83].

2 Regular position

Solving a problem in Computational and Combinatorial Geometry normally requires to distinguish between several different cases: *Line segment intersection*, e.g., must bear in mind that an endpoint of one segment may be on the other segment, that the two endpoints of one segment may be identical, etc. [Sedg88]. Since algorithms considering all possible cases often become quite complicated, early papers in Computational Geometry often only distinguish between "the general case" and "the degenerate special cases" and exclude explicitly inputs that lead to the latter ones [Sham78]. Recently, more attention has been paid to algorithms which admit all possible inputs, either by considering carefully all distinct cases of the specific problem [HiNS90] or by using a general technique which allows to establish the general case for any kind of input [EdMü90].

Space–sweep algorithms, too, have to deal with degeneracies. As for the sweep scheme (2), there is one case which is normally considered a degeneracy, namely the case where for some value of t the sweep-plane $G(t)$ meets more than one event point. In [NeSc90] a sweep-plane which never meets more than one event point at the same time is called a sweep-plane *in regular position*. In the first part of [NeSc90] it is assumed that all event points are already known at the beginning of the sweep, and an algorithm is developed which for any such input constructs *numerically* a sweep plane in regular position. More precisely, for a given finite set $S \subset I\!\!R^d$ of event points a sweep function g is constructed such that $g(p_i) \neq g(p_j)$ for $i \neq j$. The order induced by g on S is the *lexicographic order*: $x = (\xi_1, \ldots, \xi_d) <_L y = (\eta_1, \ldots, \eta_d)$ iff an index $i \in \{1, \ldots, d\}$ exists such that $\xi_j = \eta_j$ for all $j < i$ and $\xi_i < \eta_i$. That means

$$g(p_1) < g(p_2) < \cdots < g(p_n) \quad \text{iff} \quad p_1 <_L p_2 <_L \cdots <_L p_n. \tag{3}$$

The second part of [NeSc90] presents a *symbolic* approach. Any sweep using this method starts with a sweep function g whose coefficients according to (1) are not numerically fixed. During the sweep these coefficients are adjusted in such a way that all degeneracies due to the input vanish and no new degeneracies are created. More precisely, the coefficients of g are kept variable and must only fulfill condition (3) which, in general, induces further relations between these coefficients during the sweep.

The applicability of the numeric method is rather limited, of course, since with many problems new event points may occur during the sweep [BeOt79, HiNS90]. The symbolic method is much more general but it requires a system for symbolic computations.

The present paper generalizes the numeric approach in [NeSc90]. It develops an algorithm SWEEPPLANE which is meant to be used as an *auxiliary algorithm* by space–sweep algorithms which (for reasons of simplicity) assume that the sweep plane is always in regular position. SWEEPPLANE is an *on-line algorithm* [Sham78, PrSh85]: Whenever the space–sweep algorithm detects a new event point according to step (e) of scheme (2), it calls SWEEPPLANE. SWEEPPLANE provides a sweep plane in regular position with respect to all event points met so far, including the new one. This sweep plane is determined without having to reconsider the former event points and, depending on the new event point, may be identical to the sweep plane used before the call. As in [NeSc90] the induced order is always the lexicographic one, which means that a new sweep plane never changes the order of the event points already swept. This is required, of course, by many applications. SWEEPPLANE may be regarded as a numeric concretisation of the symbolic approach in [NeSc90]. It is still less general, since it does not allow to perform the whole sweep without ever changing its direction, as it is required e.g. in [BiNe83]. A typical application is certainly [BiNe85] where a sweep plane is explicitly used, but not necessarily a sweep plane whose direction never changes.

3 Mathematical preparations

We start from a point set $S = \{p_1, \ldots, p_n\} \subset \mathbb{R}^d$ which we assume to be lexicographically ordered, i.e. $p_1 <_L p_2 <_L \cdots <_L p_n$. An equivalent assumption is that all differences $z_i := p_{i+1} - p_i$ $(i = 1, \ldots, n-1)$ are lexicographically greater than $0 \in \mathbb{R}^d$. We are interested in sweep functions g with the property $g(p_1) < g(p_2) < \ldots < g(p_n)$ (cf. (3)) or, equivalently, with the property

$$g(z_i) > 0 \text{ for } i = 1, \ldots, n-1. \tag{4}$$

In [NeSc90] it is shown that there exist sweep functions satisfying (4) whose coefficients λ_i in the representation (1) are all positive. Therefore we may presuppose

$$\lambda_i \geq 1 \text{ for } i = 1, \ldots, d-1, \ \lambda_d = 1 \tag{5}$$

As in [NeSc90] we make use of the set $H := \{x \in \mathbb{R}^d; x >_L 0\}$ of all lexicographically positive elements of $\mathbb{R}^d$. H is the union of the pairwise disjoint sets $H_1, \ldots, H_d$ defined by

$$H_k := \{x \in \mathbb{R}^d; \xi_i = 0 \text{ for } i < k, \xi_k > 0\}. \tag{6}$$

Rather than with the sweep function g itself our algorithm will work with a (nonlinear) *associated function* $h : H \to \mathbb{R}$ defined as follows:

$$h(x) := \lambda_k \xi_k - \sum_{i>k} \lambda_i |\xi_i| \text{ for } x \in H_k. \tag{7}$$

We have $h(x) \leq g(x)$ for all $x \in H$. Therefore, in order to find a sweep function g satisfying (4) it will be sufficient to construct an associated function h with the property

$$h(z_i) > 0 \text{ for } i = 1, \ldots, n-1. \tag{8}$$

Now, we can formulate the main idea on which our on-line algorithm SWEEPPLANE will be based:

Every time when the plane-sweep algorithm making use of SWEEPPLANE arrives at a new event point, SWEEPPLANE solves the following computational problem:

> Given differences $z_1, \ldots, z_{n-1} \in H$, an associated function h of the form (7) which satisfies (8), and an additional difference z_n: Derive an associated function h_0 of the form (7) and satisfying (8) for $i = 1, \ldots, n$. (9)

Having found h_0, a sweep function g_0 satisfying (4) can be found trivially by means of $\lambda_1, \ldots, \lambda_d$ taken from the representation (7). Please note that we do not demand $h_0(z_i) = h(z_i)$ for $i = 1, \ldots, n-1$, i.e. h_0 is not an extension of h in general. An analogous procedure will be applied before the start of the actual sweep to construct a sweep plane in regular position with respect to the event points already known.

In order to solve (9) we first represent the coefficients $\lambda_1, \ldots, \lambda_d$ in (7) in the form

$$\lambda_m = \varphi_0 \varphi_1 \ldots \varphi_{d-m} \quad (m = 1, \ldots, d) \tag{10}$$

which implies $\varphi_0 = 1$. Combining (7) and (10) we get

$$h(x) = \xi_k \prod_{j=0}^{d-k} \varphi_j - \sum_{i>k} |\xi_i| \prod_{j=0}^{d-i} \varphi_j \quad \text{for } x \in H_k. \tag{11}$$

Lemma 1 Let $\varphi_0^*, \ldots, \varphi_{d-1}^*$ be factors satisfying $\varphi_0^* = 1$ and $\varphi_j^* \geq \varphi_j$ $(j = 1, \ldots, d-1)$, and let h^* be the corresponding associated function. It follows $h^*(x) \geq h(x)$ for all $x \in H$.

Proof: Let $k \in \{1, \ldots, d\}$ and $x \in H_k$.

1. For $m > d - k$ the factor φ_m does not occur in (11), therefore replacing φ_m by φ_m* does not influence $h(x)$.

2. Replacing φ_{d-k} by φ_{d-k}^* only influences the first term of (11) resulting in a function h_{d-k} which satisfies $h_{d-k}(x) \geq h(x)$.

3. Now let us assume that for $1 < m \leq d-k$ the factors $\varphi_m, \ldots, \varphi_{d-k}$ have already been replaced by $\varphi_m^*, \ldots, \varphi_{d-k}^*$ resulting in a function h_m with $h_m(x) \geq h(x)$. Replacing φ_{m-1} by φ_{m-1}^* leads to a function h_{m-1} which may be represented in the following way:

$$\begin{aligned} h_{m-1}(x) &= \frac{\varphi_{m-1}^*}{\varphi_{m-1}} \xi_k \varphi_0 \ldots \varphi_{m-1} \varphi_m^* \ldots \varphi_{d-k}^* \\ &\quad - \frac{\varphi_{m-1}^*}{\varphi_{m-1}} \sum_{k < i \leq d-m+1} |\xi_i| \varphi_0 \ldots \varphi_{m-1} \varphi_m^* \ldots \varphi_{d-i}^* \\ &\quad - \sum_{d-m+1 < i \leq d} |\xi_i| \varphi_0 \ldots \varphi_{d-i} \end{aligned}$$

We obtain the desired inequality $h_{m-1}(x) \geq \frac{\varphi_{m-1}^*}{\varphi_{m-1}} h_m(x) \geq h_m(x)$.

As $h^* = h_1$ we arrive at $h^*(x) \geq h(x)$.

Lemma 2 Let $Z \subset H$ be a finite set of differences such that $h(z) > 0$ for all $z \in Z$. Let difference $z_0 = (\zeta_1, \ldots, \zeta_d) \in H_k$. Then there exist factors $\varphi_0^* = 1$, $\varphi_j^* \geq 1$ $(j = 1, \ldots, d-1)$ such that $h^*(z) > 0$ for all $z \in Z \cup \{z_0\}$.

Proof: For $j \neq d - k$ we set $\varphi_j^* = \varphi_j$. If $h(z_0) > 0$ then $\varphi_{d-k}^* = \varphi_{d-k}$, too. Otherwise we replace φ_{d-k} by a larger value φ_{d-k}^* such that

$$h^*(z_0) = \zeta_k \varphi_0 \ldots \varphi_{d-k-1} \varphi_{d-k}^* - \sum_{k < i \leq d} |\zeta_i| \varphi_0 \ldots \varphi_{d-i} > 0$$

which is possible because of $\zeta_k > 0$. Now, Lemma 1 guarantees that $h^*(z) > 0$ for all $z \in Z \cup \{z_0\}$.

Remark: In case $S = \{p_1\}$ or $Z = \emptyset$, respectively, valid factors (cf. (5)) are easily defined by setting $\varphi_j := 1$ $(j = 0, \ldots, d-1)$.

4 The algorithm SWEEPPLANE

The preparations of the last section allow us to design an algorithm SWEEPPLANE for constructing and maintaining a sweep plane in general position. As already mentioned, SWEEPPLANE is meant to be an auxiliary algorithm at the disposal of a class of space-sweep algorithms (cf. (2)) which are not designed to take care of the regular position of the sweep plane themselves. Whenever a new event point appears during the execution of such a space-sweep algorithm, SWEEPPLANE may be called and executed in order to ensure the sweep plane to be in regular position.

Every time SWEEPPLANE is called its *input* consists of

- event points $p_1, \ldots, p_n \in I\!R^d$ in lexicographic order
- factors $\varphi_0 = 1, \varphi_i \geq 1 (i = 1, \ldots, d-1)$ satisfying $h(p_{i+1} - p_i) > 0$ for $i = 1, \ldots, n-1$ (if $n > 1$)
- a new event point $p_{n+1} \in I\!R^d$.

The resulting *output* consists of

- event points $p_1, \ldots, p_{n+1}$ in lexicographic order
- factors $\overline{\varphi}_0 = 1, \overline{\varphi}_i \geq 1 (i = 1, \ldots, d-1)$ satisfying $\overline{h}(p_{i+1} - p_i) > 0$ for $i = 1, \ldots, n$
- the sweep function

$$g(x) = \sum_{i=1}^{d} \xi_i \prod_{j=1}^{d-i} \overline{\varphi}_j \quad \text{for } x = (\xi_1, \ldots, \xi_d).$$

In pseudocode notation SWEEPPLANE consists of the following steps:

1. Insert p_{n+1} into the lexicographically ordered set $\{p_1, \ldots, p_n\}$.

2. Determine the (nonnegative) differences z of p_{n+1} to its lexicographic predecessor and successor in $\{p_1, \ldots, p_n\}$.

3. **if** one of these differences z is equal to 0

   then remove p_{n+1}

   else

4.   **for each** difference z **do**

5.   Determine k for which $z \in H_k$;

6.   **if** $k < d$ **then** ensure that $h(z) > 0$

            endif

        endfor

   endif.

Remarks:

- Step 2: If p_{n+1} turns out to be the lexicographically smallest or largest element of $\{p_1, \ldots, p_{n+1}\}$ then only one difference is found, of course.

- Step 6: Lemma 2 guarantees that an associated function h satisfying $h(p_{i+1} - p_i) > 0$ for $i = 1, \ldots, n$ can always be constructed. For $k = d$ we always have $h(z) > 0$, i.e. it is never necessary to update h.

- In order to find suitable functions g and h with respect to the event points initially known, we first put these event points in lexicographic order. Then we apply SWEEPPLANE successively to all of them, starting with $\varphi_0 = \ldots = \varphi_{d-1} := 1$, i.e.

$$
\begin{aligned}
g(x) &= \xi_1 + \ldots + \xi_d, \\
h(x) &= \xi_k - (|\xi_{k+1}| + \ldots + |\xi_d|) \text{ for } x \in H_k \ (k = 1, \ldots, d).
\end{aligned}
$$

This provides the input required by SWEEPPLANE when the first new event point appears during the actual sweep.

The *time complexity* for one execution of SWEEPPLANE is $O(\log n)$ under the assumption that INSERT(p_{n+1}) in step 1 needs $O(\log n)$ time, which is normally justified. The remaining steps can all be performed in constant time. If SWEEPPLANE is just used to construct a sweep plane in regular position for a fixed set of n points, it needs $O(n \log n)$ time which is optimal [NeSc90].

The numeric behavior of SWEEPPLANE depends on the coordinates of the occurring event points, of course. [Schm91] assumes that every such coordinate ξ can be represented as a binary integer of m digits at most, i.e. $0 \le \xi < 2^m$, and that $\lambda_1, \ldots, \lambda_d \in I\!N$. With these assumptions it is easy to show that $\varphi_0, \ldots, \varphi_{d-1}$ can always be found in such a way that all of them are bounded by 2^m.

5 A small illustrative example

We choose $d = 2$ and assume that a space-sweep algorithm **A** which we do not specify more precisely but which we suppose to conform to the sweep scheme (2), makes use of SWEEPPLANE. Let us assume that before starting the sweep, the 3 event points $p_1 = (1, 1)$, $p_2 = (3, 2)$ and $p_3 = (5, 3)$ are already known.

First, p_1, p_2 and p_3 are lexicographically ordered, which results in $p_1 <_L p_2 <_L p_3$. Hence Q is initialized to (p_1, p_2, p_3). SWEEPPLANE is first called for p_1 and initializes the sweep function g and the associated function h as follows: $\varphi_0 = \varphi_1 := 1$, therefore

$$
\begin{aligned}
g(x) &= \varphi_0 \varphi_1 \xi_1 + \varphi_0 \xi_2 &&= \xi_1 + \xi_2, \\
h(x) &= \begin{cases} \xi_1 \varphi_0 \varphi_1 - |\xi_2| \varphi_0 &= \xi_1 - |\xi_2| &\text{for } x \in H_1 \\ \xi_2 \varphi_0 &= \xi_2 &\text{for } x \in H_2. \end{cases}
\end{aligned}
$$

At event point p_2 the positive difference to p_1 is $z = (2, 1) \in H_1$ and we get $h(z) > 0$ which ensures that the initial sweep plane is in regular position with respect to $\{p_1, p_2\}$. g and h are not updated by SWEEPPLANE. At p_3 SWEEPPLANE is called for the third time, and the situation is analogous. With that, steps (a) and (b) of scheme (2) are completed (preprocessing).

Now algorithm A starts the actual sweep, i.e. the while-loop (c) – (g) of scheme (2). Let $p_4 = (2,3)$ be the first new event point, detected at p_1. Since $g(p_4) = g(p_2) = 5$, the (initial) sweep plane is no longer in regular position. We find $p_1 <_L p_4 <_L p_2 <_L p_3$, i.e. p_1 and p_2 are the lexicographic neighbors of p_4. The two corresponding positive differences are $p_4 - p_1 = (1,2) \in H_1$ and $p_2 - p_4 = (1,-1) \in H_1$. $h(p_4 - p_1) = -1$ indicates that SWEEPPLANE will update h and g. $h(p_4 - p_1) = h((1,2)) = \xi_1 \varphi_0 \varphi_1 - |\xi_2| \varphi_0 = \varphi_1 - 2 > 0$ implies $\varphi_1 > 2$. Choosing $\varphi_1 = 3$ we obtain the following updates of h and g:

$$\overline{h}(x) = \begin{cases} 3\xi_1 - |\xi_2| & \text{for } x \in H_1 \\ h(x) = \xi_2 & \text{for } x \in H_2, \end{cases}$$

$$\overline{g}(x) = 3\xi_1 + \xi_2.$$

As also $\overline{h}(p_2 - p_4) = \overline{h}((1,-1)) = 2 > 0$, the sweep plane defined by $\overline{g}$ is in regular position with respect to $\{p_1, p_2, p_3, p_4\}$. Let $p_5 = (4,0)$ be a further new event point, detected at p_2. A small computation shows that the sweep plane will again be updated by SWEEPPLANE although it still is in regular position. The reason is, of course, that $g(z) > 0$ does not imply $h(z) > 0$.

Our algorithm SWEEPPLANE has been implemented in Pascal and C++, and its correctness and usefulness have been tested by a number of executions using the well-known plane–sweep algorithm of Bentley–Ottmann for reporting all intersections of finitely many line segments as the calling algorithm.

The authors thank K. Hinrichs and W. Nef for helpful suggestions.

References

[BeOt79] Bentley, J.L., Ottmann, T.A.: *Algorithms for reporting and counting geometric intersections*. IEEE Transactions on Computers C-28, 643-647 (1979).

[BeMe86] Beretta, G., Meier, A.: *Scan Converting Polygons Based on Plane–Sweep*. Technical Report 68, Institut für Informatik, ETHZ (1986).

[BiNe83] Bieri, H., Nef, W.: *A sweep-plane algorithm for computing the volume of polyhedra represented in Boolean form*. Linear Algebra Appl. 52/53,69-97 (1983).

[BiNe85] Bieri, H., Nef, W.: *A sweep-plane algorithm for computing the Euler–characteristic of polyhedra represented in Boolean form*. Computing 34, 287-302 (1985).

[EdMü90] Edelsbrunner, H., Mücke, E.P.: *Simulation of Simplicity: A technique to cope with degenerate cases in geometric algorithms*. ACM Transactions on Graphics 9, 66–104 (1990).

[Fort87] Fortune, S.: *A sweepline algorithm for Voronoi diagrams*. Algorithmica 2, 153–174 (1987).

[Hadw55] Hadwiger, H.: *Eulers Charakteristik und kombinatorische Geometrie*. J. reine angew. Math. 194, 101–110 (1955).

[Hadw68] Hadwiger, H.: *Eine Schnittrekursion für die Eulersche Charakteristik euklidischer Polyeder mit Anwendungen innerhalb der kombinatorischen Geometrie.* Elem. Math. 23, 121–132 (1968).

[HiNS88] Hinrichs, K., Nievergelt, J., Schorn, P.: *Plane-sweep solves the closest pair problem elegantly.* Information Processing Letters 26, 255–261 (1988).

[HiNS90] Hinrichs, K., Nievergelt, J., Schorn, P.: *An all-round sweep algorithm for 2-dimensional nearest-neighbor problems.* To appear.

[KeWe78] Kerr, J.W., Wetzel, J.E.: *Platonic divisions of space.* Mathematical Magazine 51, 229-234 (1978).

[LePr84] Lee, D.T.,Preparata, F.P.: *Computational geometry – a survey.* IEEE Transactions on Computers C-33, 1072-1101 (1984).

[Mehl84] Mehlhorn, K.: *Data Structures and Algorithms 3. Multi-dimensional Searching and Computational Geometry.* Springer-Verlag 1984.

[Müll88] Müller, H.: *Realistische Computergraphik. Algorithmen, Datenstrukturen und Maschinen.* Informatik-Fachberichte 163. Springer-Verlag 1988.

[NeSc90] Nef, W., Schmidt, P.-M.: *Computing a sweeping-plane in regular ("general") position: A numerical and a symbolic solution.* J. Symbolic Computation 10, 633-646 (1990).

[PrSh85] Preparata, F.P., Shamos, M.I.: *Computational Geometry – An Introduction.* Springer-Verlag 1985.

[Schm91] Schmidt, P.-M.: *About the Precision of the Coefficients of a Sweeping-Plane.* In U. Eckhardt et al. (Eds.): Geometrical Problems of Image Processing. Research in Informatics, Vol. 4, 107-113. Akademie Verlag 1991.

[Sedg88] Sedgewick,R.: *Algorithms.* Addison-Wesley, 2nd Ed. 1988.

[Sham78] Shamos, M.I.: *Computational Geometry.* Ph.D. Thesis, Yale University, 1978. University Microfilms International.

[Wood84] Wood, D.: *Paradigms and Programming with Pascal.* Computer Science Press 1984.

Performance Analysis of Three Curve Representation Schemes

Salvador Dominguez and Oliver Günther
FAW Ulm, P.O. Box 2060
7900 Ulm, Germany

Abstract

This paper describes the results of an experimental performance comparison of three representation schemes for arbitrary curved shapes: the strip tree [Ballard 1981], a curve–fitting approach using Bezier curves [Bezier 1974, Pavlidis 1982], and the arc tree [Günther 1988, Günther and Wong 1989]. Each of these schemes represents a curved shape as a hierarchy of approximations, where higher levels in the hierarchy correspond to coarser approximations of the curve. The schemes are compared on several geometric operations including point inclusion, curve–curve intersection, curve–area intersection, and area–area intersection. It is shown that in most cases the arc tree is the most efficient representation scheme of the three evaluated.

1 Introduction

In application areas such as computer vision, computer graphics or cartography, data objects are often described using curves or curve–bounded areas in the plane. Typically, the curves are rather arbitrary in shape and can therefore not be described by closed mathematical formulas but only by some kind of approximation. In practice, such approximations are usually a sequence of curve points, as obtained by a digitizer, or a sequence of splines. An appropriate representation scheme has to support typical operations, such as curve intersection or point inclusion queries, adequately and efficiently.

This paper describes the experimental comparison of three well–known curve representation schemes: the strip tree [Ballard 1981], a curve-fitting approach using Bezier curves [Bezier 1974, Pavlidis 1982], and the arc tree [Günther 1988, Günther and Wong 1989]. Each of these schemes represents a curved shape as a hierarchy of approximations, where higher levels in the hierarchy correspond to coarser approximations of the curve. In addition, each approximation typically corresponds to a bounding area, which encloses the actual curve. When geometric operations are computed, one will initially address coarse approximations of the curves (near the root of the hierarchies), and will only proceed to finer approximation levels if necessary. Differences between the three representations are found in the choice of bounding areas, the type and amount of information stored at each approximation level, and how decisions are made to proceed to finer approximations.

In section 2, we introduce the general idea behind each of the three representation schemes and describe how to compute the representation of a given curve. These representations are then used to perform curve operations such as point inclusion, curve–curve intersection, curve–area intersection, and area–area intersection. Sections 3 through 6 describe each of these algorithms and evaluate the performance of their implementations. Finally, section 7 contains a summary and conclusions.

2 Three Curve Representation Schemes

2.1 Arc Tree

An arc tree representation divides a given curve into several arcs of equal length. The endpoints of these arcs are then connected by line segments to form a polygonal path, which serves as the approximation. For example, a first (very coarse) approximation would consist of the line segment connecting the two endpoints of the curve. The next approximation would consist of the segment connecting the first endpoint with the curve's midpoint (as defined by the arc length of the curve) and the segment connecting the midpoint with the second endpoint, and so on.

More formally, given a curve C with length l, the k–th arc tree approximation of C is a polygonal path with 2^k edges, each associated with an arc of length $l/2^k$ ($k \geq 0$). With E^2 denoting the two–dimensional Euclidean space, C can be defined as a function

$$C : [0,1] \to E^2$$

such that the length of the curve from $C(0)$ to $C(t_0)$ is $t_0 \cdot l$. Then the common endpoints of edge e_{ki} ($i = 1, \ldots, 2^k$) and its associated arc a_{ki} are $C((i-1)/2^k)$ and $C(i/2^k)$. The approximation is further refined by increasing k. It is important to note that the vertices of the k–th approximation form a true subset of the vertices of the $(k+1)$–th approximation. See Figure 1 for an example.

Ellipses are used as the bounding areas of an arc tree approximation: it can be shown [Günther 1988] that if E_{ki} denotes the ellipse whose major axis has length $l/2^k$ and whose focal points are the two endpoints of the edge e_{ki}, then arc a_{ki} is internal to E_{ki} (Fig. 2).

An arc tree is a balanced binary tree that stores the family of approximations of a given curve. Its root node, considered at tree level 1, contains the points $C(0), C(1/2)$ and $C(1)$. Every other node contains a single curve point defined as follows: if a node on level i contains the point $C(x/2^i)$ ($x = 1 \ldots 2^i - 1$), then its left descendant contains point $C((2x-1)/2^{i+1})$, and its right descendant contains point $C((2x+1)/2^{i+1})$. The resolution r of an arc tree is defined as the number of levels actually stored. See Figure 3 for an example.

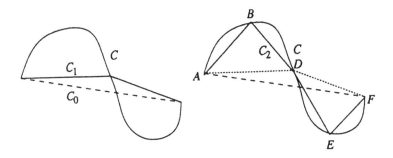

Figure 1: 0th, 1st and second arc tree approximation of curve C. Note that the subcurves AB, BD, DE, and EF all have the same arc length, viz., one quarter of the arc length of curve C.

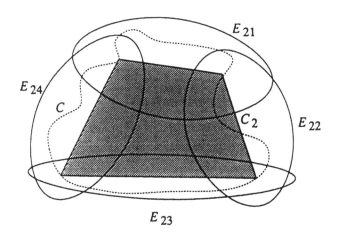

Figure 2: A curve C with its 2nd approximation C_2 and corresponding ellipses E_{2i}.

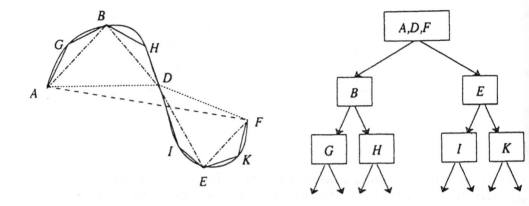

Figure 3: An arc tree approximation with corresponding arc tree of resolution 3.

To construct an arc tree from a sequence of curve points, one first passes through the points to compute the length l of the entire polygonal path connecting the curve points. l is considered to be the length of the curve. From there, each arc tree approximation can be computed by recursive subdivision [Günther 1988]. Higher approximations are computed until their accuracy (i.e., the maximal difference between a curve point and its approximation) is below a given threshold. Note that this guarantees that the resulting arc tree is balanced. The partitioning points obtained in this process are written into a file. Finally, the nodes of the arc tree are filled by reading the points $C(i/2^r)(i = 0..2^r)$ from the file and inserting them into the appropriate arc tree nodes while performing a depth–first inorder traversal of the tree.

2.2   Strip Tree

A strip tree representation approximates a curve by bounding it with rectangles *(strips)* of varying size. These strips can be subdivided further to produce smaller bounding strips and therefore a finer approximation. A strip is defined by two endpoints, (x_b, y_b) and (x_e, y_e), and two widths, w_l and w_r. The endpoints are curve points defining the beginning and the end of the strip, where the segment $((x_b, y_b), (x_e, y_e))$ is parallel to two sides of the strip . The distance from that segment to each of these two sides is designated by the two widths w_l and w_r. Initially, a curve can be approximated as one strip covering the entire curve tightly. This strip can be subdivided into two strips by picking a curve point which lies on the strip boundary. This point, along with the endpoints of the initial strip segment, is used to define two new segments and thus two new strips. Each of these two strips can in turn be subdivided, and so forth.

A strip tree is a binary tree that represents the resulting hierarchy of strips. The root node corresponds to the strip that encloses the entire curve. The curve is subdivided by picking a curve point on the strip boundary and creating two new strips. Nodes containing the information for these new strips are the children of the root node, and so on.

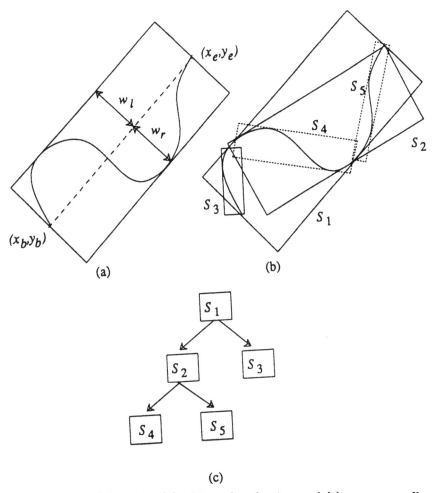

Figure 4: A curve with (a) a strip, (b) a hierarchy of strips, and (c) a corresponding strip tree.

Strips are subdivided recursively until the accuracy (i.e., the maximum distance between a subcurve and its approximating edge segment $((x_b, y_b), (x_e, y_e))$, which is $max(w_l, w_r))$ is below a given threshold. Note that proceeding in this fashion does not guarantee a balanced binary tree. See Figure 4 for an example.

2.3 Bezier Approach

Concatenations of Bezier curves are often used in interactive graphics to solve curve–fitting problems. An m-th-order Bezier curve is an m-th-order polynomial defined by $m + 1$ guiding points, such that the curve goes through the first and last guiding points and passes near the remaining ones in a well–defined manner [Bezier 1974]. The guiding points are also the vertices of a convex polygon that contains the curve (the *characteristic*

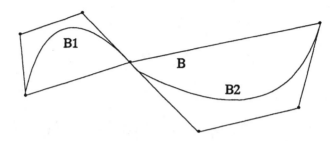

a) curve B is a concatenation of two third-order Bezier curves B1 and B2. B is enclosed within the two characteristic polygons corresponding to B1 and B2

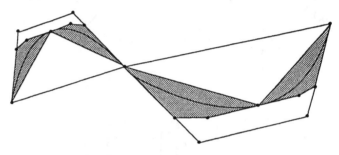

b) curve B with each initial characteristic polygon subdivided into two characteristic polygons

Figure 5: A Bezier curve approximation.

polygon). A curve representation scheme using Bezier curves is similar to the strip tree scheme except that the characteristic polygons, and not the strips, are used as bounding areas. Further refinement is achieved by subdividing the Bezier curves into shorter Bezier curves of the same degree [Pavlidis 1982]. If a Bezier curve B is subdivided into two Bezier curves $B1$ and $B2$ then the characteristic polygons of $B1$ and $B2$ are disjoint subsets of B's characteristic polygon (Fig. 5).

In our comparative study each curve is represented by a sequence of third-order Bezier curves. Each Bezier curve is defined by four guiding points. The data structure for the Bezier approach is simply a linked list of binary trees. Each tree in the list corresponds to a Bezier curve. The root node of each tree contains the four guiding points that define its curve and are also the vertices of its characteristic polygon. We can subdivide each Bezier curve into two (third–order) Bezier curves by subdividing its characteristic polygon as described in [Pavlidis 1982]. The children of each node contain the guiding points for each of the two Bezier curves obtained from the subdivision (Fig. 6). Each Bezier curve is subdivided recursively until the accuracy (i.e., the maximum of the perpendicular distance of the second and the third guiding point to the line passing through the first and fourth

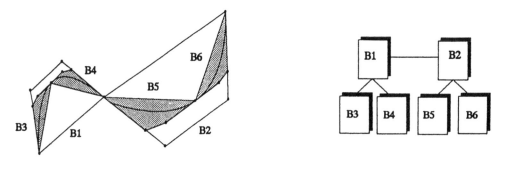

Figure 6: A Bezier curve approximation with its linked list representation. The tree nodes contain the vertices of the corresponding characteristic polygons.

guiding points) is below the given threshold. Note that the resulting binary trees do not have to be balanced.

2.4 Construction of the Representations

In our implementation we assume that the initial curve data for all three schemes are Bezier guiding points describing a sequence of third–order Bezier curves. We made this choice for three reasons: 1) Bezier guiding points can define complex curves in a more compact manner than a sequence of curve points; 2) there are numerous applications where curves are defined through the interactive placement of guiding points; 3) a sequence of Bezier curves can easily be converted into a sequence of curve points, whereas the reverse computation (from a sequence of curve points to a sequence of - possibly few - Bezier curves) is significantly more complicated. This input data allows the consideration of the Bezier approach as a promising representation scheme. However, whereas the Bezier approach can use the guiding point data directly, we need to do some precomputation for the arc and strip tree schemes.

The idea behind this preprocessing is to divide the curve recursively into two subcurves of equal length and to use the partitioning points as input for the tree representations. In order to find the length of the curve, we process the guiding points four at a time. Each 4–tuple defines a third–order Bezier curve, with the fourth guiding point coinciding with the first guiding point of the next Bezier curve. We multiply this vector of four guiding points with the Bezier geometry matrix, as described in [Foley and Van Dam 1982], to produce the coefficients of the third–order parameterization, $x(t)$ and $y(t)$, of each Bezier curve. The arc length l of each Bezier curve is now

$$l = \int_0^1 \sqrt{x'^2(t) + y'^2(t)}\,dt$$

From there, the partitioning points for the recursive subdivision can be obtained using

Figure 7: Curves $zp1$ and $zp2$.

numerical integration with Newton's approximation. The recursion terminates if the accuracy (i.e., the maximum distance between a subcurve and its approximating edge segment) is below the given threshold.

The resulting sequence of partitioning points can be read directly into an arc tree, as described in section 2.1. Concerning the strip tree, one may decide whether one wants to construct it from this sequence of curve points or directly from the Bezier curves. We chose the former option because this facilitates a direct comparison with the arc tree construction and because it leads to a simpler implementation. While the direct computation of the strip tree from the Bezier curves is more accurate in theory, it may also cause numerical problems and is extremely CPU-intensive.

2.5 Performance Comparison

We implemented the three representation schemes on a SUN 3/50 using the C programming language. For our comparisons, we created six curves: two closed curves ($zp1$ and $zp2$, Fig. 7) for testing point inclusion; one pair of open curves ($z1/z2$, Fig. 8) for testing curve–curve intersection; and two more pairs of closed curves ($zc1/zc2$, $zc3/zc4$, Fig. 8) for testing curve–curve, curve–area, and area–area intersection. All curve representations were constructed with the same accuracy, as defined above. As previously stated, the input data was converted from Bezier guiding points to curve points for the arc tree and strip tree schemes. For each curve, the number of guiding points defining the original Bezier curves, the number of curve points after conversion, and the CPU time required to perform the conversion are listed in Table 1. All CPU times in this paper are given in 1/50th of a second, which was the smallest unit of measurement available on the chosen hardware platform.

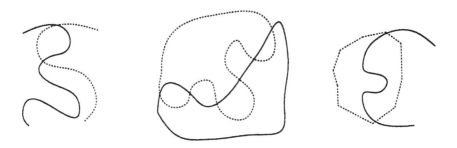

Figure 8: Curve pairs $z1/z2$, $zc1/zc2$ and $zc3/zc4$.

Conversion	No. Guiding Pts.	No. Curve Pts.	CPU time
$z1$	13	33	1025
$z2$	7	17	425
$zp1$	13	65	2376
$zp2$	31	129	4460
$zc1$	19	65	1894
$zc2$	19	65	1607
$zc3$	13	33	1086
$zc4$	10	33	844

Table 1 : Conversion into a sequence of curve points.

The times for constructing the representation of each scheme after the appropriate conversion for the arc tree and strip tree schemes has been performed are listed in Table 2. The fact that the sequence of curve points obtained from the conversion is evenly spaced was not used for the construction, i.e. the sequence was considered to be some arbitrary sequence of curve points. This way the arc tree did not gain an unwarranted advantage.

Construction	Arc	Strip	Bezier
$z1$	6	20	6
$z2$	4	7	3
$zp1$	14	32	6
$zp2$	24	77	11
$zc1$	13	33	8
$zc2$	11	28	5
$zc3$	8	13	4
$zc4$	8	12	2

Table 2 : Construction of the representations.

In the arc tree and strip tree schemes the conversion of the Bezier input into curve points is performed at a considerable cost. If we had used a sequence of curve points as input then this cost would not have occurred, but it would have been much more difficult and more costly in turn to consider the Bezier approach.

After the appropriate conversions in the arc tree and strip tree schemes, the times to construct a Bezier representation are still faster than the times for the other two schemes. The Bezier approach initially handles less data than the arc tree and the strip tree. In addition, there are very efficient algorithms for the subdivision of characteristic polygons in the Bezier scheme [Pavlidis 1982], whereas it is much more complicated to find subdivision points for the arc tree and strip tree schemes. The times for strip tree construction are even slower than those for the arc tree because, aside from finding subdivision points, the strip tree scheme also requires strip widths, which are difficult to compute.

3 Point Inclusion

All three schemes support the query whether a given point A is contained in a given closed curve. In each representation scheme, the query initially addresses a coarse approximation of the curve, near the root(s) of the representation. By testing the bounding areas of the approximation, we decide whether the query can be answered at this resolution. If that is not possible, the approximation is refined recursively until the query can be answered.

3.1 Arc Tree

The arc tree approximation of a closed curve is a closed polygon. To answer a point inclusion query, we start with a simple approximation of the curve. We then check every edge e_{ki} of the approximation to see whether its replacement with arc a_{ki} may affect the internal/external classification of A, the point tested for inclusion.

It can be shown that the replacement of a_{ki} by e_{ki} may only affect the internal/external classification of the tested point if A is internal to the bounding ellipse E_{ki} [Günther 1988].

Recall that the focal points of E_{ki} are the two endpoints of the edge e_{ki}, $C((i-1)/2^k)$ and $C(i/2^k)$, and that its major axis has length $l/2^k$. To determine whether a point is internal to an ellipse, we add up the distances between the point and each of the two focal points of the ellipse. If the sum is no more than $l/2^k$, then the point is internal, and the replacement of e_{ki} by a_{ki} may affect the classification of A.

If such edges e_{ki} exist, then we replace each of these edges by the two edges $e_{k+1,2i-1}$ and $e_{k+1,2i}$ to obtain a closer approximation of the curve. We recursively refine this polygon until (i) no more edges are found whose replacement may affect the classification of A, or (ii) the maximum resolution of the arc tree has been reached. In the latter case, we assume that the query cannot be solved at the resolution of the arc tree. On the other hand, if we do obtain a final polygon, then the initial query is addressed to this final approximation and solved using Shamos' point inclusion algorithm [Preparata and Shamos 1985]: we construct a horizontal line through A and count the number of intersections between this line and the edges of the final polygon that lie to the left of A. If this number is odd, then A is internal, otherwise it is external.

3.2 Strip Tree

Before discussing point inclusion and similar operations on strip trees, we need to define the types of intersections that can occur with strips. Two strips have a *null* intersection if none of their respective edges intersect each other. A *clear* intersection exists if the two edges of the first strip that are parallel to the line segment $((x_b, y_b)_1, (x_e, y_e)_1)$ each intersect both edges of the second strip that are parallel to the line segment $((x_b, y_b)_2, (x_e, y_e)_2)$. Lastly, two strips have a *possible* intersection if they have neither a clear nor null intersection. A possible intersection implies that one strip encloses the other strip or that some intersection exists, but not as in the case of a clear intersection (Fig. 9). From these definitions, it can be shown that two curves intersect if their corresponding strips have a clear intersection [Ballard 1981]. On the other hand, it is obvious that two curves do not intersect if their corresponding strips have a null intersection.

The idea behind point inclusion in the strip tree scheme is similar to the arc tree approach. We define a semi-infinite strip with zero width emanating from A, the test point, and compare it with the strip corresponding to the root node of the strip tree. If they have a null intersection, then the procedure returns 0. If a clear intersection exists, then a 1 is returned. If a possible intersection exists, then the procedure is recursively called on the left and right subtree, and the sum of their return values is returned. If the leaves of the strip tree are reached without obtaining a null or clear intersection then the query cannot be solved at the given resolution. Otherwise, the result depends on whether the sum returned is odd or even. The point is enclosed in the curve represented by the strip tree if and only if the sum returned is odd.

3.3 Bezier Approach

The idea behind point inclusion in the Bezier scheme is analogous to the strip tree scheme. In the Bezier approach, third–order Bezier characteristic polygons, the scheme's bounding

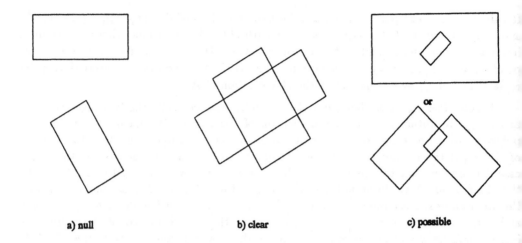

a) null   b) clear   c) possible

Figure 9: Types of strip intersection.

areas, are used in the same manner as strips in the strip tree scheme. Therefore, the same
types of intersections are defined for the characteristic polygons as were for strips. If
a clear intersection exists between two characteristic polygons then their corresponding
curves intersect.

We construct a semi–infinite polygon with zero width emanating from A and determine its intersections with each root node of the linked list of trees. As with the strip
tree scheme, if a null intersection exists, then a 0 is returned, a clear intersection returns
a 1, and a possible intersection recursively returns the sum of the intersection values of
the left and right subtrees of each root node. If the sum returned from adding up the
values from each root node is odd, then the point tested for inclusion is enclosed in the
curve represented by the sequence of Bezier curves.

3.4   Performance Comparison

We used the two closed curves $zp1$ and $zp2$ (Fig. 7) for comparing the point inclusion
operations of the three schemes. The algorithms have been implemented on a SUN 3/50.
For each curve, we tested four points for inclusion (Fig. 10, 11). Table 3 shows the depth
r of the traversal of each of the respective scheme representations and the CPU time
t needed to determine the classification of each tested point (measured in 1/50th of a
second).

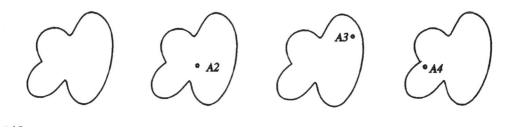

Figure 10: Point inclusion on $zp1$.

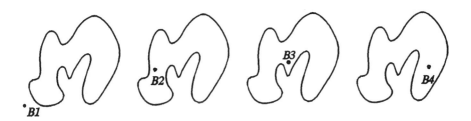

Figure 11: Point inclusion on $zp2$.

Point Inclusion		Arc Tree	Strip Tree	Bezier Approach
$zp1$	A1	$r=2, t<1$	$r=1, t=1$	$r=1, t=2$
	A2	$r=3, t=1$	$r=3, t=2$	$r=3, t=4$
	A3	$r=3, t=1$	$r=3, t=2$	$r=2, t=2$
	A4	$r=4, t=1$	$r=4, t=3$	$r=3, t=4$
$zp2$	B1	$r=4, t=1$	$r=1, t=1$	$r=1, t=7$
	B2	$r=4, t=1$	$r=3, t=2$	$r=3, t=8$
	B3	$r=6, t=1$	$r=6, t=8$	$r=3, t=7$
	B4	$r=4, t=1$	$r=9, t=7$	$r=3, t=8$

Table 3 : Point inclusion.

For the first curve, $zp1$, although we see that all three schemes solved each query with approximately the same resolution, the times for the arc tree appear to be slightly faster than for the other two schemes. Since the queries were resolved at the same level, this suggests two things. First, the arc tree provides a better localization of those parts of the curve which are of critical importance in determining the internal / external classification of A. This means that the fraction of the tree nodes that is investigated is smaller in the case of the arc tree than for the other two representations. Second, the tests performed in the arc tree scheme are faster than the tests performed in the other two approaches. The arc tree scheme proceeds to finer approximations if the tested point is enclosed within the bounding ellipse. Hence, the only significant work done is computing the distances from the tested point to each of the two foci of the ellipse, which are found in the nodes of the arc tree. On the other hand, the strip tree and Bezier schemes have the more time–consuming task of determining the types of intersection between strips and polygons in order to refine their approximations.

The times for the strip tree approach appear to be slightly faster than for the Bezier scheme. This is due to the fact that the Bezier approach is hampered by the fact that it must process every tree in its linked list, regardless of the proximity of the tested point to the curve. In addition, it is of course more complicated to test intersections between the characteristic polygons compared to strips.

For the second curve, $zp2$, on the average the times for the arc tree are considerably faster than for the other two schemes although one often needs to go deeper into the tree. Here, not only is the strip tree scheme slowed down by its more time–consuming tests, but the values of r suggest that the unbalanced representation of the strip tree contributed to its relatively poor performance. In resolving one query, the traversal of the strip tree reached 9 levels. The fairly consistent but slow times of the Bezier scheme show that, once again, this approach is hindered by the fact that it must process every tree in its linked list. In addition, since the second curve is more complicated, the time to traverse the extra trees in the Bezier scheme had a significant effect on its performance.

4   Curve–Curve Intersection

The detection and computation of the intersection of two curves C and D is the second operation supported by the three curve representation schemes. In order to detect whether two curves intersect, the algorithms begin with coarse approximations for both curves. Tests are performed on the bounding areas of the approximations to determine if the query can be answered at the current resolution. If that is not the case, the approximations for one or both curves are refined. A possible intersection area can be *computed* in a similar manner.

4.1   Arc Tree

To *test (detect)* whether the curves C and D intersect, we start with simple approximations, C_{app} and D_{app}, and continue with higher resolutions where necessary. For each pair of edges, e_{ki} of C_{app} and f_{kj} of D_{app} $(i, j = 1..2^k)$, we examine whether their corresponding arcs *may* intersect by checking whether the corresponding ellipses E_{kj} and F_{kj} intersect.

f they do, then we tag these edges and check whether the corresponding arcs *must* intersect, as described in [Günther 1988]. If the arcs intersect, then we report a curve–curve
intersection and stop. Otherwise, we replace all tagged edges by the corresponding edges
of the next higher resolution, and proceed recursively on the refined approximations until
the maximum resolution is reached.

We implemented this algorithm essentially as described above except for a couple of
modifications to speed up execution. First, we replaced the test whether two ellipses
E_{ki} and F_{kj} intersect by the test whether the two *circumscribing circles* of E_{ki} and F_{kj}
intersect. Although testing ellipses for intersection is more accurate than testing circles,
the latter test is significantly faster at a cost of a few more tagged edges. Second, we
maintain a matrix to keep track which pairs of ellipses (E_{ki}, F_{kj}) intersect. Then we
process a pair of edges (e_{ki}, f_{kj}) only if the ellipses $E_{k-1,i/2}$ and $F_{k-1,j/2}$, corresponding to
their parent edges, intersect. If these ellipses do not intersect, then we know that E_{ki} and
F_{kj} do not intersect either.

The algorithm to actually *compute* the intersection of two curves is a variation of the
above detection algorithm. In the computation algorithm, we never check whether two
arcs *must* intersect. We continue refining the curves recursively until (i) there are no
more tagged edges, or (ii) we reach the maximum resolution of the arc trees involved. In
the first case, the curves C and D do not intersect. In the second case, we intersect each
tagged edge of C_{app} with each tagged edge of D_{app} and return the intersection points.

4.2   Strip Tree

Two curves intersect if their corresponding strips have a clear intersection. Therefore, in
order to *detect* a curve–curve intersection in the strip tree scheme, we simply test pairs
of strips from both curves until a clear intersection is found.

We start with the root nodes of the strip trees corresponding to the given curves C
and D. If the strips corresponding to these nodes have a null intersection, then we report
no intersection and stop. If they have a clear intersection, then we report an intersection
and stop. If they have a possible intersection, then we recursively test the children of
these root nodes. If the area of the strip represented by the first root node T_1 is greater
than the area of the strip of the second node T_2, then the algorithm is recursively called
with (i) T_2 and the left child of T_1 and (ii) T_2 and the right child of T_1, and vice versa.

The algorithm to *compute* the actual intersection of the curves is a variation of the detection algorithm. In case of a clear intersection we keep calling the algorithm recursively
on the children of the nodes involved, as described above, until a null intersection or the
maximum resolution has been reached. For each remaning clear intersection at maximum
resolution we return the center point of the strip intersection as the approximation of an
intersection point.

4.3   Bezier Approach

The algorithm to *detect* a curve intersection for the Bezier scheme is identical to the
strip tree algorithm. Here, polygons from both curves are tested until we find a clear

intersection. However, whereas in the strip tree scheme, we perform the initial call on the root nodes of the two strip trees, in the Bezier approach, each curve is represented by a linked list of root nodes. Therefore, the initial call involves a nested loop: the inner loop traverses the linked list of the second curve, while the outer loop traverses the linked list of the first curve. The algorithm stops when we find the first clear intersection. The modifications to convert the above detection algorithm to one that returns actual intersection points are identical to those made in the strip tree scheme.

4.4   Performance Comparison

These algorithms have been tested on a SUN 3/50 with three different pairs of curves: $z1/z2$, $zc1/zc2$, and $zc3/zc4$. Table 4 shows for each scheme the CPU times needed to detect and to compute an intersection between those curves, measured in 1/50th of a second.

Curve-Curve Intersection		Arc Tree	Strip Tree	Bezier Approach
Detection	$z1 / z2$	$t = 4$	$t = 20$	$t = 13$
	$zc1/zc2$	$t = 9$	$t = 21$	$t = 19$
	$zc3/zc4$	$t = 2$	$t = 7$	$t = 20$
Computation	$z1 / z2$	$t = 5$	$t = 36$	$t = 26$
	$zc1/zc2$	$t = 18$	$t = 60$	$t = 95$
	$zc3/zc4$	$t = 8$	$t = 23$	$t = 99$

Table 4 : Curve–curve intersection.

In detecting curve–curve intersections, the times for the Bezier scheme are usually faster than for the strip tree approach. We can report a curve–curve intersection once we find just one clear intersection between a pair of polygons. Therefore, the Bezier scheme needs not necessarily process every tree in the two linked lists, as was necessary in the point inclusion operation.

When actually *computing* all curve–curve intersections, then the Bezier scheme needs to process every tree in the two linked lists. If the curves are not very complicated, as in the $z1/z2$ example, then the times for the Bezier approach remain competitive compared to the times for the strip tree. However, if the curves are complicated, as in the $zc1/zc2$ example, then the time used by the Bezier scheme to perform the nested loop on the two linked lists takes its toll on the scheme's performance.

However, for both detection and computation, the times for the arc tree scheme are consistently faster than the times for the other two approaches. This indicates that the arc tree scheme is more efficient in localizing possible intersections, which means that the fractions of the trees that are traversed are smaller than for the other two representations. In addition, we attribute this advantage of the arc tree to two facts: first, it is a balanced representation, and second, the tests to proceed to finer approximations are simpler and aster in the arc tree scheme than in the other two approaches. When finding curve–curve intersections in the arc tree scheme, we need to determine whether two circles intersect. On the other hand, in both the strip tree and Bezier approaches, one has to determine intersection types, which is significantly more complex.

5 Curve–Area Intersection

Each of the three schemes also supports the detection and computation of curve–area intersections. For all three schemes, the algorithm to *detect* a curve–area intersection is very similar to the one that detects curve–curve intersections. However, the idea behind *computing* a curve–area intersection for the strip tree and the Bezier approach involves pruning the original curve representation - a strategy substantially different from the idea behind curve–curve intersection.

5.1 Arc Tree

In the arc tree scheme, the algorithm to compute the intersection of a curve C and an area $P(D)$ that is bounded by a curve D is an extension of its curve–curve counterpart. As was done for the curve–curve operation, the algorithm begins by recursively processing edges until either (i) there are no more tagged edges, or (ii) the maximum resolution has been reached for C and D. In case (i), the curve C could be enclosed by the area $P(D)$. We decide this by checking whether some point of C is enclosed by D. In case (ii), let C_{app} and D_{app} denote the approximations reached. We intersect the tagged edges of C_{app} with the tagged edges of D_{app} to obtain intersection points. These intersection points are used to subdivide C_{app} into disjoint edge segments. Each edge segment of C_{app} is either internal or external to $P(D)$. This classification usually changes at each intersection point (for the treatment of some special cases see [Günther 1988]). Therefore, we start at a certain segment, determine its classification, and alternately label each segment internal or external. We then replace all internal untagged edges by their corresponding edges of maximum resolution. Finally, we return a sorted list of the internal edges and edge segments.

5.2 Strip Tree

The idea behind computing a curve–area intersection is to create a new strip tree which represents those subcurves of C that are contained in the area $P(D)$. This can be done by pruning the original strip tree of curve C. First we test if the strip S_C corresponding to the root of C's strip tree has a null intersection with the strip S_D corresponding to the root of D's strip tree. If that is the case, then C and $P(D)$ have no intersection. Otherwise a variation of the point inclusion algorithm is used to determine if S_C or any of its substrips are enclosed by D. As in the case of curve–curve intersection (see section 4.2), the algorithm proceeds in a recursive manner to traverse the strip trees of C and D until all strips are found or until the maximum resolution has been reached. The strips that have been found form a new strip tree, which represents those part(s) of C that are contained in $P(D)$.

5.3 Bezier Approach

To compute a curve–area intersection, one proceeds similarly as in the case of the strip tree. The algorithm for the strip tree scheme essentially works by pruning the original strip tree of the curve. However, it is not feasible to prune across the linked list of trees in the Bezier representation because if we trim any root nodes we lose the links to subsequent root nodes. Therefore, our calls to the procedure are made in a nested loop. The outside loop traverses the linked list for the curve representation and the inner loop traverses the linked list for the area representation. If any curve–area node pair contributes to the curve–area intersection, then we report the appropriate strips of the curve and continue. Other than this change in our initial calls, the rest of the implementation is virtually identical to the strip tree method.

5.4 Performance Comparison

These algorithms have been tested on a SUN 3/50 with two different pairs of curves, $zc1/zc2$ and $zc3/zc4$, Table 5 shows for each scheme the CPU time t needed to detect and compute a curve–area intersection for curve C and area $P(D)$, measured in 1/50th of a second.

Curve–Area Computation	Arc Tree	Strip Tree	Bezier Approach
$zc1/zc2$	$t = 19$	$t = 288$	$t = 236$
$zc3/zc4$	$t = 11$	$t = 78$	$t = 143$

Table 5 : Curve–area computation.

The times to compute a curve–area intersection are noticeably slower in the strip tree and Bezier schemes than in the arc tree approach. In fact, the time to compute the intersection in the arc tree scheme is just a little higher than the time to compute a curve–curve intersection. This is because the curve area operation is simply an extension of the curve–curve procedure: once we have obtained the intersection points of two curves C and D, it is a relatively simple matter to find those segments of C that are contained in $P(D)$.

On the other hand, the strategy used by the strip tree and, to a certain extent, the Bezier scheme is to obtain the intersection by pruning the original curve approximation. We have already seen in the strip tree approach that processing an unbalanced representation can lead to poor performance. Not only does the strip tree need to prune a possibly nefficient structure, it also must maintain the construction of a new strip tree. Similarly, since the Bezier procedure calls are made in a nested loop, processing complicated curves can lead to poor results.

However, the strip tree scheme does have the advantage of returning the intersection in a new strip tree. (Note that it is possible to construct a strip tree whose corresponding curve is disconnected.) This allows further queries to be made on the resultant strip tree. The same advantage holds for the Bezier scheme, but not for the arc tree, where the result is not represented as an arc tree but rather as a linked list of edges.

6   Area–Area Intersection

Each scheme also provides the detection and computation of area–area intersections. The algorithms to detect and compute area–area intersections are essentially identical for all three schemes and are simply extensions of other operations.

We tested area–area intersections with two pairs of curves $zc1/zc2$ and $zc3/zc4$. Table 6 shows the CPU time t needed to compute the intersection of the two enclosed areas, measured in 1/50th of a second.

Area–Area Computation	Arc Tree	Strip Tree	Bezier Approach
$zc1/zc2$	$t = 43$	$t = 501$	$t = 478$
$zc3/zc4$	$t = 24$	$t = 162$	$t = 295$

Table 6 : Area–area computation.

The times for computing the area–area intersections simply reflect the performance of the curve–area computation operation of each scheme. Not surprisingly, because of its balanced representation and its faster test of bounding areas, the arc tree operation outperformed the area–area intersection operations of the other two schemes.

7   Conclusions

In this paper, we introduced and described three curve representation schemes. We discussed the algorithms for various curve operations, analyzed the performances of their implementations, and compared the results.

Aside from the preprocessing of the initial Bezier data, the arc tree scheme outperformed the strip tree and the Bezier approach in practically all test cases. The arc tree scheme had the advantage of having faster tests on bounding areas: its tests on ellipses and circles were faster than the strip and polygon intersection tests of the strip tree and Bezier schemes. Furthermore, the single balanced tree representation of the arc tree proved to be more time–efficient than the possibly unbalanced tree representation of the strip tree and the linked list of trees of the Bezier scheme. Finally, the arc tree provides a more efficient localization of those parts of the curve(s) that are critical with regard to the given query. In other words, the decision where to proceed to finer approximations is fastest for the arc tree scheme, and it is also more selective, i.e. the average number of nodes visited is smallest.

In the arc tree scheme, the conversion from the initial Bezier data to curve points was very time–consuming. Nevertheless, if we use Bezier curve data as input, then we can perform this conversion once per curve and save the converted data. With repeated queries on the curve, the time spent in data conversion is amortized in the time saved using the arc tree scheme over the Bezier approach. If we use *curve points* as the initial data, then we no longer need to perform this time–intensive task, and the times for arc tree construction are even somewhat faster than those for the strip tree construction. Once constructed, the arc tree proves to be the most efficient curve representation among the three studied.

References

Ballard, D. H., Strip trees: A hierarchical representation for curves, *Communications of the ACM*, May 1981.

Bezier, P. E., Mathematical and practical possibilities of UNISURF, *Computer Aided Geometric Design*, Academic Press, New York, NY, 1974.

Foley, J. and Van Dam, A., *Fundamentals of Interactive Computer Graphics*, Addison–Wesley, Reading, Mass., 1982.

Günther, O., *Efficient structures for geometric data management*, Lecture Notes in Computer Science No. 337, Springer–Verlag, 1988.

Günther, O. and Wong, E., The arc tree: an approximation scheme to represent arbitrary curved shapes, *Computer Vision, Graphics, and Image Processing*, Vol. 51, 313–337, 1990.

Pavlidis, T., *Algorithms for graphics and image processing*, Computer Science Press, Rockville, MD, 1982.

Preparata, F.P. and Shamos, M.I., *Computational Geometry*, Springer–Verlag, 1985.

Preclassification and Delayed Classification of Boundary Entities in Arbitrary Dimensions

Martin J. Dürst

Institut für Informatik der Universität Zürich
Winterthurerstrasse 190, CH-8057 Zürich-Irchel
email: mduerst@ifi.unizh.ch

Abstract

The preclassification of the vertices of a polyhedron is the central part of a recently proposed point-in-polyhedron algorithm [HT89]. We show that this preclassification is related to the convex hull on a sphere. This allows an unified approach for very general polyhedra of arbitrary dimensions and for all boundary entities (vertices, edges, faces). It also reduces the time complexity of preclassification in 3D. Replacing preclassification by delayed classification leads to an additional reduction of time complexity.

1. Introduction

1.1 The Point-in-Polyhedron Problem

The point-in-polyhedron problem, i.e. deciding whether a given point (or several points) is contained in a polyhedron or not, is one of the most basic operations in a wide area of application fields such as computer graphics, solid modeling, and image analysis.

The probably best known algorithm to solve this problem counts the number of intersections of the boundary of the polyhedron with a ray (a semi-infinite line) from the point in question [Kal82]. Other solutions include to find the first intersection of a ray with the polyhedron boundary, and to reduce the dimension by intersecting the polyhedron with a plane containing the point [Kal82]. Another possibility is to calculate the net sum of the angles under which the boundary of the polyhedron is seen from the test point, using the Gauss-Bonnet integral [LMR84].

All these algorithms assume that the polyhedron is defined by its boundary, a form that is called Boundary Representation (B-Rep) [Req80] in solid modeling. The problem becomes trivial if the polyhedron is given in the form of a set-theoretic combination of primitives [Nef78, BN88], often called Constructive Solid Geometry (CSG) [Req80]. Great speedups are possible with any form of space subdivision, such as the bintree or the octree [TKM84, Sam90]. However, for the individual cells of the subdivision, the point-in-polyhedron problem still has to be solved in some form. A particularly elegant solution is available for 2D [GRS83, DGHS88]; unfortunately it does not generalize to higher dimensions.

The above mentioned algorithms are usually described for one given number of dimensions (polygons in 2D or polyhedra in 3D), with additional restrictions regarding the form of the allowed polyhedra such as boundedness and connectedness. An extension to higher dimensions and to less restricted polyhedra is sometimes possible in theory, but mostly difficult in practice because of the many cases that have to be considered to maintain consistency. This is especially true for the popular ray-casting approach.

1.2 The Closest Boundary Entity Algorithm

The closest boundary entity algorithm is another solution to the point-in-polyhedron problem, which is unfortunately not yet well known. It has been described for simple polygons in 2D by Nordbeck and Rystedt [NR67] and extended to bounded and connected 2-manifold polyhedra in 3D by Horn and Taylor [HT89]. As there are very few degenerate cases, which are easy to handle, the algorithm is highly robust [HT89]. Also, the closest boundary entity algorithm can easily be generalized to arbitrary dimensions and a very general class of polyhedra, as we will show in this paper. This makes it the algorithm of choice for many future implementations of the point-in-polyhedron problem. The main steps of the algorithm are as follows:

Step 0: Preprocessing: Determine the classification of each boundary entity (vertex, edge, face,...) of the polyhedron based on the form of the neighborhood of the boundary entity.

Step 1: Calculate the distance from the point in question to all the boundary entities of the polyhedron.

Step 2: Select the boundary entity with the smallest distance to the point in question (this entity is called the closest boundary entity).

Step 3: Determine whether the point lies inside or outside the polyhedron based on the classification of the neighborhood of the closest boundary entity.

Choosing the boundary entity of the polyhedron (vertex, edge, face) with the smallest distance to the test point (Step 2) reduces the global problem to a local problem. The most crucial point of the algorithm is the fact that each boundary entity can be classified in such a way that the final result is obtained immediately once the closest boundary entity is known. This classification can be done in two ways: Preclassifying all boundary entities in a preprocessing step, as proposed in [HT89], or delaying classification of a boundary entity until a point is found that lies closest to this boundary entity, as proposed newly in this paper. In both cases, a classification once obtained can be used for all points that lie closest to the respective boundary entity.

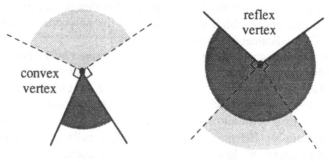

reflex vertex

convex vertex

inside of polygon   area where points are closest to the vertex

Figure 1. Convex and reflex vertices of a polygon
and the location of the points closest to the vertex

That such a classification is possible for any boundary entity in any dimension will be shown in Section 4. To give a first, intuitive idea of how the classification works, we can use the example of a vertex of a 2D polygon (see Figure 1). Here, classification amounts to deciding whether the vertex is convex or reflex (concave). A point closest to the vertex can only lie outside the polygon if the vertex is convex, and only lie inside if the vertex is reflex.

The paper is structured as follows: The next section introduces the necessary definitions for polyhedra and for the sphere. The mapping of the neighborhood of an entity of the polyhedron boundary (called a boundary feature) to the sphere is then explained in Section 3. Section 4 relates the configuration on the sphere to the closestness of a point to a boundary entity. Based on this, a classification of the boundary entities of the polyhedron is developed. The two ways to implement classification, preclassification and delayed classification, are discussed in Section 5. To give new, nontrivial upper bounds for the preclassification of boundary entities, the convex hull on the sphere is used. Section 6 concludes the paper.

This paper concentrates on the geometrical aspects of the algorithm, showing why and how a general and efficient classification of the boundary entities can be obtained. Numerical aspects, such as the calculation of the distance to each of the boundary entities, and implementation details, such as the actual data structures used (cf. [PS86]), are currently under study and will be reported later.

2. Definitions

This section contains the basic definitions for polyhedra and for the sphere. In general, capital letters are used for objects (point sets), and lower case letters for points. By $\Omega(X)$, we denote the embedding space of the object X, by $\dim(X)$ the dimension of X, and by $\text{dist}(X,Y)$ the distance between X and Y, i.e. the greatest lower bound (infimum) of the distances between any point of X and of Y. The letters H and p are used throughout the paper for the polyhedron and the point to be classified.

2.1 Polyhedra

The polyhedron definition that is introduced here is much more general than usual. We do not require boundedness, connectedness, manifoldness (i.e. the boundary of a k-dimensional polyhedron is a $(k-1)$-manifold), or regularity (i.e. the polyhedron is the closure of its interior [Req80]). Even the notions of interior and exterior are discarded, which is at first sight contradictory to the very aim of developing a point-in-polyhedron algorithm.

Such a general polyhedron definition has been chosen mainly because it turned out that the applicability of the closed boundary entity algorithm is much wider than its description in [HT89] let assume. Also, the generality of the definition simplifies the argumentation by eliminating special cases. Our polyhedron definition is at the moment not intended as the base of a mathematical theory of polyhedra. Without a notion of inside and outside, set operations and other important operations on polyhedra are difficult to define. However, it should be noted that some algorithms for set operations are also based on a general space partition [CR91].

The reader not familiar with higher dimensions or with non-manifold and non-regular objects should not get confused. It is at any point of the paper sufficient to imagine the discussed objects in 2D or 3D. A good introduction to the special cases and the terminology for non-manifold objects in 3D can be found in [CR91].

Definition 1: A k-dimensional polyhedron H is a partitioning of a k-dimensional space (denoted $\Omega(H)$) into a finite number of entities so that:
- Each entity B of dimension i ($0 \le i \le k$) is a (relatively) open subset of an i-dimensional subspace of $\Omega(H)$. This subspace is denoted $\Omega(B)$.
- Each i-dimensional entity's boundary is the union of entities of dimension $j < i$.
- A boundary entity B of dimension i that bounds an entity C of dimension $i+1$ does so with uniform orientation (i.e. C is always on one and the same side, or on both sides, of B).

Entities B of H for which $\dim(B)=k$ are called *volume entities*, and those with $\dim(B)<k$ are called *boundary entities*. Volume entities never touch each other directly, they are always separated by some boundary entities. The *boundary* of the polyhedron H, denoted $bd\ H$, is the union of the boundary entities of H, whereas the boundary of any individual entity of H is defined in the usual topological sense.

Definition 1 refrains from specifying the embedding space of the polyhedron and its entities. This allows to capture the common properties of both planar and spherical polyhedra. Nevertheless, we will later omit the word "planar" for simplicity. Spherical polygons, discussed in the next subsection, will always be denoted as such.

Definition 2: A *planar k-dimensional polyhedron* is a k-dimensional polyhedron according to Definition 1 where $\Omega(H)$ is the Euclidian space $\mathbf{R}^k$, and the subspaces $\Omega(B)$ are the linear manifolds (points, lines, planes, hyperplanes) of $\mathbf{R}^k$.

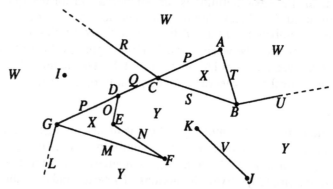

Figure 2. A 2-dimensional polyhedron conforming to Definition 2.

An example of a polyhedron is depicted in Figure 2. The letters A-K denote vertices, L-V edges, and W-Y volume entities. Note that the two line segments identified with P could be identified differently, resulting in a (slightly) different polyhedron. The same is true for the two parts of W separated by R, or the two parts of X. To conform with Definition 1, this would enforce different identifications also for the two parts of P.

The point-in-polyhedron problem now amounts to finding the entity B of the polyhedron H that contains p. We also call this to "classify p against H" and use the notation $p?H$ (pronounced "where is p in H"), so that $p?H=B \Leftrightarrow p \in B$.

Traditionally, solving the point-in-polyhedron problem means obtaining an answer from the set {inside, outside} or {inside, outside, boundary}. Any traditional polyhedron can be modeled as a polyhedron according to Definition 1 by regarding the interior and the exterior as volume entities, regarding the vertices, edges, and faces as boundary entities, and associating with each entity one of the answers from the set {inside, outside, boundary}. Then, the answer to the traditional form of the point-in-polyhedron problem is easily obtained from $p?H=B$ by a simple table lookup.

A boundary entity B of H is said to be *closest* to a point p (and vice versa) if $dist(p,B)>0$ and for all boundary entities C of H, $dist(p,B)\leq dist(p,C)$. If $dist(p,B)=0$, then there may be other boundary entities C of H with $dist(p,C)=0$. To guarantee that $p?H=B$ $\Leftrightarrow p\in B$, we have to select from the boundary entities with zero distance to p the boundary entity of lowest dimension, which is unique.

In the following sections, the discussion will concentrate on the case where p is closest to B, which implies $dist(p,B)>0$. The case $dist(p,B)=0$ will not be discussed. To show that the closest boundary entity algorithm terminates, a third case has to be considered. If the polyhedron consists of a single volume entity, the answer to $p?H$ is trivial.

2.2 The Sphere

The k-dimensional sphere S^k ($k\geq 0$) is defined in $\mathbf{R}^{k+1}$ as the set of points that have one and the same distance from its center. S^0 is a set of two points on a line, S^1 a circle, and S^2 the familiar sphere in 3D. If we do not want to specify the dimension of the sphere, we simply write S. Intersecting S^k with a $j+1$-dimensional plane of $\mathbf{R}^{k+1}$ passing through the center results in a sphere of dimension j, called a j-subsphere (cf. [Ber80], p. 257). A set of $j+2$ points in $\mathbf{R}^{k+1}$ define a $j+1$-dimensional plane of $\mathbf{R}^{k+1}$. Thus facts involving $j+1$ points in general position on S^k, plus the center as the $(j+2)$th point, can be proved using S^j instead of S^k.

A *line* on S is defined as a great circle, which is a 1-subsphere of S. The *line segment* between two points p and q in general position, denoted pq, is the (open) minor arc of the great circle through p and q. The line segment from p to itself, pp, is p, and the line from p to its antipodal point, denoted $-p$, consists of p and $-p$. These definitions are modeled after [PJ79].

Distance on S is the distance along a line segment and corresponds to the angle from the origin. Although there is no corresponding line segment, $dist(p,-p)=\pi$. The distance $\pi/2$ plays a special role in this paper, as it relates both to the convex hull and to the closestness of a point to a boundary entity of a polyhedron. An *angle* on S is measured in the obvious way.

A *hemisphere* is a part of S^k delimited by a $(k-1)$-subsphere. To every point $p\in S$, there is an open hemisphere $oHem(p) = \{x|x\in S, dist(p,x)<\pi/2\}$, and a closed hemisphere $cHem(p) = \{x|x\in S, dist(p,x)\leq\pi/2\}$. The sphere is partitioned by $oHem(p)$ and $cHem(-p)$, as well as by $cHem(p)$ and $oHem(-p)$. The *antipodal hemispheres* of p are the hemispheres $cHem(-p)$ and $oHem(-p)$.

One of the central concepts in (computational) geometry is convexity. In this paper, we use convexity on the sphere to obtain lower bounds for the classification of boundary entities. A set of points X is *convex* if and only if for any two points $p, q \in X$, the line segment pq is contained in X [PJ79,p.71]. The *convex hull* of a set Y, denoted *conv* Y, is defined as the smallest convex set containing Y. In this form, convexity is easily applied to the sphere.

The following lemma about convex hulls on the sphere will be needed in Section 5.1:

Lemma 3: For any set $X \subset S$ and any point $p \in S$, $conv\ X \subset cHem(-p) \Leftrightarrow X \subset cHem(-p)$.
  Proof: In any case, $X \subset conv\ X$ and so $conv\ X \subset cHem(-p) \Rightarrow X \subset cHem(-p)$. On the other hand, $cHem(-p)$ itself is convex, so that $X \subset cHem(-p) \Rightarrow$
$\Rightarrow conv\ X \subset cHem(-p)$ as the least convex set is contained in all other convex sets containing X.   Q.E.D.

Polyhedra can also be defined on the sphere:

Definition 4: A *spherical k*-dimensional polyhedron is a polyhedron according to Definition 1 where $\Omega(H)$ is S^k, and the subspaces $\Omega(B)$ are the subspheres of S^k.

3. Boundary Features

To solve $p?H$ for a point p closest to a boundary entity B, we have to consider the neighborhood of B while discarding all the parts and aspects of H that are not relevant. We do this by abstracting H in the neighborhood of the boundary entity B in the form of a so-called boundary feature, which is itself a polyhedron. The dimension of the problem is then reduced by mapping the boundary feature to a sphere, resulting in a spherical polyhedron.

  For a boundary entity B of H, the *boundary feature* H_B is constructed as follows: For each open ray R starting at a point $q \in H_B$, there exists an ε such that all points $r \in R$ with $dist(q,r) \leq \varepsilon$ lie in one and the same entity C of H. Then all points on R belong to the entity C_B of H_B. The conditions for a polyhedron in Definition 1 assure that this construction is consistent for all rays. C_B retains the supporting space and the dimensionality of C. The boundary feature H_B is itself a polyhedron (see Fig. 3). The entities of H that have a corresponding entity in H_B are B itself and those entities that are bounded by B.

  The "locally adjoined pyramid" of Nef and Bieri [Nef78, BN88] is basically similar to our boundary feature. The differences are due to the differences in the original polyhedron definition. Expressed in our language, they allow only two volume entities, the interior and the exterior. Also, they use the locally adjoined pyramid to define the faces (boundary entities) of the polyhedron, whereas here, two boundary entities can be different even if their boundary features are identical.

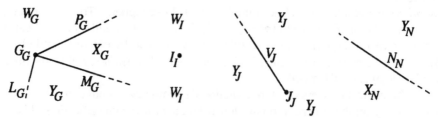

Figure 3. The boundary features H_G, H_I, H_J, and H_N
of the polyhedron of Figure 2.

The following theorem shows the relevance of the boundary feature for the point-in-polyhedron problem:

Theorem 5: For a point p closest to a boundary entity B of H, $p?H=C \Leftrightarrow p?H_B=C_B$.

Proof: That p is closest to B means that there is an open ray R through p starting at the point q of B which is closest to p. All points $r \in R$ with $dist(q,r) \leq \varepsilon \leq 2 \cdot dist(q,r)$ have to lie in one and the same entity C of H, because otherwise p would not be closest to B.   Q.E.D.

Compared with H, H_B exhibits a high degree of regularity and can be simplified as follows: We can consider two points as equivalent if they lie on two open rays used in the construction of H_B and these rays contain a common point. One of the resulting equivalence classes, B_B itself, can be ignored since the case $dist(p,B)=0$ has been treated at the end of Section 2.1. Each of the remaining equivalence classes is an open halfplane of dimension $\dim(B)+1$ bounded by B_B and subset of a single entity C_B of H_B.

The quotient space of this equivalence relation can be represented as a sphere, denoted S_B, with $\dim(S_B)=k-\dim(B)-1$. This sphere not only captures the topology of the equivalence sets, but if centered at a point on B_B, also has a geometric meaning. The representant on S_B will be one of the points of the equivalence class it represents; the other points can be mapped to it by a projection, composed of a projection parallel to B_B onto a (hyper)plane of dimension $k-\dim(B)$ orthogonal to B_B and a radial projection.

The projection maps linear subspaces to subspheres. Therefore H_B is reduced to the spherical polyhedron $H_B°$, each entity C_B of H_B (with the exception of B_B) is reduced to the corresponding entity $C_B°$ of $H_B°$, and the point p is projected to the point $p_B°$ on S_B. Obviously, a point-in-polyhedron problem can also be defined on the sphere, and so the relation of $H_B°$ to $p?H_B$ and $p?H$ is expressed as follows:

Theorem 6: $p?H_B=C_B \Leftrightarrow p_B°?H_B°=C_B°$.

Proof: $H_B°$ is independent of the size and position of S_B (besides it being centered on B_B). Thus we can choose S_B so that $p=p_B°$. In any position of S_B, $C_B° \subset C_B$, and the theorem follows.   Q.E.D.

Theorems 5 and 6 can be combined in the following corollary:

Corollary 7: For a point p closest to a boundary entity B of H, $p?H=C \Leftrightarrow p_B°?H_B°=C_B°$.

The spherical polyhedron $H_B°$ has at least one dimension less than H_B and H, so that Corollary 7 reduces the point-in-polyhedron problem to a lower dimension. An additional simplification is due to the fact that boundary features and their corresponding spherical polyhedra can be classified such that the answer to $p?H$ can be found easily once we know that p is closest to B. This classification is developed in the next section.

4. Classification

4.1 Closestness and the Sphere

To show how boundary features can be classified, we have to investigate the relation between the existence of points closest to a boundary entity B and the form of the corresponding spherical polyhedron $H_B°$. This relation can be expressed with the following theorem:

Theorem 8: For points closest to a boundary entity H, the following holds:
(a) If a point p is closest to the boundary entity B of H, then $bd\ H_B° \subset cHem(-p_B°)$.
(b) If $bd\ H_B°$ is a subset of a closed hemisphere of S_B, then there exist points closest to B.

Proof: (a) The proof works by contradiction. We assume $bd\ H_B° \not\subset cHem(-p_B°)$ and show that then p cannot be closest to B. There will be a point $r_B°$ so that $r_B° \in bd\ H_B°$, or concretely $r_B° \in C_B°$, where $C_B°$ is a boundary entity of $H_B°$, but $r_B° \notin cHem(-p_B°)$, which means $r_B° \in oHem(p_B°)$ and thus $dist(p_B°,r_B°)<\pi$. There will thus always be a point r on the corresponding boundary entity C of H so that $dist(p,C) \leq dist(p,r) < dist(p,B)$ and thus p is not closest to B. The situation is depicted in Figure 4.a; without loss of generality it can be assumed that S_B is positioned so that $p=p_B°$.

(b) This is the (slightly weakened) reversal of (a), and so we can reverse the above argument, using Figure 4.b and 4.c. Let the center of the closed hemisphere that contains $bd\ H_B°$ be called $-p_B°$. There will be no entity C of H that is bounded by B for which $dist(p_B°,B)>dist(p_B°,C)$. But the boundary entities of H bounded by B are the only boundary entities of H that can come arbitrarily close to B. Therefore if there is a point p corresponding to $p_B°$, but which is not closest to B, then we can move this point towards B until it gets closer to B than to D.   Q.E.D.

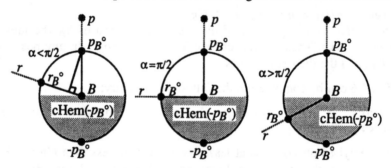

Figure 4. Illustration of Theorem 8.

The following corollary is a stronger version of part (a) of Theorem 8:

Corollary 9: A point p can be closest to a boundary entity B of H, with $p?H=C$, only if $S_B \backslash C_B° \subset cHem(-p_B°)$.

Proof: The difference to Theorem 8 is that $bd\ H_B°$ is replaced by $S_B \backslash C_B°$, i.e. the theorem says that not only the boundary of $H_B°$, but also all volume entities of $H_B°$ besides $C_B°$ are contained in the antipodal hemisphere of $p_B°$. But in $oHem(p_B°)$ there cannot be another volume entity besides $C_B°$, as otherwise some part of $bd\ H_B°$ is necessary to separate it from $C_B°$.   Q.E.D.

4.2 The Classification

We are now ready to develop the classification of the boundary features of H. Primarily, classification is based on the number of entities a point closest to the corresponding boundary entity can possibly lie in. This number is limited by the following theorem:

> **Theorem 10**: Given a boundary entity B, there are at most two entities of H bounded by B that can contain a point p closest to B.
>
> Proof: According to Corollary 9, for a point p to be closest to B and contained in an entity C, p_B° must be surrounded by an open hemisphere completely contained in C_B° (oHem(p_B°)$\subset C_B^\circ$). But at most two disjoint open hemispheres, separated by a $(k-1)$-subsphere, fit on a k-sphere.   Q.E.D.

The three classes of boundary features resulting from Theorem 10 are called FLAT, CONVEXOID and COMPLEX. The following paragraphs explain these classes, the reasons for their names, and the additional information that is part of the classification and that is necessary for each class to allow the fast evaluation of $p?H$.

FLAT: A boundary feature H_B with *two entities* C and D that contain points closest to B is called FLAT because the boundary entities that have to separate C and D all lie on a plane of dimension $k-1$ through B. In the simplest case, B itself is of dimension $k-1$ (a face in 3D), and $S_B=S^0$. Otherwise, some of the boundary entities separating C and D will be of dimension $k-1$. Whether the result of $p?H$ is C or D has to be decided for every p individually by using the orientation of the normal vector of such a boundary entity of dimension $k-1$.

CONVEXOID: A boundary feature H_B with *a single entity* C that contains points closest to B is called CONVEXOID because it is a generalization of the 2D convex vertex case presented in Figure 1 in Section 1.2. Once C has been identified and stored, the result $p?H=C$ is immediate when we know that p is closest to B.

COMPLEX: A boundary feature H_B with *no entity* C that can contain points closest to B (i.e. no points closest to B at all) is called COMPLEX. B bounds many different entities in a complex way so that no entity occupies a full hemisphere on S_B by itself.

In Figure 2 in Section 2.1, the boundary feature H_C is COMPLEX. The boundary features of all other vertices are CONVEXOID, whereas those of all edges are FLAT.

5. Implementing Classification

The classification outlined in Section 4.2 is obviously very valuable when solving $p?H$. This classification can be obtained mainly in two ways. Preclassification fully analyses and classifies each boundary feature of H during preprocessing. Delayed classification only classifies a boundary feature if a point p is found closest to it.

5.1 Preclassification

Theorem 8 and Corollary 9 give enough information to decide the class of each boundary feature. If $bd\ H_B^\circ$ has the form of a $(\dim(S_B)-1)$-subsphere, then H_B is FLAT, if $bd\ H_B^\circ$ fits into a closed hemisphere, then H_B is CONVEXOID, and if neither of this is the case, then H_B is COMPLEX.

Checking for FLATness can be implemented easily, examining the boundary entities of H_B° one by one and stopping as soon as they do not fit into a $(\dim(S_B)-1)$-subsphere any more. On the other hand, the distinction between CONVEXOID and COMPLEX does not directly suggest an implementation. For this, we have to use the convex hull. Part (a) of Theorem 8 can be strengthened as follows:

> **Theorem 11**: If a point p is closest to the boundary entity B of H, then $conv\ bd\ H_B° \subset cHem(-p_B°)$.
>
> Proof: Using Lemma 3. Equally, part (b) of Theorem 8 and Corollary 9 can be strengthened by using the convex hull operator.   Q.E.D.

A convex set on the sphere that does not fit into a closed hemisphere will automatically fill out the whole sphere. Thus to distinguish CONVEXOID and COMPLEX boundary features, we can calculate $conv\ bd\ H_B°$. If it is equal to the whole sphere S_B, then B is COMPLEX, otherwise it is CONVEXOID (or FLAT).

There are many algorithms that compute the convex hull in some dimension of Euclidean space, and most of them can be adapted to compute the convex hull on the sphere with the same dimension. Also, the detection of FLAT boundary entities and the calculation of the additional information necessary for each class can easily be incorporated into such an algorithm.

To give an example of how using the convex hull leads to an efficient classification of boundary entities, we examine the case of a vertex V of a 2-manifold polyhedron in 3D. For this, Horn and Taylor [HT89] introduced the concept of a discrimination plane and obtained a time complexity of $O(m^3)$, where m is the number of edges connected to V (the complexity of $O(m!)$ given in [HT89] is a misprint [DK89]).

In this case, the edges and faces connected to V will form the simple spherical polygon $H_V°$ with m arcs and m vertices on a 2-sphere. The convex hull of $bd\ H_V°$ can be calculated in time $O(m)$ adapting the convex hull algorithm by Melkman [Mel87] to the sphere [DK89]. The algorithm of Melkman has the advantage that it does not need a point on the convex hull to start with; if such a point were known already, then computing the convex hull would not be necessary any more.

The time complexity of $O(m)$ to classify a vertex of a 2-manifold polyhedron in 3D is optimal in the worst case. For non-manifold polyhedra and higher dimensions, it is not clear whether constructing $conv\ bd\ H_B°$ is the optimal way to find the class of H_B, as it can no more be done in linear time. However, even for these cases, using the convex hull an upper bound can be obtained that is much lower than that of a naive approach.

Calculating $conv\ bd\ H_B°$ not only produces the classification of H_B, it also allows to derive the set of locations on S_B that corresponds to points closest to B. This set is a sort of dual of $conv\ bd\ H_B°$. Each of the entities of its boundary corresponds directly to an entity of the boundary of $conv\ bd\ H_B°$, with a distance of $\pi/2$ and with the dimensions of the two corresponding entities adding up to $\dim(S_B)$.

Although the details of the form of the convex hull are not needed in the basic form of the algorithm, it is important when considering incremental or parallel preprocessing. When only part of the boundary entities of $H_B°$ are known, their convex hull is exactly the information that is needed for further processing.

5.2 Delayed Classification

Preclassifying all boundary features without knowing whether this classification will ever be used is obviously not very efficient. Instead, it is possible to delay classification until there is an actual point p closest to the corresponding boundary entity. Nevertheless, if a second point is found closest to the same boundary entity, the classification already obtained can be used.

Delayed classification proceeds as follows: We start with all boundary features of H classified as UNKNOWN. A COMPLEX boundary feature will never have a point closest to it, so that it will stay UNKNOWN. If a point p is found to be closest to a boundary entity B, then a spherical version of the closest boundary entity algorithm is applied to $H_B°$ to find $C_B°=p_B°?H_B°$. This means that we have to find the boundary entity $D_B°$ of $H_B°$ that is closest to $p_B°$ on S_B. Because we already know that p is closest to B, the following theorem simplifies finding $D_B°$:

Theorem 12: For any point p closest to a boundary entity B of H, either $bd\ H_B°=\{\}$, or there is a boundary entity $D_B°$ of $H_B°$ closest to $p_B°$ with $\dim(D_B°)=0$, and thus $\dim(D)=\dim(B)-1$, or there is a boundary entity $E_B°$ closest to $p_B°$ that is a subsphere of S_B, and $\text{dist}(p_B°,E_B°)=\pi/2$.

Proof: On a sphere, for any closed line segment connecting two points in $c\text{Hem}(-p_B°)$, one of its endpoints will be at least as close to $p_B°$ as any other points in the segment. As only three points are involved, the reader can check this on S^2, and we omit a more detailed proof.

Now, according to Theorem 8, if $bd\ H_B°\neq\{\}$, any points of $bd\ H_B°$ are in $c\text{Hem}(-p°)$. Assume a point r of a boundary entity $E_B°$ of $H_B°$ with $\dim(E_B°)>0$, where $E_B°$ is not a subsphere of $H_B°$. This point r is always contained in a line segment $\subset bd\ H_B°$ whose endpoints are in boundary entities of $H_B°$ with dimension $<\dim(E_B°)$. One of these endpoints will be closer to $p_B°$ than r. By induction, there is a $D_B°$ with $\dim(D_B°)=0$ so that $\text{dist}(p_B°,D_B°)\leq\text{dist}(p_B°,r)$, or there will be a boundary entity $E_B°$ that is a subsphere of S_B. If there is a boundary entity $E_B°$ that is a subsphere of S_B, then we have $\text{dist}(p_B°,E_B°)=\pi/2$ because subspheres always contain antipodal points. Otherwise, of all $D_B°$ with $\dim(D_B°)=0$ there will always be at least one closest to $p_B°$.   Q.E.D.

Obviously, to find the boundary entity of $H_B°$ that is closest to $p_B°$, only the boundary entities of dimension 0 and those that are subspheres of S_B have to be considered. If all these boundary entities have a distance of $\pi/2$ to $p_B°$, then there exists a possibility that $H_B°$ is FLAT. This can be checked immediately, or the check can be delayed until a second point closest to B is found. If there is a boundary entity $D_B°$ of $H_B°$ with $\dim(D_B°)=0$ and $\text{dist}(p_B°,D_B°)>\pi/2$, then H_B is definitively CONVEX.

To find $C_B°=p_B°?H_B°$, we have to define and classify the boundary feature $(H_B°)_{D_B°}$. It is not possible to define this boundary feature directly on S_B. Rays starting on $D_B°$ would meet on the 'back side' of S_B. It is however possible to define a boundary feature for $D_B°$ by extending the neighborhood of a point of $D_B°$ on a (hyper)plane tangential to S_B in this point. If we choose the point of $D_B°$ closest to $p_B°$, $p_B°$ can also be mapped to this plane in a natural way. We can then again map $p_B°$ and $(H_B°)_{D_B°}$ to the sphere $S_{D_B°}$, with $\dim(S_{D_B°})\leq\dim(S_B)-1$. Continuing this recursion, we finally arrive at a sphere of dimension 0, with the corresponding boundary entity being FLAT, and so obtain the necessary information to classify H_B and to answer $p?H$.

5.3 Comparing Preclassification and Delayed Classification

Comparing preclassification and direct classification, it turns out that there is only one case where preclassification performs better: The on-line case, where the polyhedron is known in advance and can be fully preprocessed, but the query points arrive one by one and have to be answered as fast as possible.

In all other cases, delayed classification is faster, and this for several reasons. First, the classification of each boundary feature is faster, because the point closest to the corresponding boundary entity serves to focus calculations. Second, there is never any time spent to classify COMPLEX boundary features. Third, only a few of the boundary features are classified if the number of query points is small or they are locally concentrated. Forth, no time is spent preprocessing the polyhedron, so that the answer for the first point query is obtained faster.

6. Conclusions and Future Work

This paper has analyzed the geometric aspects of the closest boundary entity algorithm for the point-in-polyhedron problem. This was done by mapping the neighborhood of a boundary entity to a sphere. The very general polyhedron definition showed the general applicability of the algorithm. Indeed, a method using a semi-infinite ray or the Gauss-Bonnet integral fails already for non-bounded polyhedra.

Compared with the description of the algorithm given by Horn and Taylor, the following improvements have been made: First, the algorithm was extended to arbitrary dimensions and a much more general class of polyhedra. Second, the time to classify a 2-manifold vertex in 3D was reduced from $O(m^3)$ to $O(m)$. Third, boundary entities of any dimension are all treated in a uniform way; this will simplify implementation. Forth, the introduction of delayed classification made the algorithm considerably more efficient.

The numeric aspects of the algorithm, especially the calculation of the distances, are currently under study. Also, the exact data structures that are necessary during the algorithm require some additional consideration. The optimality of the algorithm for preclassification and the extension of the closest boundary entity approach to polyhedra in R^k with nonplanar boundaries at present remain open problems. Also, the relation of the boundary feature classification to the problem of converting boundary representations to CSG, solved only for 2D [DGHS88], and the variant of this problem for spherical polyhedra, even in 2D, are of deep interest.

Acknowledgements

I am grateful to acknowledge the support of Prof. T. L. Kunii of the University of Tokyo and Prof. P. Stucki of the University of Zürich, and the interesting discussions with Prof. W. Nef, Prof. H. Imai, and many of my friends from the University of Tokyo and the University of Zürich.

References

[Ber80]   Berger, M. *Geometry II*, Springer-Verlag, Berlin, 1987.

[BN88]   Bieri, H., and Nef, W. Elementary set operations with d-dimensional polyhedra. *Proc. 4th Workshop for Computational Geometry*, March 1988.

[CR91]   Crocker, Gary A., and Reinke, William A. An editable nonmanifold boundary representation. *IEEE Comp. Graph. and Appl.* **11**, 2 (March 1991), 39-50.

[DGHS88] Dobkin, D., Guibas, L., Hershberger, J., and Snoeyink, J. An efficient algorithm for finding the CSG representation of a simple polygon. *Computer Graphics (Proc. SIGGRAPH)* **22**, 4 (1988), 31-40.

[DK89]   Dürst, Martin J., and Kunii, Tosiyasu L. Vertex classification using the convex hull on a sphere. *Proc. Intern. Workshop on Discrete Algorithms and Complexity*, H. Imai, Ed. IPSJ, Tokyo, 1989, pp. 25-32.

[GRS83]   Guibas, L., Ramshaw, L., and Stolfi, J. A kinetic framework for computational geometry. *Proc. 24th Annual IEEE Symposium on the Foundations of Computer Science.* IEEE, 1983, pp. 100-111.

[HT89]   Horn, William P., and Taylor, Dean L. A theorem to determine the spatial containement of a point in a planar polyhedron. *Comput. Vision Graphics Image Process.* **45** (1989), 106-116.

[Kal82]   Kalay, Y. E. Determining the spatial containment of a point in general polyhedra. *Comput. Graphics Image Process.* **19** (1982), 303-334.

[LMR84]   Lane, J., Magedson, B., and Rarick, M. An efficient point in polyhedron algorithm. *Comput. Vision Graphics Image Process.* **26** (1984), 118-125.

[Mel87]   Melkman, A.A. On-line construction of the convex hull of a simple polyline. *Information Processing Letters* **25** (April 1987), 11-12.

[Nef78]   Nef, W. *Beiträge zur Theorie der Polyeder, mit Anwendungen in der Computergraphik.* Herbert Lang, Berne, Switzerland, 1979.

[NR67]   Nordbeck, S., and Rystedt, B. Computer cartography point-in-polygon programs. *BIT* **7** (1967), 39-64.

[PJ79]   Prenowitz, W., and Jantosciak, J. *Join Geometries: A Theory of Convex Sets and Linear Geometry.* Springer-Verlag, New York, 1979.

[PS86]   Putnam, L.K., and Subrahmanyam, P.A. Boolean operations on n-dimensional objects. *IEEE Comp. Graph. and Appl.* **6**, 6 (June 1986), 43-51.

[Req80]   Requicha, Aristides A. Representations for rigid solids: theory, methods, and systems. *Comp. Surv.* **12**, 4 (Dec. 1980), 437-464.

[Sam90]   Samet, Hanan. *The Design and Analysis of Spatial Data Structures.* Addison-Wesley, Reading, 1990.

[TKM84]   Tamminen, Markku, Karonen, Olli, and Mäntylä, Martti. Ray-casting and block model conversion using a spatial index. *Computer-Aided Design* **16**, 4 (July 1984), 203-208.

On the morphology of polytopes in $\mathbf{R}^d$

Peter Engel

Department of crystallography, University of Bern,
Freiestrasse 3, 3012 Bern, Switzerland

Abstract

An algorithm to enumerate the combinatorial types of d–spheres is described. A purely geometric condition is given to determine the non–polytopal spheres. New results are presented for d=4 and 5.

1. Introduction

The characterization of the facial structure of polytopes has a long tradition in combinatorial geometry. Its beginnings go back to Euler [13]. His well–known theorem gives the relation between the numbers n_0, n_1, and n_2 of vertices, edges, and facets of a 3–dimensional convex polytope:

$$n_0 - n_1 + n_2 = 2.$$

The facial structure of 3–polytopes is completely characterized by Steinitz' [25] fundamental theorem that *every 2–sphere can be realized as a 3–polytope*. Reformulated in the language of graph theory, Steinitz' theorem asserts that *every 3–connected planar graph with n_0 nodes and n_1 edges is isomorphic to the edge graph of a 3–polytope.*

The investigation of higher dimensional polytopes started in 1852 with Schläfli's [23] generalization of Euler's theorem to arbitrary dimensions:

$$\sum_{i=0}^{d-1}(-1)^i n_i = 1 - (-1)^d.$$

The analogous higher dimensional Steinitz problem has found great interest in recent years. According to Bokowski and Sturmfels [7] it seems likely that in higher dimensions, d≥4, there exists no intrinsic characterization for d–polytopes. As a partial result a theorem due to Mani [20] for simplicial spheres, and, in general, to Kleinschmidt [19] states that *every (d–1)–sphere with at most d+3 vertices is polytopal.* However, correcting an earlier result of Brückner [8], Grünbaum and Sreedharan [16] proved that *there exists a 3–sphere with eight vertices that is not isomorphous to the boundary complex of any 4–polytope.* In general, it holds that *for each dimension d≥4 there exist non–polytopal (d–1)–spheres with equal or more than d+4 vertices.*

Our particular interest is concerned with the enumeration of (d–1)–spheres and to decide their polytopality. A systematic investigation of the combinatorial types of 3–polytopes was started by Euler who determined the respective numbers 1, and 2 of combinatorial types having four, and five facets. The corresponding Schlegel diagrams are shown in Fig.1. In 1829 Steiner [24] stated the number 7 of combinatorial types having six facets and he asked for the rate of increase of the number of combinatorial types with respect to the number of facets. Further results on 3–polytopes, summarized in Table 1, were obtained by Hermes [17], Duivestijn and Federico [9], and Engel [10].

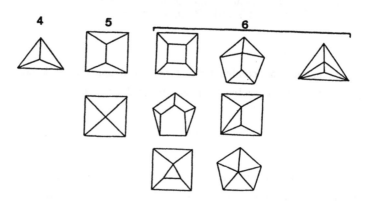

Fig. 1 Schlegel diagrams for the combinatorial types of 3–polytopes having 4, 5, and 6 facets

The enumeration of higher dimensional polytopes was started by Brückner [8] in 1905. He derived the simple 4–polytopes with eight facets without, however, realizing that one of his constructions corresponds to a non–polytopal 3–sphere. A complete enumeration of the simplicial 4–polytopes with eight vertices was performed by Grünbaum and Sreedharan [16]. These authors also proved the existence of a non–polytopal 3–sphere which gave a great impetus to the study of the analogue of Steinitz' theorem in higher dimensions. Altshuler and coworkers, in a series of papers [2,3], enumerated the general 3–spheres with eight vertices and the simplicial ones with nine vertices and decided their polytopality. Engel [12] determined the general spheres with nine facets and gave lower bounds for the numbers of non–polytopal ones. Perles [22] enumerated the combinatorial types of d–polytopes with up and including d+3 vertices using Gale diagrams for d=4, 5, and 6. Goodman and Pollack [14] gave an upper bound $c(d,n_0)$ for the number of combinatorial types of simplicial d–polytopes and Kalai [18] gave a lower bound $s(d,n_0)$ for the number of combinatorial types of simplicial (d–1)–spheres. Combining these bounds Kalai obtained, for every d≥5, *that* $\lim_{n_0 \to \infty} (c(d,n_0)/s(d,n_0))=0$.

We have developed an algorithm to derive the (d–1)–spheres from the d–simplex. The main processes used are the halfspace–intersection, the edge–reduction, and combinations thereof. In order to classify the spheres into combinatorial types, we determine for each sphere the unified polytope–scheme. All spheres of the same combinatorial type have an identical unified polytope–scheme.

2. Derivation of polytopes through halfspace–intersections

For the proposed algorithm it is most convenient to consider a convex d–polytope $P_n \subset E^d$ having n facets as the intersection of n closed halfspaces H_i,

$$P_n := \bigcap_{i=1}^{n} H_i .$$

Equivalently, in dual notation, a d–polytope is the convex hull of a finite set of points in E^d. Results on the facial structure of polytopes can be directly transformed to the dual notation and vice versa. The 0–, 1–, and (d–1)–faces of a d–polytope are called the vertices, edges, and facets of the polytope. Among the faces of a polytope exists a partial order with respect to inclusion which is revealed by its hierarchical structure as shown in Fig. 2. By $\mathcal{L}(P)$ we denote

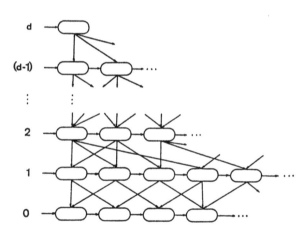

Fig. 2. The hierarchical structure of a d–polytope

the face–lattice given by the hierarchical structure of P. Two polytopes are isomorphic and belong to the same combinatorial type if they have isomorphous face–lattices. We obtain a polytope P_n from P_{n-1} by its intersection with a further halfspace H_n which cuts off certain vertices v_j of P_{n-1}. We denote by v^+, v^0, v^- a vertex which lies inside, on the boundary, or outside of H_n respectively. By identifying the vertices of P_{n-1} by +, 0, or −, we can calculate the transformation $\mathcal{L}(P_{n-1}) \to \mathcal{L}(P_n)$ using Nef's [21] halfspace–intersection algorithm. The operations which have to be performed on the higer level elements of the hierarchical structure depend only on the operations performed on the lower level elements (for details see Engel [11]). It follows that the combinatorial type of a sphere S_n is completely determined if we identify the vertices of S_{n-1} in a consistent way by +, 0, or − and then make the transformation $\mathcal{L}(S_{n-1}) \to \mathcal{L}(S_n)$. We call such a transformation a cut. We have to be aware that not every possible cut can be realized as a halfspace–intersection. The realization strongly depends on the actual shape of P_{n-1}. However, for 3–polytopes Steinitz' theorem asserts that the shape of P_{n-1} can always be chosen such that any cut can be realized. Barnette and Grünbaum [4] proved that

the shape of any facet of a 3–polytope can be preassigned.

This remains no longer true in dimensions $d \geq 4$ as was shown by Barnette [5].

Of particular interest are those cuts which can always be realized by a halfspace–intersection, independently on the particular shape of P_{n-1}. We call them free cuts, and the resulting P_n we call a freely generated polytope. They will allow us to separate closely the set of polytopal spheres from the non–polytopal spheres. We conject that

almost all polytopes in $\mathbb{R}^d$ are freely generated.

It is easy to see that the cutting off of any part of a 2–face, of a k–simplex, $k=0,\cdots, d$, or of a complete h–face, $h=0,\cdots, d-1$ are free cuts. We denote by $\mathcal{V}^-$, $\mathcal{V}^0$, and $\mathcal{V}^+$ the sets of vertices identified by −, 0, or + respectively. We say that a cut γ operates on a k–face F^k if the sets $\mathcal{V}^-$ and $\mathcal{V}^0$ belong to F^k but γ does not operate on some $F^h \subset F^k$. For any $F^l \supseteq F^k$ we call $\bar\gamma$ the complement cut in F^l which is defined by interchanging + and − for all the vertices of F^l. Clearly, with γ beeing free, also the complement cut $\bar\gamma$ in F^l is free. For simple spheres we observed that if $l=d-1$ then both cuts γ and $\bar\gamma$ result in spheres of the same combinatorial type.

3. Derivation of non–simple spheres by edge–reduction

Edge–reductions are only considered for dimensions $d \geq 3$. Given all simple spheres S_n with n facets, we can derive in a systematic way the non–simple spheres with n facets by the process of edge–reduction. An edge E_{ij} of S_n is shrunk until both adjacent vertices v_i and v_j coincide. Thereby, the number of edges is reduced by one. The operation of reduction corresponds to a transformation of the face–lattice into the derived face–lattice, $\mathcal{L}(S_n) \to \mathcal{L}(S_n')$. We require that an edge–reduction does not diminish the number of facets, thus putting a well–defined end to the reduction process.

We say that the reduction of an edge E_{ij} results in a bridge if there exist two faces which meet in $v_i \equiv v_j$ whose intersection is not a face of $\mathcal{L}(S_n')$. By reducing further edges, in general, we can remove the bridge. We denote this a combined reduction. We note that after a edge–reduction, we cannot make any statement about the polytopality of S_n'.

Because of the condition that the number of facets has to be conserved during the process of edge–reduction, it follows that for a $(d-1)$–sphere S_n at most a $(d-2)$–face can completely be removed by edge–reductions. Thus, with increasing d, the possibilities for combined edge–reductions becomes more and more complicated.

For 2–spheres we have to consider only single edge–reductions. All cases where bridges occur are discarded. For 3–spheres a combined reduction may occur if the edge to be reduced belongs to a trigonal 2–face $F^2 := \{v_1, v_2, v_3\}$. If one edge is reduced, say E_{12}, then F^2 collapses into E_{13}. Thereby, two edges disappear and the valence of v_3 is decreased by one. In the case that the valence of v_3 becomes less than four, a bridge occurs and we have also to reduce the edge E_{13}. For 4–spheres, at most a 3–face may be removed and new, non–trivial cases occur. Two

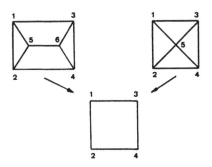

Fig. 3. Possible combined edge–reductions for 3–faces

examples are represented in Fig. 3. In Fig. 3a), the Schlegel diagram of the trigonal prism is shown. By a combined reduction of the edges E_{15} and E_{36} the prism collapses into the quadrilateral $\{v_1, v_2, v_3, v_4\}$. Thereby, five edges are removed and the valences of v_3 and v_4 are decreased by one. This would require that the valences of v_3 and v_4 are greater than five. In Fig. 3b) is shown the Schlegel diagram of the four–sided pyramid. By the reduction of the edge E_{15}, the pyramide collapses to the quadrilateral $\{v_1, v_2, v_3, v_4\}$. Thereby, four edges are removed and the valences of v_2, v_3, and v_4 are decreased by one. Again this would require that the valences of these vertices are greater than five.

4. Characterization of the combinatorial type

The algorithm to enumerate the combinatorial types of (d–1)–spheres requires a unique characterization of the face–lattice $\mathcal{L}(S_n)$. We call a subseries of successively subordinated faces of $\mathcal{L}(S_n)$, $F^0 \subset F^1 \subset \cdots \subset F^d \equiv S_n$, a d–flag. Given an arbitrary d–flag we can number all faces of $\mathcal{L}(S_n)$ in a unique way. For each level we use a separate numbering scheme. The numbering starts at vertex $v_1 := F^0$. Next, we ascend the d–flag to F^1 and descend to the vertex v_2 adjacent to v_1 relative to F^1. Thereby, the d–flag is changed to $F^0 := v_2$. In general, if all faces subordinated to F^k are numbered through, we assign a successive number to F^k and ascend along the d–flag to F^{k+1}. We now descend in the face–lattice along a path of relative minimal adjacency until we reach a vertex, or a h–face whose subordinated faces all are numbered. One step, descending from level k+1 to level k, is as follows: among all faces F_i^k not yet numbered and subordinated to F^{k+1} we find a face F_m^k which has subordinated itself a face F^{k-1} with relative minimal number. Thereby we change the d–flag to $F^k := F_m^k$. Thus, ascending and descending within the face–lattice, we can number through all faces. Next we set up the polytope–scheme by writing down for each facet the numbers of its vertices in increasing order, separating the numbers of each facets by a slash as is shown in Fig. 4. A sequence of at least three successive numbers from n_1 to n_2 we write, by abbreviation, as n_1–n_2. This polytope–scheme depends only on the arbitrarily chosen initial d–flag. We set up the polytope–schemes for each possible d–flag and put them into classes of equal polytope–schemes which become lexicographically ordered. A representative of the first class we call the unified polytope–scheme and the number of equal schemes in this class is equal to the order of the combinatorial automorphism group of the face–lattice.

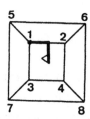

1–4/1,2,5,6/1,3,5,7/2,4,6,8/3,4,7,8/5–8

Fig. 4. The Schlegel diagram of a 3–polytope and its associated unified polytope–scheme

The unified polytope–scheme gives a unique characterization of the face–lattice of $\mathcal{L}(S_n)$, and hence, of its combinatorial type. Vice versa, from the polytope–scheme we can always rege–nerate the complete face lattice $\mathcal{L}(S_n)$.

5. Determination of the non–polytopal spheres

In what follows, we only consider $(d-1)$–spheres with $d\geq4$. Among the cuts which are not free we have to find those which cannot be realized through a halfspace–intersection. We say that a set of three 2–faces $\{F_1^2, F_2^2, F_3^2\} \subset F^3$ which share no common vertex form a trihedron if there exists a trigonal section through the three 2–faces. From elementary geometry it is known that the carrier planes of the three 2–faces which form a trihedron intersect in a common point (which may be at infinity). For a 4–face F^4 it holds that

if there exist an edge $E_{ij} \subset F^4$ which is common to two trihedra of F^4 and a consistent cut which involves the edge E_{ij} and two 2–faces which meet in E_{ij}, then there exists such a cut which is not realizable through a halfspace–intersection.

As an example consider a 4–polytope of combinatorial type 7.14–1. The Schlegel diagrams of

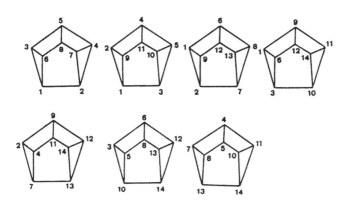

1–8/1–5,9–11/1,2,6–9,12,13/1,3,6,9–12,14/2,4,7,9,11–14/3,5,6,8,10,12–14/4,5,7,8,10,11,13,14

Fig. 5. The facial structure and the unified polytope–scheme of the combinatorial type 7.14–1

its facets, and its unified polytope–scheme are shown in Fig. 5. We consider the following three 2–faces $F_1^2 := \{3,5,6,8\}$, $F_2^2 := \{4,5,7,8\}$, and $F_3^2 := \{5,8,10,13,14\}$ which meet in a common edge E_{58}. The 2–faces F_1^2 and F_3^2 belong to a trihedron having edges E_{58}, E_{36}, and E_{1014}. The straight lines defined by these edges intersect in a common point Q, as it is shown in Fig. 6. Similarly, F_2^2 and F_3^2 belong to a second trihedron having edges E_{58}, E_{47}, and E_{1314}, with the corresponding straight lines intersecting in the common point R. Because of the convexity of F_3^2 the point R lies outside of the closed intervall $[Q,v_8]$, either beyond v_8, or else beyond Q. Now, assume we want to perform the cut γ defined by the set $\mathcal{V}^- := \{v_4^-, v_5^-, v_6^-, v_8^-\}$. It implies that the trace $e_1 := H \cap aff(F_1^2)$ intersects the edge E_{36} and intersects $aff(E_{58})$ in P. Thus the point P is within the open intervall (Q,v_5). The trace $e_2 := H \cap aff(F_2^2)$ meets P but avoids v_4. Since the point R is outside of the closed intervall $[Q,v_8]$, it follows that the halfspace H necessarily cuts off also the vertex v_7. Hence, the cut γ cannot be realized through a halfspace–intersection. Clearly, if $\mathcal{V}^-$ defines a non–realizable cut γ which operates on a certain 3–face F^3 then also the

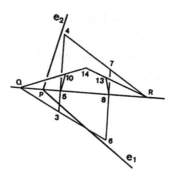

Fig. 6. A non–realizable cut

complement cut $\bar{\gamma}$ in any 1–face $F^1 \supseteq F^k$ determines a non–realizable cut. On the other hand, if S_n is polytopal and the cut γ is non–realizable then necessarily the counter cut $\hat{\gamma}$ defined by the set of vertices $\hat{\mathcal{V}} := \{v_3^-, v_5^-, v_7^-, v_8^-\}$ is always realizable. Some of the counter cuts $\hat{\gamma}$ provide the polytopes which are not freely generated.

The combination of edge–reduction followed by cutting off complete facets has proved to be very efficient to derive the non–polytopal spheres. Only facets which contain at least one vertex with valence greater than d have to by cutt off. Thereby, the number of facets is preserved. We use this combination also to decide the polytopality for most of the non–simple spheres. Given a derived sphere $S_n{}'$, we can cut off a complete facet. The sphere S_n thus generated has the same number of facets but, in general, is of a different combinatorial type. Now, if there exists a generated sphere S_n which is non–polytopal, then necessarily $S_n{}'$ is also non–polytopal. However, there exist non–polytopal spheres $S_n{}'$ which are assotiated with no generated sphere S_n. We call them initial spheres. Clearly, the simple spheres are initial, but there exist others.

For dimensions $d \geq 5$ we can find non–polytopal spheres by inspection of their faces. From the bservation that we had to use only cuts which operate on 3–faces (together with the corresponding complement cuts) we conject that

most of the non–polytopal (d–1)–spheres with d≥5 have a non–polytopal 4–face.

. Results

We have written a computer program POLYTOPE–C in C programing language. In [12] we had calculated the 3–spheres with up to nine facets. The corresponding numbers of combinatorial types are given in Table 2. Using these results we calculated the simple 3–spheres with ten facets. The freely generated polytopes were readily obtained from the free cuts. The complete list of simple 3–spheres with ten facets was obtained by edge–reduction followed by cutting off complete facets. The numbers of combinatorial types of simple 3–spheres with up to en facets are stated in Table 3. Upper bounds for the numbers of non–polytopal spheres are btained by taking the difference of the numbers of spheres and the numbers of freely generated polytopes. A sphere S_n is called neighborly, if the intersection of each $\lfloor d/2 \rfloor$ facets is a face of t. Our number of neighborly spheres –they have 35 vertices– agrees with the number stated by Altshuler [1]. By results of Altshuler [1] and Bokowski and Sturmfels [5] it is known that there exist 431 combinatorial types of neighborly polytopes with ten vertices. Among those, 333 are reely generated.

Table 1. Numbers of combinatorial types of 3–polytopes with n_2 facets and n_0 vertices

n_2 / n_0	4	5	6	7	8	9	10	11
4	1							
5		1	1					
6		1	2	2	2			
7			2	8	11	8	5	
8			2	11	42	74	76	38
9				8	74	296	633	768
10				5	76	633	2635	6134
11					38	768	6134	25626
12					14	558	8822	64439
13						219	7916	104213
14						50	4442	112082
15							1404	79773
16							233	36528
17								9714
18								1249
Total	1	2	7	34	257	2606	32300	440564

We also calculated the combinatorial types of simple 4–spheres with up to nine facets. The corresponding numbers are given in Table 4. Those with six, seven, and eight facets are polytopal and were determined by Perles [22]. There exist exactly 337 combinatorial types of simple 4–spheres with nine facets. At least 15 types are non–polytopal. They are easily recognized by having at least one non–polytopal 4–face. Four more cases still are doubtful.

Table 2. Numbers of combinatorial types of 3–spheres with n_3 facets and n_0 vertices. The numbers of non–polytopal 3–spheres are given within parentheses

n_0 \ n_3	5	6	7	8	9
5	1				
6		1	1	1	1
7		1	3	5	7
8		1	5	27	76
9		1	7	76	467(≥1)
10			6	138(1)	1908(≥6)
11			4	209(4)	5411(≥57)
12			3	231(6)	11974(≥260)
13			1	226(8)	21129(≥778)
14			1	173(7)	31234(≥1706)
15				122(5)	39875(≥3046)
16				70(5)	44461(≥4488)
17				33(2)	43870(≥5529)
18				16(2)	38493(≥5836)
19				5(1)	30216(≥5408)
20				4(1)	21089(≥4313)
21					13231(≥3154)
22					7181(≥1872)
23					3604(≥1133)
24					1390(≥444)
25					567(236)
26					121(45)
27					50(27)
Total	1	4	31	1336(42)	316355(>38339)

Table 3. Numbers of combinatorial types of simple 3–spheres with n_0 vertices and n_3 facets. The numbers of non–polytopal spheres are given within parentheses

n_3	5	6	7	8	9	10
n_0						
5	1					
6						
7						
8		1				
9		1				
10						
11			1			
12			2			
13			1			
14			1	3		
15				5		
16				8		
17				8	7	
18				6	23	
19				5(1)	45	
20				4(1)	84	30
21					128	124
22					175(3)	385
23					223(11)	952
24					231(22)	2142
25					209(46)	4340(≤28)
26					121(45)	8106(≤152)
27					50(27)	13853(≤583)
28						21702(≤1863)
29						30526(≤4565)
30						38552(≤9337)
31						42498(≤16475)
32						39299(≤22207)
33						28087(≤20689)
34						13542(≤11353)
35						3540(3109)
Total	1	2	5	39(2)	1296(154)	247678(<90361)

Table 4. Numbers of combinatorial types of simple 4–spheres with n_0 vertices and n_4 facets. The numbers of non–polytopal spheres are given within parentheses

n_0 \ n_4	6	7	8	9
6	1			
8				
10		1		
12		1		
14			1	
16			2	
18			3	3
20			2	7
22				15
24				34
26				50
28				89(2)
30				139(≤17)
Total	1	2	8	337(≤19)

References

[1] Altshuler, A., Neighborly 4–polytopes and neighborly combinatorial 3–manifolds with ten vertices. Canadian J. Math. 29 (1977) 400–420

[2] Altshuler, A., Bokowski, J., Steinberg, L.,The classification of simplicial 3–spheres with nine vertices into polytopes and non–polytopes. Discrete Math. 31 (1980), 115–124

[3] Altshuler, A., Steinberg, L.,The complete enumeration of the 4–polytopes and 3–spheres with eight vertices. Pacific J. of Math. 117 (1985) 1–16

[4] Barnette, D., Grünbaum, B., Preassigning the shape of a face. Pacific J. Math. 32 (1970) 299–306

[5] Barnette, D., The triangulations of the 3–sphere with up to 8 vertices. J. Comb. Theory (A) 14 (1975) 37–52

[6] Bokowski, J., Shemer, I., Neighborly 6–polytopes with 10 vertices. Israel J. Math. 58 (1987) 103–124

[7] Bokowski, J., Sturmfels, B., Polytopal and nonpolytopal spheres an algorithmic approach. Israel J. Math. 57 (1987) 257–271

[8] Brückner, M., Ueber die Ableitung der allgemeinen Polytope und die nach Isomorphismus verschiedenen Typen der allgemeinen Achtzelle (Oktatope). Verh. Kon. Akad. v. Wetensch., Amsterdam, 1. Sec. 10 (1905) 1–27

[9] Duivestijn, A.J.W., Federico, P.D.J., The number of polyhedral (3–connected planar) graphs. Math. of Computation 37 (1981) 523–532

[10] Engel, P., On the enumeration of polyhedra. Discrete Math. 41 (1982) 215–218

[11] Engel, P., Geometric Crystallography. D. Reidel publishing company, Dordrecht (1986)

[12] Engel, P., The enumeration of four–dimensional polytopes. Discrete Math. 90 (1991) in press

[13] Euler, L., Demonstratio nonnullarum insignium proprietatum, quibus solida hedris planis inclusa sunt praedita. Novi comm. acad. sci. Petropolitanae 4 (1752/3) 140–160; Opera omnia, series prima Vol. 26, p. 94–108, Orell Füssli, Zürich (1953)

[14] Goodman, J.E., Pollack, R., Upper bounds for configurations and polytopes in R^d. Discrete Comput. Geom., 1 (1986) 219–227

[15] Grünbaum, B., Convex polytopes. Wiley–Interscience, London (1967)

[16] Grünbaum, B., Sreedharan, V.P., An enumeration of simplicial 4–polytopes with 8 vertices. J. Comb. Theory 2 (1967) 437–465

[17] Hermes, O., Die Formen der Vielflache. J. reine angew. Math. 120 (1899) 305–353

[18] Kalai, G., Many triangulated spheres. Discrete Comput. Geom., 3 (1988) 1–14

[19] Kleinschmidt, P., Sphären mit wenigen Ecken. Geom. Dedicata 5 (1976) 307–320

[20] Mani, P., Spheres with few vertices. J. Comb. Theory A13 (1972) 346–352

[21] Nef, W., Beiträge zur Theorie der Polyeder mit Anwendungen in der Computergraphik. H. Lang, Bern (1978)

[22] Perles, M.A., (results stated in Grünbaum [14], p. 424)

[23] Schläfli, J., Theorie der vielfachen Kontinuität. (1852, unpublished); Gesammelte mathematische Abhandlungen Vol. 1, p. 169–287, Verlag Birkhäuser, Basel (1950)

[24] Steiner, J., Théorèmes à démontrer et problèmes à résoudre. Ann. de Math. 28 (1829) 302–304. Gesammelte Werke Bd I, p. 227. Reimer, Berlin (1881)

[25] Steinitz, E., Rademacher, H., Vorlesung über die Theorie der Polyeder. Springer–Verlag, Berlin (1934); Reprint: Springer–Verlag, Berlin, Heidelberg, New York (1976)

Robustness in Geometric Modeling
– Tolerance-Based Methods

Shiaofen Fang and Beat Brüderlin

Computer Science Department
University of Utah
Salt Lake City, UT 84112

Abstract: Two tolerance-based methods are presented: the *linear model method* and the *curved model method*, both of which make geometric algorithms robust by testing for ambiguous situations and correcting them. The linear model method only applies to linear objects. It faithfully preserves the original meaning of the problem but may detect too many ambiguous situations and fail. The curved model method can be used for both linear and curved objects and creates fewer ambiguities, but it does not necessarily preserve all the properties of linear objects because it uses a curved model to approximate linear object. Both methods are implemented and applied for 3D Boolean operations on polyhedra.

1 Introduction

Geometric modeling algorithms often fail due to inconsistent decisions based on tolerances. The robustness of geometric operations is considered a key factor in successful application of geometric modeling and computational geometry in areas such as CAD/CAM, robotics and computer animation[7][1][10].

Errors in geometric operations can occur for the following reasons:

- Inaccurate input data: When data are from scientific experiments, measurements, or generated by of other inaccurate computations, they may have errors.

- Floating-point arithmetic errors: The floating-point representation of real numbers in a computer is only an approximation. The cumulative effect of round-off errors may cause disastrous results.

- Approximation errors: An accurate treatment of computations of curved objects is sometimes too complicated or impossible. Usually, an approximation of a curved object by a piecewise linear object is used to obtain an approximated solution. This approximation may lead to inconsistencies.

In geometric algorithms, decisions are based on the result of numerical computations of geometric relations between objects. When the numerical data and their computation are inaccurate, tolerances have to be introduced. Unfortunately, the direct utilization of tolerances may result in inconsistent decisions, ambiguous representations of geometric objects and even failure of an algorithm. For example (in figure 1) we want to determine the incidence relationships among three points P_1, P_2 and P_3 with circular regions as their tolerances. First we determine that $P_1 \neq P_3$

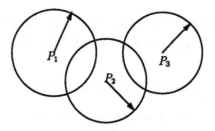

Figure 1: Comparing Three Points

because their tolerances do not intersect. Then we find $P_1 = P_2$ and $P_2 = P_3$ because of their intersecting tolerance regions. This leads to the conclusion $P_1 = P_3$ which contradicts the first decision. More examples can be found in [7] regarding topological invalidity as a result of inconsistent decisions.

Various approaches have been developed for geometric robustness[12][5][8][9][4][13][11][6][7]. Brüderlin[2] presented a method of detecting ambiguities in geometric relations (e.g coincidence and incidence) of linear geometric entities using tolerances for robust of geometric computation. Its application on Boolean operations of 3D objects and a brief description of the approach can be found in [3].

Inspired by [2] and [3], we developed two new tolerance-based robustness methods – the linear model method and the curved model method. We will show that the linear model method is an improvement over [2] and [3]. The curved model method generalizes and revises the idea of the linear model method, so that it can be applied to both, linear objects and curved objects uniformly, and does not create as many ambiguities as the linear model method does.

We first introduce our basic idea (in section 2) with the simple example of computing the point coincidence relation for a set of 2D points. In section 3, the linear model method is presented. Its improvements over the method of [2] and [3] will be discussed at the end of this section. Section 4 discusses the curved model method. In section 5, we present an application of both, the linear model method and the curved model method on 3D Boolean operations of polyhedra.

2 Point Coincidence Relation

We first describe the basic ideas with the example of finding 2D point-point coincidence relations.

We define the r *region* of a point to be the circular region with the point as center and radius r. Assume τ is the error bound of a point (we call it tolerance) and ν is the error bound of computing this tolerance (we call it secondary error) then we can define a tolerance environment of a point as three regions: ε, δ and Δ regions. Initially, $\varepsilon = \tau - \nu$, $\delta = \tau + \nu$ and $\Delta = +\infty$.

Definition 1 *(coincident, apart, ambiguous, equivalence class, consistent)*
   Two points P_1, P_2 are coincident ($P_1 = P_2$): iff $\varepsilon_1 \cap \varepsilon_2 \neq \emptyset$, $\delta_1 \subset \Delta_2$ and $\delta_2 \subset \Delta_1$.
   Two points P_1, P_2 are apart ($P_1 \neq P_2$): iff $\delta_1 \cap \delta_2 = \emptyset$.
   Two points P_1, P_2 are in an ambiguous relation iff they are neither coincident nor apart.
   The points $P_1, ..., P_n$ build an equivalence class iff $\bigcap_i \varepsilon_i \neq \emptyset$ and $\bigcup_i \delta_i \subset \bigcap_i \Delta_i$.
   A set of points S is consistent iff $\forall P_i, P_j \in S$, they are not in an ambiguous relation. □

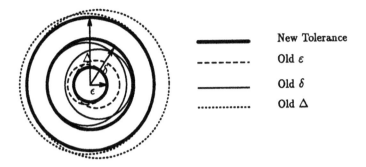

Figure 2: Merging two coincident points

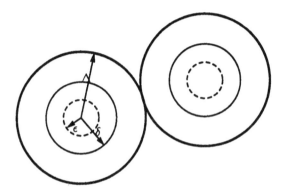

Figure 3: Δ regions of two apart points

If the ε regions of two points overlap, they are declared to be coincident and the data of these two points are merged. The new ε region is the maximal circular region inside the intersection of the two old ε regions, the new δ region is the minimal circular region that has the same center as the new ε region and encloses the union of the two old δ regions. The new Δ region is the maximal region that has the same center as the new ε region and is inside the intersection of the two old Δ regions. (See figure 2).

When the δ regions of two points do not intersect, we say they are apart and their distance is computed. The Δ of each point has to be updated if their distance is smaller than 2Δ (See figure 3).

When the δ regions of two points intersect but their ε regions do not intersect, or when the δ region of a point grows out of its Δ region, we detect an ambiguity. The situation in figure 1 will be detected as an ambiguity regardless of the order in which we compare these three points. With following definition and theorem, we will show that the coincidence relation defined with tolerances is an equivalence relation and so truly simulates the coincidence relation in Euclidean space.

In the following, we will use ε, δ and Δ to denote the ε, δ and Δ regions of a point if no confusion is created.

Theorem 1 *If a set of points S is consistent then coincidence of points is an equivalence relation for this set.*

Proof: To be an equivalence relation, it must be reflexive, symmetric and transitive. Reflexivity and symmetry apparently follow from the definition. We now prove transitivity. Assume $P_1 = P_2$ and $P_2 = P_3$. So P_1, P_2 are in an equivalence class E_1 and P_2, P_3 are in an equivalence class E_2. If $E_1 \neq E_2$, the unions of their δ regions do not intersect. But because the δ region of E_1 and E_2 both include the δ region of P_2 which conflicts the statement that their unions of δ regions are apart. So, $E_1 = E_2$ and therefore $P_1 = P_3$. □

When we find an ambiguity, it means that the initial tolerances (τ) of these points do not allow us to make consistent decisions. We can try to adjust the tolerances such that the relations become consistent. The following lemma deals with operations on the tolerance regions that preserve previously made decisions. The diameters of the error regions are either uniformly increased or uniformly decreased by the amount d.

Lemma 1 *If the r_1 region of point P_1 is inside the r_2 region of point P_2, then*

- *The $(r_1 - d)$ region of P_1 is inside the $(r_2 - d)$ region of P_2.*
- *The $(r_1 + d)$ region of P_1 is inside the $(r_2 + d)$ region of P_2.*

Proof: Regions of 2D points are circles. The lemma is apparently true for any two circles. □

Theorem 2 *Assume $P_1, ... P_n$ are coincident and merged to a single point P. If we increase the ε and δ regions of all P_i and P by the same amount or decrease the ε and δ regions of all P_i and P by the same amount, then we still have $P_1 = ... = P_n = P$ as long as no new ambiguity is detected.*

Proof: After increasing or decreasing the ε and δ regions of P_i and P by the same amount, if no ε is zero or negative and no δ is greater than it's Δ, from the lemma the ε region of P is still inside all the ε regions of P_i and the δ region of P still contains all the δ regions of P_i. So their coincidence relations still hold. □

For each point, if we increase or decrease its tolerance τ, its ε and δ will also be increased or decreased by the same amount. With above theorem, relations among points will not be affected by changing the tolerances of all points that are coincident as long as no new ambiguity is found and the amount of tolerance changes are the same for all the coincident points. Now, if two points (both can be the results of merging a set of coincident points) do not have a consistent relation, we can dynamically adjust their ε regions and δ regions to seek a consistent relation between them.

This adjustment is not always successful because we may be in the situation that any adjustment will create a new ambiguity. In these cases, a global adjustment has to be done and the algorithm has to be re-run with a new set of tolerances. But eventually, this ambiguity handling process will stop and ambiguity will be solved because in the worst case with a very big tolerance every point will be included, so no ambiguity can result.

3 The Linear Model Method

3.1 Representations, Models and Tolerances

In [7], C. Hoffmann introduced the concept of separating representation and model of geometric objects. A representation is a data structure intended to describe a geometric object in Euclidean space (the model of the representation). Representation consists of two parts: symbolic constraints, describing relationships among different geometric entities, and numerical data, containing the position and orientation information of the object. However when we work with inaccurate data and computations the numerical data do not generally satisfy the symbolic constraints and so the representation may have some internal inconsistencies. We say a representation has a model if there exists an object in Euclidean space (the model) satisfying all symbolic constraints of the representation.

For tolerance-based methods, we also require that a model of a geometric object is inside the tolerance of this object. Since a geometric algorithm can create new symbolic constraints (e.g. detect relations among objects) and numerical data (e.g. intersections), tolerances need to be updated and computed to guarantee the existence of a model satisfying those new constraints. For instance, in the point coincidence relation of section 2, when an ambiguity occurs, there is a possibility of inconsistent decisions which create symbolic constraints that are impossible to be satisfied. So the model's existence is not guaranteed. The tolerance adjustment for solving ambiguities can be interpreted as follows: the tolerances are either increased to try to include a possible model or shrunk to release some symbolic constraints.

We next present the linear model method. Our strategy is to try to update and compute the tolerances so that for each linear object there is a linear model inside its ε region satisfying all the constraints (relations). This method only works for linear objects.

3.2 Relations Between Simple Objects

A simple object in 3D is either a point, line or a plane. A complex object is a set of simple objects, connected by some incidence and/or coincidence relations (e.g. a polygon). We will first generalize the approach in section 2 to decide the relation of two simple objects, and then discuss complex objects. In the rest of this section, objects always mean simple objects unless indicated.

Since lines and planes are infinite objects whose tolerances are difficult to define and manipulate (e.g. two planes should always intersect when considering errors), we will cut them with a big enough bounding sphere (our working space) and only consider those parts of the lines and planes that are inside the bounding sphere when defining their tolerances. Two simple objects of different dimensions can be coincident, incident, intersecting or apart. The tolerance of an object is defined similarly to the tolerance of a point. An r *region* of an object is a volume in our working space in which each point has a distance to this object of less than r (r is sometimes directly used to denote a region if no confusion is possible).

The tolerance of an object consists of three regions: the ε, the δ and the Δ region. With an initial tolerance τ and a secondary error bound ν, these regions can be initialized similarly to the region initializations in section 2, that is $\varepsilon = \tau - \nu$, $\delta = \tau + \nu$ and $\Delta = +\infty$.

With this tolerance definition, we require that a model of an object must be inside the ε region of this object. The strategy to guarantee the existence of model is that we dynamically update tolerances so that the ε region of an object consists of a set of currently valid models. So a model always exists as long as no ε region is empty and no ambiguity is detected

Definition 2 *(linear model)*
  A *linear model of an object O is an linear object inside the ε region of O with the same dimension as O.* □

Definition 3 *(apart, coincident, incident, intersect, ambiguous, consistent)*

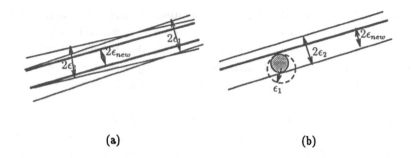

Figure 4: ε region updating for (a) coincidence (b) incidence.

Two objects O_1, O_2 are apart iff $\delta_1 \cap \delta_2 = \emptyset$.

Objects O_1 and O_2 are coincident : $O_1 = O_2$ iff they have the same dimension, and there exists a model inside the ε region of O_1 and O_2, and $\delta_1 \subset \Delta_2$, $\delta_2 \subset \Delta_1$.

Object O_1 is incident on object O_2 : $O_1 \subset O_2$ iff O_1 is a lower dimensional object than O_2 and O_1 is coincident with its projection on O_2 with the same tolerance values (ε, δ and Δ) as O_2.

Two objects O_1, O_2 intersect iff $\varepsilon_1 \cap \varepsilon_2 \neq \emptyset$ and there is no incidence or coincidence relation between them.

Two objects O_1, O_2 are in ambiguous relation iff none of the above relations hold.

A set of objects S is consistent iff for all pairs $O_i, O_j \in S$ they are not in an ambiguous relation.

$\square$

Tolerance updating for general linear objects is similar to that for points. When the δ regions of two objects do not intersect, they are apart and their Δ regions may need to be updated according to their minimal distance. When two objects are coincident, i.e. they have a common model, these two objects then will be merged and the intersection of the two ε regions will be approximated to become an ε region of the merged object and consists of all models of new object. The new δ and Δ regions are approximated from the union and intersections of the two old δ and Δ regions, respectively (see figure 4(a)).

For an incidence relation $O_1 \subset O_2$, the coincidence of O_1 and its projection on O_2 will update the tolerances of both O_1 and its projection on O_2. The tolerance of O_2 then will be updated along with the projection object of O_1 in order to make sure that every instance of O_2 satisfies the constraint: $O_1 \subset O_2$ (figure 4(b)).

When the δ regions of two objects intersect but their ε regions do not intersect (there will be an empty ε region) or when $\delta > \Delta$ an ambiguity is detected (the decisions are likely to be inconsistent). The tolerances have to be adjusted in order to resolve the ambiguity (see section 3.5).

For an intersection relation, if I is the intersection of O_1 and O_2, their tolerances should satisfy the relations $I \subset O_1$ and $I \subset O_2$. The tolerance of I can be defined as follows: $\varepsilon = min(\varepsilon_1, \varepsilon_2)$; $\delta = max(\delta_1, \delta_2)$; $\Delta = min(\Delta_1, \Delta_2)$. And then tolerances of O_1 and O_2 must be updated to satisfy the incidence relations as depicted in figure 5.

Theorem 3 *If a set of objects is consistent, then coincidence is an equivalence relation and incidence relation is a partial order relation.*

The proof of this theorem is similar to the proof of theorem 1.

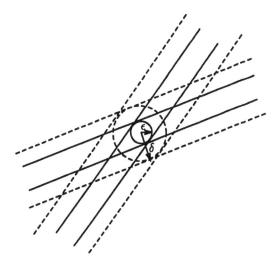

Figure 5: intersection of two lines

3.3 Constraint Propagation

A complex object is a set of simple objects connected by symbolic constraints such as incidence and coincidence. For instance, a triangle in 2D space is a complex object with three lines, three points and six point-line incidence constraints (each point is incident to two lines). When the tolerance of a simple object in a complex object is updated, the tolerances of those simple objects that are directly connected with this object by some constraints (e.g. incidence relation) may have to be updated as well in order to preserve all these constraints. This kind of updating will eventually be propagated to the whole complex object. For example, if the ε of a point in a triangle is decreased, all the ε's in this triangle have to be decreased in order to preserve this triangle relation. In other words, we have to re-satisfy all the constraints in the complex object by updating the tolerances in the complex object. We call this procedure the constraint propagation. When the relations of simple objects in a complex object are very complicated, the constraint propagation problem can sometimes be very difficult.

Tolerance updating may be done in two possible ways. One way is to move the object itself and to adapt the ε, δ and Δ values such that the corresponding regions overlap afterwards (called the perturbation method) for instance, when merging two coincident objects. The other possibility is to locally reevaluate the ε, δ or Δ regions as in the triangle example before.

Let's assume two simple objects O_1, O_2 are detected to have a certain relation (e.g. coincidence, incidence) and their tolerances need to be updated. Two cases are considered. The first case is when O_1, O_2 are not in a common complex object. Then a simple solution is to carry out the same perturbation for O_1 as well as for the whole complex object to which O_1 belongs, and to carry out another perturbation for O_2 as well as for the complex object to which O_2 belongs such that O_1 and O_2 overlap. We call this the global propagation approach. To speed up this error propagation we have complex objects share a data record for ε, δ and Δ (see [2]).

The second case occurs when O_1 and O_2 are part of the same complex object, i.e. they are already connected by constraints. A simple example is in figure 6 where the points P_1, P_2 are found

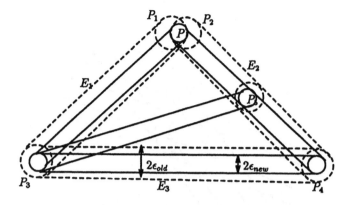

Figure 6: Localize perturbation in a complex object

coincident, but they are already connected by the three edges. The global propagation approach doesn't work here because the perturbation operations for P_1 and P_2 would be different. However, for some simple cases it is possible to localize the constraint propagation. The basic idea is to try to perturb simple objects in a way that the perturbation will not be propagated further. For example in figure 6, after P_1, P_2 are merged to P and their coordinates are perturbed to be the coordinates of P, tolerances of line E_1 and E_2 can be updated simply by connecting P_3, P and P_4, P. So the edge E_3 need not to be perturbed.

In some cases, re-intersection might be necessary when trying to localize perturbation as in the case of finding the new position for point P_5 in figure 6. But for linear objects, this re-intersection may still be affordable. This second approach is called local propagation approach.

3.4 Zero Perturbation Approach

Although in many cases the constraint propagation problem can be solved by global and local propagation approaches, there still exist some situations that cannot be handled by these two approaches because some of the simple objects in a complex object are over-constrained. So, the localization of perturbation is not possible, in general. An example is shown in figure 7. Eight of the nine points can be constructed from other points and lines, but the 9th intersection is dependent on the other eight (through Pascal's theorem).

The following approach is called a zero perturbation approach since it does tolerance updating without perturbation of the objects.

For the coincidence relation, the two coincident objects will not be merged, instead they are simply linked together with the knowledge that they are coincident. Their ε and Δ values are decreased so that the new ε regions and Δ regions are inside the two old ε regions and Δ regions, their δ values are increased so that the new δ regions enclose both the two old δ regions, as depicted in figure 8. So the model for the two coincident objects exists in both of their ε regions.

The incidence relation can be represented by the coincidence of one object with its projection on the other object, and the intersection relation can be represented by two incidence relations, their zero perturbation tolerance updating can be obtained similarly.

The problem with this zero perturbation approach is that the tolerance updating operation is

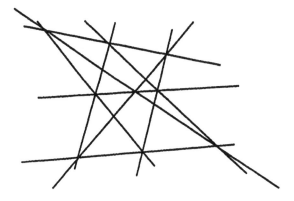

Figure 7: Over-constrained object (PASCAL's theorem)

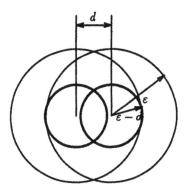

Figure 8: Zero Perturbation of two ε regions

not idempotent. i.e. when we perform the same operation (e.g. comparison) on these two objects again, their tolerances do not remain the same – the ε and Δ become smaller and δ becomes bigger. Once more, since all simple objects in a complex object share a common set of tolerance values and therefore this non-idempotence effect will be propagated to the whole complex object. It means that all the errors in zero perturbation tolerance updating on any simple object of a common complex object will accumulate on the same set of ε, δ and Δ values. If there are many incidences or coincidences between simple objects of a common complex object the likelihood of detecting an ambiguity increases and the performance of the algorithm will be greatly affected. This problem will be solved in the curved model method.

3.5 Ambiguity Handling

A similar theorem to theorem 2 can be proved for handling ambiguity with general linear objects, so that we can cure some cases of ambiguous relations.

Lemma 2 *If the r_1 region of a point P is inside the r_2 region of a linear object O, then*

- *The $(r_1 - d)$ region of P is inside the $(r_2 - d)$ region of O.*
- *The $(r_1 + d)$ region of P is inside the $(r_2 + d)$ region of O.*

Proof: Take the projection point of P on O. The lemma then can be derived easily from lemma 1.

Lemma 3 *If the r_1 region of an object O_1 is inside the r_2 region of object O_2, then*

- *The $(r_1 - d)$ region of O_1 is inside the $(r_2 - d)$ region of O_2.*
- *The $(r_1 + d)$ region of O_1 is inside the $(r_2 + d)$ region of O_2.*

Proof: We can see the r_1 region of O_1 as the union of all the r_1 regions of points of O_1. For each such point's region, from lemma 2, an increasing or decreasing of the regions does not affect the containment relation of the point's region and the region of O_2. So the region of O_1, the union of all the regions of its points, also has the same relation as the regions of its points. □

Theorem 4 *For a complex object formed by some simple objects with coincidence and incidence relations, if we increase or decrease the tolerance τ of each simple object in the complex object, then as long as no new ambiguity is detected, all current relations of objects are unchanged.*

Proof: From the definition of tolerances, increasing or decreasing τ results in the increasing or decreasing of ε and δ regions by the same amount. From lemma 3, these changes of tolerances do not affect the subset relations among regions. So if no ε is zero or negative after the changes of tolerances, the coincidence and incidence relations among simple objects in the complex object are not changed. For apart objects, as long as δ is not greater than its Δ, they are still apart. □

With above theorem, we can now dynamically adjust the tolerances of a complex object in which the ambiguity occurs to solve the ambiguity and continue the algorithm with new tolerances. When this adjustment is not successful (always create new ambiguities), re-running the algorithm with a different set of tolerances is necessary.

4 The Curved Model Method

There are two reasons why we need a curved model method. Firstly, with the linear model method, the constraint propagation often causes too many ambiguities, especially when we have to use the zero perturbation approach. Frequent ambiguity handling greatly affects the performance of the algorithm. Secondly, we want to apply geometric operations to curved objects (e.g. sculptured solid modeling). The generalization of linear model method to problems with curved objects is not straightforward. Some conceptual adjustments are needed.

The curved model method modifies the model definition of the linear model method by relaxing some constraints on models.

4.1 Tolerances and Curved Models

The definitions of tolerance and regions in a curved model method are the same as for the linear model method. The difference is the definition of models.

Definition 4 *(curved model)*
  The curved model of a point is any point inside its ε region.
  a curved model of an object O is an object inside the ε region of O. □

This model definition is less strict than the one for the linear model method and therefore allows a much broader range of objects to be considered as models. For example, a model M of a curve C (including lines) is any curve in ε region of C that satisfies all the constraints of C but not necessarily C's mathematical form. The model of a surface (including plane) can be defined similarly.

In the linear model method, a model must be linear (the same mathematical form of the original object). This requirement is dropped in curved model method. So, the model of a line could be any curve within the tolerance region of that line, a model of a plane could be any curved surface, a model of a curve or a surface could be a curve or a surface of other form. It turns out that this concept adjustment of models is significant because it essentially eliminates the constraint propagation problem and makes the method applicable uniformly for both linear and curved objects.

4.2 Tolerance Updating

Let us examine the impact the curved model has on tolerance updating. We still use the same definition of relations of simple objects the linear model method used except we now consider curved models instead of linear models. For instance, two objects are coincident if they have common curved models.

The tolerance updating for coincidence relation is basically the same as for the linear model method. i.e. merge coincidence objects and compute intersections of ε and Δ regions and union of δ regions to form new ε, δ and Δ regions of the new object. But because we have to represent the ε, δ and Δ regions as some system recognized object type, approximations have to be made in most cases.

For the incidence relation, tolerance updating is different. Assume $O_1 \subset O_2$. Then O_1 and P (O_1's projection on O_2) are coincident and their tolerances are updated. With the linear model method, the tolerance of O_2 would also be updated along with P's tolerance. The reason for doing this is to make sure that O_2's models are all linear objects. But in curved model method, it is no

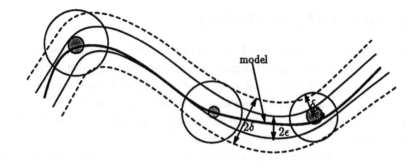

Figure 9: incidence relations of curved objects

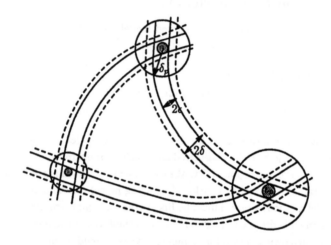

Figure 10: Localized tolerance updating of a Complex Object

longer necessary because of the flexibility of curved model. In other words, the tolerance of O_2 will not be updated along with O_1's projection object. Note that the ε and Δ regions of O_1 must be inside the ε and Δ regions of O_2 and the δ region of O_1 must enclose the δ region of O_1's projection on O_2 as depicted in figure 9. Also when O_2's tolerance is updated for some reason (e.g. coincidence relation or O_2 is detected to be incident on another object), tolerances of O_1 have to be updated as well to preserve their incidence relations.

It is now much easier to handle complex objects with the curved model method because the tolerance updating of lower dimensional objects does not affect the tolerances of higher dimensional objects, therefore tolerances will not be propagated to the whole complex object. For example, in the configuration of figure 10, the changes of tolerances of points do not affect the tolerances of the edges, so the ε and Δ values of points can be smaller than those of the edges (Δ regions are not illustrated in the figure), and the δ values of points can be bigger than those of incident edges.

Ambiguity handling for the curved model method works the same way as for the linear model method.

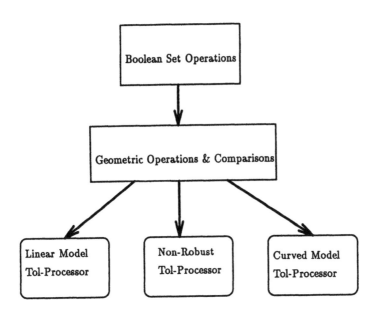

Figure 12: Structure of the Boolean operations with tolerance Processors

will refer to appropriate tolerance processing module depending on which robustness method (linear model, curved model or without robustness) it is using (see figure 12).

Tolerance definition, computation and updating, ambiguity handling and relation detection are all done inside the individual tolerance processing modules. The implementations of the robustness methods in these modules use the definitions and rules we described in section 3 and section 4. For the details of the Boolean operation algorithm itself, refer to [3].

The different models can be switched during a modeling session which allows us to experiment with different sets of data and different methods, and thus to compare the effectiveness of the different approaches. Table 1 shows the results if computing the union of two unit cubes, one of which has been rotated by a small θ angle about the x-axis. The curved model method creates ambiguities for a much smaller range. When using the simple tolerance approach invalid results (dangling edges and faces) are created in the range $0.0025° \geq \theta \geq 10^{-4°}$. Figure 13 shows some examples of our test case. Figure 13(a) shows the positions of the two cubes before Boolean operation. Figure 13(b) is the result of the Boolean operation with curved model method. Figure 13(c) is the result of the same operation with a much smaller angle. Figure 13(d) shows an invalid result when using simple tolerance approach (without considering robustness) with the same small angle figure 13(c) used.

6 Conclusions

We present two new tolerance-based methods, the linear model method and the curved model method, for robustness in geometric modeling. The linear model method only applies to linear objects. It faithfully preserves all properties of the objects but may detect too many ambiguous situations for the reasons explained in the paper. The curved model method applies to both, linear and curved objects uniformly, and creates fewer ambiguities than the linear model method does.

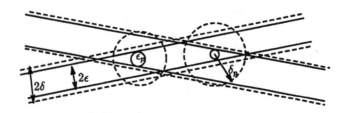

Figure 11: Inconsistent relation of two points on two lines

4.3 Preserving Geometric Properties

Because an object and its curved models do not necessarily have the same mathematical form, they also don't necessarily have the same geometric properties. However, in algorithms and their implementation for certain geometric problems, the algorithm designers or the programmers often assume geometric properties of objects to deduce results without explicit computation. This could results in unexpected and incorrect results when we use curved models because these properties may not be preserved by the curved model. For instance, in figure 11, two lines are detected to be intersecting. Then we find that two non-coincident points are incident on both lines which contradicts the fact that two lines can only intersecting once.

It is not feasible for curved models to preserve all properties of each kind of objects. But some common sense properties, as "two lines only intersect at one point", are more likely to be used in reasoning and should be preserved in curved models. To preserve such properties, new constraints must be added to curved models. These constraints might not be explicitly specified. For instance, they can be in the form of new tolerance updating rules. We only briefly outline an approach of tolerance updating to preserve the property of "two linear objects only intersect once".

In order to preserve this property, we must make sure that every object that is incident on two or more linear objects has a big enough δ to prevent it from separating from other intersection objects (e.g. They all should enclose some common point). When an object O is incident on object P, we say P is a parent object of O. A list of each object's parent objects is kept. This list is updated when incidence relations are detected. When the list is updated by adding a new linear object or changing the tolerance of a linear object, we must check the δ of the child object to make sure that it's big enough to contain the intersection of the δ regions of both intersecting lines. So in figure 11, two points that are incident to both lines would be either coincident or ambiguous, but never apart.

5 Robust Boolean Operations of Polyhedra

We have implemented a robust Boolean set operation algorithm on polyhedra with both the linear model method and the curved model method in an experimental solid modeler based on half-space representation. The Boolean operation algorithm is designed to be independent of the implementation of the low level operations such as intersection and comparisons. The details of the relation computation and are hidden from the main algorithm and are implemented in three independent tolerance processing modules, using a) the linear model method, b) the curved model method or c) a fixed ε method without any robustness consideration. Relation detection in Boolean operation

Angle	Curved-Model	Linear-Model
$\theta > 0.0025°$	intersect	intersect
$0.0025° \geq \theta \geq 10^{-4°}$	ambiguous	ambiguous
$10^{-4°} > \theta > 10^{-7°}$	coincident	ambiguous
$\theta \leq 10^{-7°}$	coincident	coincident

Table 1: Results of union of two unit cubes with $\tau = 10^{-3}, \nu = 10^{-4}$

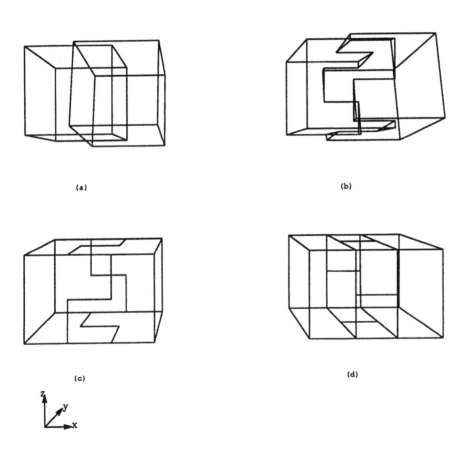

(a)

(b)

(c)

(d)

Figure 13: Some examples of our test case

However it does not preserve all the properties of the original objects because it uses curved models to simulate these objects. To preserve some of these properties, more computation and book keeping is needed which slows down the algorithm.

It turns out that to achieve robustness we have to pay the costs in other aspects of the algorithm. The algorithm can be significantly slower because we need to always compute and update tolerances, sometimes even re-run the algorithm to achieving a correct result.

Future research includes finding an appropriate set of geometric properties to be preserved, and developing applications of the two robustness methods for geometric modeling problems especially those with curved objects such as sculptured solid modeling.

7 Acknowledgment

This work was supported in part by the National Science Foundation, grants DDM-89 10229 and STC 89-20219. All opinions, findings, conclusions or recommendations expressed in this document are those of the authors, and do not necessarily reflect the views of the sponsoring agencies.

References

[1] Allen, G. Testing the accuracy of solid modellers. *Computer Aided Engineering* (June 1985), 50–54.

[2] Bruderlin, B. Detecting ambiguities: an optimistic approach to robustness problems in computational geometry. Tech. Rep. UUCS 90-003 (submitted), Computer Science Department, University of Utah, April 1990.

[3] Bruderlin, B. Robust regularized set operations on polyhedra. In *Proc. of Hawaii International Conference on System Science* (January 1991).

[4] Edelsbrunner, H., and Mucke, E. Simulation of simlicity: a technique to cope with degenerate cases in geometric algorithms. In *Proc. of 4th ACM Symposium on Comp. Geometry* (June 1988), pp. 118–133.

[5] Greene, D., and Yao, F. Finite resolution computational geometry. In *Proc. 27th IEEE Symp. Fundations of Computer Science* (1986), pp. 143–152.

[6] Guibas, L., Salesin, D., and Stolfi, J. Epsilon geometry: building robust algorithms from imprecise computations. In *Proc. of 5th ACM Symposium on Computational Geometry* (1989), p. .

[7] Hoffmann, C. M. The problems of accuracy and robustness in geometric computation. *IEEE Computer 22*, 3 (March 1989), 31–41.

[8] Hoffmann, C. M., Hopcroft, J. E., and Karasick, M. S. Robust set operations on polyhedral solids. *IEEE Computer Graphics and Application 9* (November 1989).

[9] Milenkovic, V. Verifiable implementations of geometric algorithm using finite precision arithmetic. *Artificial Intelligence 37* (1988), 377–401.

[10] Requicha, A. A. G. Solid modeling - a 1988 update. In *CAD Based Programming for Sensory Robots* (1988), B. R. ed., Ed., Springer Verlag, New York, pp. 3–22.

[11] Segal, M. Using tolerances to guarantee valid polyhedral modeling results. *Computer Graphics* *24*, 4 (1990), 105–114.

[12] Sugihara, K., and Iri, M. Geometric algorithms in finite precision arithmetic. Res. Mem. 88-14, Math. Eng. and Information Physicas, University of Tokyo, 1988.

[13] Yap, C. K. A geometric consistency theorem for a symbolic perturbation theorem. In *Proc. of 4th ACM Symposium on Comp. Geometry* (June 1988), pp. 134–142.

On Shortest Networks for Classes of Points in the Plane

Edmund Ihler   Gabriele Reich   Peter Widmayer [1]

Abstract

We are given a set P of points in the plane, together with a partition of P into *classes* of points; i.e., each point of P belongs to exactly one class. For a given network optimization problem, such as finding a minimum spanning tree or finding a minimum diameter spanning tree, we study the problem of choosing a subset P' of P that contains at least one point of each class and solving the network optimization problem for P', such that the solution is optimal among all possible choices for P'. We show that solving the minimum spanning tree problem for classes of points in the plane is NP-complete, where the distance between points is defined by any of the Minkowski metrics L_p, $1 \leq p \leq \infty$. In contrast, a class solution for the minimum diameter spanning tree problem can be computed in time $O(|P|^3)$.

By proving the NP-completeness of the minimum spanning tree class problem we also get some results for distance graphs. Here, computing a class solution for the minimum spanning tree problem is NP-complete, even under several restrictions, e.g., if the graph is part of a unit grid and is a tree, where the vertex degree and the number of vertices per class are both bounded by three. This is true even if the graph is a minimum spanning tree for its set of vertices.

1 Introduction

Consider the problem of connecting a number of existing local networks into a global network; for example, several local area computer networks might have to be connected into a metropolitan area or wide area network. Due to capacity, reliability and other technological considerations, we want the new links that need to be added to form a connected network of their own; most backbone networks in university campusses or companies are of this type. It is clearly sufficient for the new network to reach one of the computers in each local network. In this paper, we look at the problem of minimizing the total length of the new network (no Steiner points allowed), and at the problem of minimizing the length of the connection between any two local networks. In the plane, the length of a direct link is the distance between both points; we extend our considerations to other cost functions and the problem of identifying minimum subgraphs for given graphs.

In this paper, we consider network optimization problems for sets of points as well as for graphs. For our purposes, a set of n points $P = p_1, \ldots, p_n$ in the plane is identified with its undirected, complete *Euclidean graph* $G = (V, E)$, where $V = P$, i.e., each point is a vertex in the graph, $E = V \times V$, and the length of edge (p_i, p_j) is the Euclidean distance between p_i and p_j. In fact, our results hold for any of the Minkowski distance functions instead of the Euclidean distance. In

[1] Authors' address: Institut für Informatik, Universität Freiburg, Rheinstraße 10-12, W-7800 Freiburg, Germany
e-mail:gabriele.reich@informatik.uni-freiburg.dbp.de

general, we are given a vertex set V and a partition of V into *classes* $V_1, \ldots, V_c$, i.e. $\bigcup_{i=1}^{c} V_i = V$, $V_i \cap V_j = \emptyset$ for $i \neq j$, and a connected, undirected distance graph $G = (V, E)$. The objective is to compute a tree that contains at least one vertex of each class, a *class tree*. We consider two types of class trees: those that have minimum total length, the *minimum class trees MCT*, and those where the diameter is minimum, *minimum diameter class trees MDCT*. As usual, the diameter of a graph G is defined as the longest of the minimum paths among all pairs of vertices of G. If each class consists of a single vertex, then the problems described above reduce to the well-known problems of computing a minimum spanning tree and a minimum diameter spanning tree. Both problems can be solved in polynomial time for points in the plane, see [2, 5].

In sharp contrast to the minimum spanning tree problem, the problem of computing a minimum class tree is NP-complete. In Section 2, we will show that this holds for points in the plane as well as for arbitrary distance graphs. For specific types of distance graphs, problems that are NP-hard in general become solvable in polynomial — even linear — time, but minimum class tree remains NP-complete. In this sense, the latter is harder than Steiner's problem if the given graph is a tree with edge lengths restricted to 1, or if the tree is part of a unit grid. In each case the vertex degree and the number of vertices per class both can be restricted to at most three. The same result holds even if the input graph is not an arbitrary tree, but a minimum spanning tree for the given set of vertices.

In contrast to the NP-completeness of computing a minimum class tree, in Section 3 we will show that a minimum diameter class tree in the plane can be computed in polynomial time. To this end, we will present an algorithm that computes a minimum diameter class tree for a set of n points in the plane in time $O(n^3)$.

2 Minimum class tree is hard

In this section, we prove the NP-completeness of computing a minimum class tree for a finite set of points in the plane and for a connected, undirected distance graph. The former follows directly from the latter.

2.1 The minimum class tree of a graph

Let us now prove that the problem of computing a minimum class tree for a graph is NP-hard, even if the given graph is a tree.

Problem: MINIMUM CLASS TREE

Instance: A class graph $G = (V, E)$ with positive integer c, sets $V_1, \ldots, V_c$, such that $V_i \cap V_j = \emptyset$ for $i \neq j$, $1 \leq i, j \leq c$, and $\bigcup_{i=1}^{c} V_i = V$, and a positive integer k.

Question: Is there a class tree CT of G whose total length does not exceed k?

We will show the NP-completeness (in the strong sense) of MINIMUM CLASS TREE by reducing 3-satisfiability (3SAT) to it, where 3SAT is defined as follows (see [1, p. 46]):

Problem: 3SAT

Instance: Collection $C = \{c_1, c_2, \ldots, c_n\}$ of clauses on a finite set $U = \{u_1, u_2, \ldots, u_m\}$ of variables such that $|c_i| = 3$ for $1 \le i \le n$.

Question: Is there a truth assignment for U that satisfies all the clauses in C?

Theorem 2.1: MINIMUM CLASS TREE is NP-complete, even if the given class graph G is a tree and each class contains at most three vertices.

Proof: Clearly, MINIMUM CLASS TREE is in NP, since a guess for CT can be verified in linear time to be a tree and a subgraph of G, contain one vertex of each class and have total length no more than k.

To see that MINIMUM CLASS TREE is NP-hard, even if G is a tree, consider the reduction of 3SAT to MINIMUM CLASS TREE as shown in Figure 1.

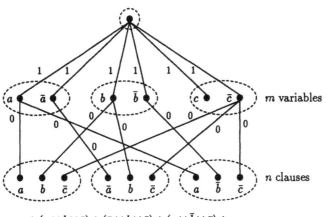

$$\ldots \wedge (a \vee b \vee \bar{c}) \wedge (\bar{a} \vee b \vee \bar{c}) \wedge (a \vee \bar{b} \vee \bar{c}) \wedge \ldots$$

Figure 1: Reduction of 3SAT to MINIMUM CLASS TREE

For each variable u_i, $1 \le i \le m$, there is a class of two vertices v_{i1} and v_{i2}, representing the assignment of *true* and *false* to u_i, respectively. For each clause c_i, $1 \le i \le n$, there is a class of three vertices v_{i1}, v_{i2}, v_{i3}, $m+1 \le i \le m+n$, representing the three literals in the clause. For each literal, the vertex is connected via an edge of length 0 to the vertex for the variable assignment that makes the literal true. All variable assignment vertices are connected via edges of length 1 to a single additional vertex v_c. Hence, there are $m + n + 1 = c$ classes in the tree and the length of a minimum class tree is m.

Clearly, the graph defined by this reduction is a tree and each class contains at most three vertices; its size and the time for its computation are both polynomials in m and n. We show that the answer to the 3SAT question is the same as the answer to the MINIMUM CLASS TREE question for $k = m$.

Claim: The answer to MINIMUM CLASS TREE is "yes" iff the answer to 3SAT is "yes".

if: If the answer to 3SAT is "yes", there exists an assignment of values *true, false* to the variables u_i, $1 \leq i \leq m$, such that each clause is satisfied. For each variable u_i, $1 \leq i \leq m$, that is true, select the vertex v_{i1} to belong to the class tree CT, for each false variable u_j, $1 \leq j \leq m$, select v_{j2}. Enlarge CT by all edges incident to vertices already in CT and by the endpoints of these edges. This yields a tree containing one vertex of each of the classes representing variables (each variable is either true or false), the single vertex v_c, and — since at least one literal of each clause is true — at least one vertex of each of the classes representing clauses. Clearly, the length of this class tree is m. Hence, the answer to MINIMUM CLASS TREE is "yes".

only if: If the answer to MINIMUM CLASS TREE is "yes", there exists a tree CT, with total length at most k, containing at least one vertex of each class. Obviously, v_c must be in CT. Because of the total length of CT and the length 1 of the edges between v_c and the variable assignment vertices v_{ij}, $1 \leq i \leq m$, $1 \leq j \leq 2$, at most m of the $2m$ vertices v_{ij} may be in CT. Moreover, each of the classes V_i must be represented by a vertex in CT; therefore, exactly one variable assignment vertex of each of the classes V_i, $1 \leq i \leq m$, is in CT. At least one vertex of each class V_i, $m+1 \leq i \leq m+n$, needs to be in T. Any such vertex v in CT implies that the variable assignment vertex v' where (v, v') is an edge of G is also in CT. Now assign u_i the value *true* if v_{i1} is in CT, and *false* if v_{i2} is in CT, $1 \leq i \leq m$. Hence, for each clause at least one literal is true, and so the answer to 3SAT is "yes".

This proves the claim and the theorem. ☐

From the proof it is clear that G is a tree and that each class contains no more than three vertices. Moreover, a straightforward modification of the proof allows to restrict the degree of each vertex to three.

2.2 The minimum class tree in the plane

Recall that the *Euclidean graph* $G = (P, E)$ is the complete graph that is induced by a set P of points $p_1, \ldots, p_n$ in the plane, where the length of an edge (p_i, p_j) is the Euclidean distance $\sqrt{(x_i - x_j)^2 + (y_i - y_j)^2}$ between p_i and p_j. A minimum class tree for a Euclidean graph is called a *geometrical minimum class tree gMCT*.

We will show that computing a minimum class tree for a Euclidean graph is NP-hard.

Problem: GEOMETRICAL MINIMUM CLASS TREE

Instance: A set P of points in the plane, a positive integer c, a partition of P into subsets $P_1, \ldots, P_c$, such that $P_i \cap P_j = \emptyset$ for $i \neq j$, $1 \leq i, j \leq c$, and $\bigcup_{i=1}^{c} P_i = P$, a positive integer k.

Question: Is there a class tree CT of the Euclidean graph $G = (P, E)$ with total length at most k?

We will show the NP-completeness of GEOMETRICAL MINIMUM CLASS TREE by a reduction of 3SAT to GEOMETRICAL MINIMUM CLASS TREE, similar to the one presented in Section 2.1. Note that NP-hardness for GEOMETRICAL MINIMUM CLASS TREE does not follow directly from NP-hardness for MINIMUM CLASS TREE, since the former is quite a special case of the latter.

Theorem 2.2: GEOMETRICAL MINIMUM CLASS TREE is NP-complete, even if the number of points per class is restricted to at most three.

Proof: By the same arguments as for MINIMUM CLASS TREE, GEOMETRICAL MINIMUM CLASS TREE is in NP. It remains to be shown that GEOMETRICAL MINIMUM CLASS TREE is NP-hard.

Consider the reduction of 3SAT from Section 2.1 and the NP-complete variant of 3SAT, where for each $u \in U$ there are most five clauses that contain either u or $\bar{u}$ [1, p. 259].

For each variable u_i create two lists l_{i1} and l_{i2} of ten vertices each, l_{i1} for the assignment of *true* to u_i, l_{i2} for u_i *false*. Connect with every other vertex of l_{i1} a vertex representing an input literal u_i, with every other vertex of l_{i2} a vertex representing an input literal $\bar{u}_i$. Since for each variable there are at most five input literals, the ten vertices of each of l_{i1} and l_{i2} will suffice for this purpose. As in the proof of Theorem 2.1, let the vertices representing literals of the same clause form a class. Also, the j-th vertices of lists l_{i1} and l_{i2}, $1 \le i \le m$, $1 \le j \le 10$, form a class. Now create a list l_c of $6m - 2$ vertices, each belonging to a new class. Connect l_{11} with the first vertex of l_c, and then connect all remaining lists l_{ij}, $1 \le i \le m$, $1 \le j \le 2$, with every third vertex of l_c. Embed the tree constructed in this way into a unit grid as it is shown in Figure 2.

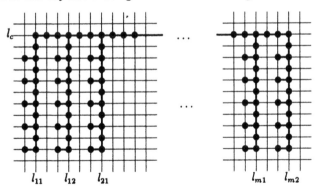

Figure 2: Reduction of 3SAT to GEOMETRICAL MINIMUM CLASS TREE

This tree consists of $n + 10m + 6m - 2$ classes; now, let $k = n + 10m + 6m - 2 - 1$.

Claim: The answer to GEOMETRICAL MINIMUM CLASS TREE is "yes" iff the answer to 3SAT is "yes".

if: Our arguments are similar to those in the proof of Theorem 2.1. There must exist an assignment of values *true*, *false* to the variables u_i, $1 \le i \le m$, satisfying each clause, since the answer to 3SAT is "yes". For each variable u_i that is true, select the corresponding list l_{i1} to belong to CT, for each false variable u_j select l_{j2}, $1 \leq i, j \leq m$. Connect these lists with the complete list l_c. Eventually, for each clause c_i connect the vertex representing the first true literal of c_i with CT by an edge of length 1. Since each clause is satisfied, exactly one vertex of each of the n classes representing clauses belongs to CT. Moreover, each of the $6m-2$ classes of l_c is in CT and one vertex of each of the $10m$ classes for the m variables is in CT, too, since either l_{i1} or l_{i2}, $1 \leq i \leq m$, completely belongs to CT. Since all edge lengths in CT are 1, the total length of CT is $n + 10m + 6m - 2 - 1$, so the answer to GEOMETRICAL MINIMUM CLASS TREE is "yes".

only if: The answer to GEOMETRICAL MINIMUM CLASS TREE is "yes", hence, a class tree CT with total length at most $n + 10m + 6m - 2 - 1$ exists. Since all edge lengths are at least 1, the number of vertices in CT is at most $n + 10m + 6m - 2$. Because the number of classes also is $n + 10m + 6m - 2$, it follows that for each class exactly one vertex belongs to CT and that each edge in CT has length 1. Therefore, for any i, $1 \leq i \leq m$, either the complete list l_{i1} or l_{i2} may be in CT, but not parts of both or even both completely. Moreover, for each of the n classes representing clauses, the single vertex v that belongs to CT must be connected with one of the lists l_{ij}, $1 \leq i \leq m$, $1 \leq j \leq 2$, that belongs to the tree, otherwise the length of the edge connecting v with CT is at least 2. Clearly, l_c must belong to CT completely. It follows that we can assign the variable u_i the value *true*, if l_{i1} belongs to CT, and *false* otherwise, $1 \leq i \leq m$, thereby satisfying each of the n clauses. Hence, the answer to 3SAT is "yes". □

Corollary 2.1: GEOMETRICAL MINIMUM CLASS TREE is NP-complete if the distance between points in the plane is defined by any of the Minkowski metrics L_p, $1 \leq p \leq \infty$.

Proof: From Figure 2 it follows that for any metric L_p, $1 \leq p \leq \infty$, the distance defined by L_p between any two vertices in the graph is at least 1. Since only edges of length 1 belong to the minimum class tree constructed in the proof of Theorem 2.2, the corollary follows. □

With the proof of Theorem 2.2 also the following corollary is self-evident.

Corollary 2.2: MINIMUM CLASS TREE is NP-complete for class graphs that are part of a unit grid and where both, the vertex degree and the number of vertices per class, is restricted to at most three.

Furthermore, it follows immediately that MINIMUM CLASS TREE is NP-complete also for trees with edge lengths restricted to 1.

2.3 The minimum class tree of a minimum spanning tree

Since computing a minimum class tree for a graph as well as for points in the plane is NP-complete, approximate rather than exact solutions are of interest [3, 4, 6]. A straightforward approach to approximate a minimum class tree is first to compute a minimum spanning tree for the vertices of the given graph or for the given set of points in the plane, and then to compute a minimum class tree *MCT* for the given minimum spanning tree *MST*. NP-completeness of *MCT* for an input tree does not immediately imply NP-completeness of *MCT* for an input *MST*. The following corollary, however, shows that *MCT* is a hard problem, even for this specific input.

Corollary 2.3: The decision whether there exists a minimum class tree of length at most k is NP-complete for the following input graphs:

1. class graphs that are minimum spanning trees for the vertex sets of complete graphs;

2. class graphs that are minimum spanning trees for arbitrary sets of points in the plane, where the distance between points is defined by any of the Minkowski metrics L_p, $1 \leq p \leq \infty$;

3. class graphs that are minimum spanning trees of graphs that are part of a unit grid.

Proof:

ad 1) Consider the proof of Theorem 2.1 and the graph G shown in Figure 1. Let G' be the complete graph of G, obtained by adding to G all missing edges and assigning to each of the new edges a length greater than the sum of all edge lengths of G. Then, G is a minimum spanning tree for the given set of vertices.

ad 2), 3) Consider the proof of Theorem 2.2 and the graph G shown in Figure 2. The number of vertices in G is $3n + 20m + 6m - 2$. Since the minimum distance between any two vertices is 1, a minimum spanning tree for G must have length $3n + 20m + 6m - 2 - 1$. But this is exactly the total length of G; hence, G is a minimum spanning tree for the given set of vertices. Obviously, this result holds for sets of points in the plane with any of the Minkowski metrics L_p, $1 \leq p \leq \infty$, as well as for graphs that originate from a unit grid. $\qquad\square$

3 Minimum diameter class tree is easy

As mentioned in Section 1, when connecting networks by a tree it sometimes will be more appropriate to minimize the length of connections between pairs of networks rather than minimizing the total length of the connecting network. Hence, the connecting network should be a minimum diameter class tree.

In this section, we will describe how to find a class tree with minimum diameter for n points and c classes in the plane, a *geometrical minimum diameter class tree gMDCT*, in time $O(n^3)$.

First, consider the problem of computing the *geometrical minimum diameter spanning tree gMDST* for a Euclidean graph without classes. In [2], it was shown that the diameter of a *gMDST* with the minimum number of edges contains at most three edges and can be computed in time $O(n^3)$ for n points in the plane.

Obviously, for class graphs also, the diameter of a *gMDCT* with the minimum number of edges contains at most three edges. Hence, the *gMDCT* has as its center either a single point, called a *monopole*, or two points, called a *dipole*. A *gMDCT* then can be determined by computing for each possible monopole and for each possible dipole a class tree with minimum diameter and selecting the best of these.

To compute a monopolar tree, first the distance of each point to the monopole is computed. In a second step, for each class the point closest to the monopole is determined. Let p be the monopole and let p' and p'' be the two points of different classes that have the largest distances to the monopole among the points determined in the second step. Then, the diameter is the sum of the distances between p and p' and between p and p''. Computing one monopolar tree clearly takes time $O(n)$.

The computation of a dipolar tree is slightly more complicated. Let the dipole be two points p and q. We must compute two circles, with centers at p and q, respectively, such that both circles together contain a point of each class and the sum of the radiusses is minimal. Then, the diameter is the sum of the radiusses of the two circles plus the length of the edge between p and q. The two circles can be found as follows: Let the points be given in two lists, one sorted by distance to p, the other sorted by distance to q. Eliminate for each class all but one point of smallest distance from the p-list and from the q-list; the resulting lists each contain c points. Now, scan the p-list in increasing order. For each point p' in the list, a class is said to be contained in the circle of p' around p if a point of that class is in that circle. Now, determine the circle with smallest radius around q that contains a point for each class not contained in the circle of p' around p. For the sequence of all these points p', the corresponding circle around q can be determined by scanning the q-list once in decreasing order, since the circle around p grows monotonically for increasing distance of p' to p, and the circle around q shrinks monotonically. During the scan, for a point q' in the q-list we check whether the class of q' is already contained in the circle of p' around p. By marking each class contained in the circle of p' around p, each check can be performed in constant time. Hence, the scan of the lists costs only $O(n)$ time for each dipole. If the class of q' is already contained in the circle around p, q' is skipped and the next point in the q-list is inspected. For the smallest circle around q containing points of all classes that are not contained in the circle of p' around p, the diameter is computed. That is, for each point in the p-list a dipolar class tree and its diameter are computed. For the dipole p, q, we keep track of the smallest of these trees.

Since there are n points that can be the center of a monopolar tree and n^2 pairs of points that can be the centers of a dipolar tree, it takes time $O(n^3)$ to compute the set of class trees in which a class tree with minimum diameter and minimum number of edges on the diameter must be contained.

We get the following theorem.

Theorem 3.1: A geometrical minimum diameter class tree for a Euclidean graph with n points can be computed in time $O(n^3)$.

4 Conclusion

We have considered the problem of computing a global network that connects local networks. The new network should either have minimum total length or minimum diameter. We modelled the problem by a more general type of problem: a given set of points in the plane or of vertices of a graph is partitioned into classes. Each class represents a property common to all points in a class. Finding a global network of minimum total length then equals computing a minimum class tree, i.e., a tree of minimum total length that contains at least one point of each class.

We showed that computing a minimum class tree is NP-hard for points in the plane, where the distance between points is defined by any of the Minkowski metrics L_p, $1 \leq p \leq \infty$. The same result holds for distance graphs, even under several restrictions, like e.g., if the given graph is a tree, where the vertex degree is at most three and where each class contains no more than three vertices.

Finding a network with minimum diameter that connects all local networks can be modelled by computing a tree with minimum diameter that contains at least one point of each of the classes. We showed that a class tree with minimum diameter can be computed in time $O(n^3)$ for n points in the plane.

There are other application areas where problems can be modelled by classes of vertices of a graph or classes of points in the plane. In the wire routing phase in physical VLSI design, for instance, components on a chip may be flipped or rotated without changing the placement of the components on the chip, generating a class of possible pin positions (see e.g. [6]).

References

[1] M.R. Garey, D.S. Johnson:
Computers and intractability: a guide to the theory of NP-completeness, Freeman, New York, 1979.

[2] J.-m. Ho, D.T. Lee, C.-H. Chang, C.K. Wong:
Bounded-diameter minimum spanning trees and related problems, Symp. on Comp. Geometry, 1989, 276–282.

[3] E. Ihler:
Bounds on the quality of approximate solutions to the Group Steiner Problem, Graph-Theoretic concepts in Computer Science, WG90, Lecture Notes in Computer Science, Springer, 1991.

[4] E. Ihler:
The complexity of approximating the class Steiner tree problem, Tech. Report, Institut für Informatik, Universität Freiburg, Germany, 1991.

[5] F.P. Preparata, M.I. Shamos:
Computational Geometry: an introduction, Springer, New York, 1985.

[6] G. Reich, P. Widmayer:
Beyond Steiner's problem: a VLSI oriented generalization, Graph-Theoretic concepts in Computer Science, WG89, Lecture Notes in Computer Science, Vol. 411, Springer, 1990, 196–210.

Determination of the symmetries of polyhedra and an application to object recognition

X. Y. Jiang,* H. Bunke
Institut für Informatik und angewandte Mathematik
Universität Bern, Länggass-Strasse 51, 3012 Bern, Switzerland

Abstract

In this paper we present a simple and efficient algorithm for determining the rotational symmetries of polyhedral objects in $O(m^2)$ time using $O(m)$ space where m represents the number of edges of the object. Our algorithm is an extension of Weinberg's algorithm for determining isomorphisms of planar triply connected graphs. The symmetry information detected by our algorithm can be utilized for various purposes in artificial intelligence, robotics, assembly planning and machine vision. In particular, an application of symmetry analysis to object recognition will be described in some detail.

1 Introduction

Information about the symmetry of solid objects can serve several purposes in artificial intelligence and related fields. Some examples are efficient model representation of symmetrical objects [4, 12], robotics [15], assembly planning [10], and machine vision where the symmetry information can be applied to speed up the process of object recognition. An example of such a vision system has been described recently [7, 8]. It is based on a matching procedure that tries to find correspondences between regions extracted from an image of a scene and regions stored in a model base. Utilizing information about regions equivalent with respect to symmetry, a number of redundant matches can be avoided. Thus the incorporation of symmetry information in the model base of a machine vision system can result in a significant speedup of the recognition process. Another system that makes explicit use of symmetry information is described in [4].

In industrial machine vision, recognition is usually based on object models. One method of model generation is the deductive approach where vision models are automatically derived from CAD-models [3, 5, 6]. In an industrial environment the deductive approach can be superior to other appoaches, particularly if CAD-models of the objects to be recognized are available, without additional costs, from the earlier stage of object design. As a shortcoming, however, CAD-models usually don't contain symmetry information. Thus this kind of information must be separately derived from the topological and geometrical data of the CAD-model if it is to be applied in a machine vision system.

In this paper we discuss the problem of determining rotational symmetries of polyhedral objects. The underlying motivation is to facilitate the automatic generation of vision models from CAD-models. We restrict our consideration to rotational symmetry and neglect mirror symmetry which cannot in general be realized by any physical movement and, thus, has little relevance to 3-D vision. In [16] optimal algorithms have been proposed for finding the symmetries of a

---

*The support of the Swiss National Science Foundation under the NFP-23 program, Grant No. 4023-027026, is gratefully acknowledged.

number of geometric entities in two and three dimensions. The algorithm for polyhedra runs in $O(n \log n)$ time where n is the number of vertices. Because of a complicated linear time graph isomorphism algorithm used therein, the algorithm for polyhedra has a rather large constant. The authors stated "While the asymptotic behavior of the algorithms is good, the 3D cases share a rather large constant because they require a graph isomorphism test. Thus, the full 3D symmetry algorithms are of primarily theoretical interest."

A polyhedron can be thought as a set of vertices, edges, and faces. Any rotational symmetry axis of the polyhedron goes through two elements of this set. An algorithm for finding rotational symmetries of polyhedra has been reported in [13]. It is based on a type of generate and test scheme. Partial descriptions of hypothetical symmetry axes are generated and then verified. A byproduct of the verification process is the completion of the symmetry axis description if the hypothetical axis is in fact a symmetry axis. The algorithm has a quadratic time complexity. In this algorithm all possible symmetry axes are classified into three different classes: axes which pass through at least one face; at least one vertex; or two edges. For each class of symmetry axes, a different treatment is needed. Two dual data structures are used. As the hypothetical symmetry axes pass through two features of the polyhedron (vertices, edges and faces), the proposed algorithm doesn't work on polyhedra with holes.

In this paper we present a simple algorithm for determining the symmetry of polyhedra. Although our algorithm has a quadratic complexity and is thus not optimal, it has a very small constant. Furthermore, our algorithm can be easily implemented in linear time on a parallel architecture. Like the algorithm in [13], our algorithm uses also a type of generate and test scheme. However, our 'generate' scheme generates 3-D rotation matrices as hypotheses. Thus, all three classes of symmetry axes can be uniformly treated, and the algorithm is much simpler. Another advantage of our algorithm over that in [13] is that it also works on polyhedra with holes.

2 Symmetry analysis

A polyhedral object comprising n vertices, m edges and h faces can be defined as a graph $G = (V, E, F)$ embedded on the surface of a solid object with

$$V = \{v_1, v_2, \ldots, v_n\}$$
$$E = \{e_k = (v_{k1}, v_{k2}) \mid k = 1, 2, \ldots, m\}$$
$$F = \{f_k = (v_{k1}, v_{k2}, \ldots, v_{kl_k}) \mid k = 1, 2, \ldots, h\}$$

where V represents the set of vertices, E is the set of edges, and F is the set of oriented faces where each face is represented by the closed chain of its vertices. A chain of vertices is oriented clockwise if we look at the corresponding face from outside of the polyhedron. In this work we only consider polyhedra whose vertex-edge-graph is connected.

A rotational symmetry $Sym(\theta, R)$ consists of a topological part and a geometrical part. The topological part is an automorphism of the embedded graph G, i.e. there exists a bijective mapping $\theta : V \to V$ such that the following conditions are fulfilled

$$(v_k, v_j) \in E \implies (\theta(v_k), \theta(v_j)) \in E.$$
$$(v_{k1}, v_{k2}, \ldots, v_{kl_k}) \in F \implies (\theta(v_{k1}), \theta(v_{k2}), \ldots, \theta(v_{kl_k})) \in F.$$

Besides these two topological conditions, an additional geometrical constraint has to be satisfied. We require that there exists a spatial rotation, represented by a 3×3 rotation matrix R, such that

$$v_k \in V \implies R \cdot v_k = \theta(v_k), \tag{1}$$

i.e. the location of the rotated version $\theta(v_k)$ of any vertex v_k must be identical to the location obtained by applying the rotation R to v_k.

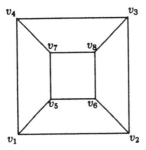

Figure 1: A finite connected graph.

Symmetry represented in the form $Sym(\theta, R)$ is sometimes not easy to use in practice. Another way of symmetry representation is to record the equivalence of vertices, edges, or faces under symmetry transformations. For this purpose we define a relation $e_V \subseteq V \times V$,

$$(v_i, v_j) \in e_V \Longleftrightarrow \exists Sym(\theta, R)(\theta(v_i) = v_j).$$

Clearly, e_V is an equivalence relation. We can thus build the equivalence classes of e_V. All vertices of some equivalence class are equivalent in the sense that a symmetry rotation transforms one into another. Similarly, two relations e_E and e_F for edges and faces, respectively, can also be defined. For the sake of description clarity, we will only consider e_V in the following.

The task considered in this paper is to determine all symmetries $Sym(\theta, R)$ and all equivalence classes of e_V of a polyhedron.

3   Basic results

In this section we give some results of graph theory on which our symmetry detection algorithm is based.

One of the basic results concerning path problems in graph theory is

Theorem 1 *In a finite connected graph it is always possible to construct a cyclic directed path passing through each edge once and only once in each direction.*

For a proof see [9, p. 41]. As an example, we consider the graph in Fig. 1. One possible path of this kind is

$$P(v_1 v_2) = v_1 v_2 v_3 v_4 v_1 v_4 v_7 v_5 v_1 v_5 v_6 v_2 v_6 v_8 v_3 v_8 v_7 v_8 v_6 v_5 v_7 v_4 v_3 v_2 v_1. \tag{2}$$

A simple way of actually finding such a path in an embedded graph was given in [14], which is a variation of a method due to Trémaux [1, 2]. We call a vertex *old* or *new* if it has or has not been reached previously, respectively. An edge being traversed in one direction is termed *old* if it has been previously traversed in the opposite direction, and *new* otherwise. In traversing an edge we go from an *initial* vertex to a *terminal* vertex. Starting with an edge traversed in one of its directions, we follow these rules:

1. When a new vertex is reached, take the right-most edge relative to the edge on which the vertex is reached.

2. When an old vertex is reached on a new path, go back in the opposite direction.

3. When an old vertex is reached on an old path, leave the vertex on the right-most edge that has not previously been traversed in that direction. (Thus, edges are traversed only once in each direction.)

For the graph in Fig. 1, if we start with the edge (v_1, v_2) in the direction from v_1 to v_2, we will get the path given in (2).

In his algorithm for determining isomorphisms of planar triply connected graphs [14], Weinberg added a numbering scheme to the path generation process above so that a code is obtained. The coding rule is as follows. As we reach a new vertex we label it with a number in natural order. Thus, the label of the initial vertex of the starting edge is 1, the label of its terminal vertex is 2, the label of the next new vertex is 3, a.s.o. We form a code consisting of the numbers of the vertices in the order in which they are visited. The code is thus given by a vector of dimension $2m + 1$ where m is the number of edges in the graph. The code of the path in (2) is, for example,

$$C(v_1 v_2) = 1234145616727838587654321.$$

In [14] Weinberg has proved

Theorem 2 *Let $G_1 = (V_1, E_1, F_1)$ and $G_2 = (V_2, E_2, F_2)$ be two triply connected plane graphs (graphs embedded in the plane), and (v_i, v_j), (v_i', v_j') be edges of G_1, G_2, respectively. There exists an isomorphism between G_1 and G_2 such that v_i and v_j are similar to v_i' and v_j', respectively, if and only if $C(v_i v_j) = C(v_i' v_j')$.*

Based on this theorem a simple algorithm for determining isomorphism was proposed. We choose arbitrarily an edge (v_i, v_j) of G_1 and compute $C(v_i v_j)$. Then, for each edge (v_i', v_j') of G_2, we compute $C(v_i' v_j')$. G_1 and G_2 are isomorphic if and only if there is some edge (v_i', v_j') of G_2 such that $C(v_i v_j) = C(v_i' v_j')$. Under an isomorphism found in this way, two vertices v_k and v_k' are similar if and only if they have the same coding number.

For our purpose of determining the symmetry of polyhedra, we extend Theorem 2 to:

Theorem 3 *Let $G = (V, E, F)$ be the graph of a polyhedron, and (v_i, v_j), (v_i', v_j') be two edges of G. There exists an automorphism of G such that the corresponding vertices of v_i and v_j are v_i' and v_j', respectively, if and only if $C(v_i v_j) = C(v_i' v_j')$.*

Proof. In proving Theorem 2 Weinberg has only used one property of triply connected plane graphs, namely that the set of edges incident to a vertex has a unique order around the vertex. As this remains true in the case of graphs of polyhedra, Theorem 2 holds also in this domain. Finally, since an automorphism is an isomorphism of a graph to itself, our theorem is thus proved. □

According to this theorem, the isomorphism detection algorithm given above can also be used to determine the automorphisms of graphs of polyhedra.

4   The algorithm

A rotational symmetry $Sym(\theta, R)$ of a polyhedron $G = (V, E, F)$ consists of an automorphism of the embedded graph G and and a rotation matrix R such that the geometrical constraint in (1) is satisfied. As we have seen in the last section, Weinberg's algorithm can be used to find all automorphisms of G. In this section we extend it to incorporate the geometrical constraint.

In order to simplify the calculation of spatial rotations it is advantageous to define a coordinate system so that the rotation axis goes through the origin. Since for a symmetrical object the axis of any symmetry rotation goes through the centroid of the vertex set V, we first transform the coordinate system in which the polyhedral object is initially defined so that the new origin corresponds to the centroid of V.

The simplest way of incorporating the geometrical constraint proceeds as follows. We choose arbitrarily an edge (v_i, v_j) of G and compute $C(v_i v_j)$. Then, for each other edge (v_i', v_j') of G, we compute $C(v_i' v_j')$. If $C(v_i' v_j') = C(v_i v_j)$, then G has an automorphism θ which maps v_i and v_j to v_i' and v_j', respectively. In this case we check the geometrical constraint by testing whether a spatial rotation can be found to transform v_k to $\theta(v_k)$, for each $v_k \in V$.

For two reasons, however, this approach is not economical. First, if some vertex number during the generation of $C(v_i' v_j')$ is not identical to the corresponding vertex number in $C(v_i v_j)$, we needn't to continue the coding process. Secondly, three vertices v_i, v_j and v_k together with their rotated versions v_i', v_j' and v_k' under the mapping θ determine the underlying rotation unambiguously. The matrix R representing the transformation from v_i, v_j, v_k to v_i', v_j', v_k' can be computed by

$$R = [v_i' \ v_j' \ v_k'] \cdot [v_i \ v_j \ v_k]^{-1}. \tag{3}$$

Thus, when the first three numbers of $C(v_i' v_j')$ have been generated, the transformation matrix R can be determined and used to check the geometrical constraint during the generation of the other numbers of $C(v_i' v_j')$. As soon as the geometrical constraint is violated by a vertex, the coding process can be stopped.

Based on the discussions above we give now a simple algorithm which incorporates the geometrical constraint in the coding process. We assume that the polyhedron is represented in some suitable data structure so that we can retrieve the right-most edge of a vertex relative to another edge in $O(1)$ time. In our implementation we used the double-connected edge list (DCEL) described in [11]. In the formulation of the algorithm we will call three neighboring vertices (v_i, v_j, v_k) of the vertex chain of some face a *wedge*. According to the path traversing rules, the first three vertices in a path build always a wedge. Our symmetry detection algorithm consists of eight steps.

1. Search for a wedge (v_i, v_j, v_k) of G such that the matrix $[v_i \ v_j \ v_k]$ is invertible. Compute

$$P(v_i v_j) = u_1 u_2 u_3 u_4 u_5 \cdots u_{2m} u_{2m+1}$$

where $u_1 = v_i$, $u_2 = v_j$, $u_3 = v_k$ and

$$C(v_i v_j) = b_1 b_2 b_3 b_4 b_5 \cdots b_{2m} b_{2m+1}$$

where $b_1 b_2 b_3 = 123$.

2. For each wedge (v_i', v_j', v_k') of G, which is different from (v_i, v_j, v_k), do steps 3 to 8.

3. Compute the transformation matrix R by (3). If R is a rotation matrix, i.e. $RR^T = I_3$ holds, continue the coding process with step 4, otherwise go to step 2 for checking the next wedge.

4. Initialization: $P(v_i' v_j') := v_i' v_j' v_k'$, $C(v_i' v_j') := 123$, $\theta(v_i) := v_i'$, $\theta(v_j) := v_j'$, $\theta(v_k) := v_k'$.

5. For $i := 4$ to $2m+1$ do steps 6 and 7.

6. Generate the i-th vertex u_i' of $P(v_i' v_j')$ and the i-th number b_i' of $C(v_i' v_j')$. If $b_i' \neq b_i$, then go to step 2 for checking the next wedge.

7. If u_i' is a new vertex (reached for the first time), then check the geometrical constraint by testing whether $R \cdot u_i = u_i'$ holds. If this is true, then register $\theta(u_i) = u_i'$, otherwise go to step 2 for checking the next wedge.

8. Now a symmetry $Sym(\theta, R)$ has been found. We incrementally update the equivalence relation e_V. (This step will be explained below.)

If the coding process and the geometric constraint test are successful, the symmetry just found is registered in step 8. Before all symmetries have been found, however, the relation e_V is only partially defined. How can we incrementally register the equivalence relation e_V? For each equivalence class

$$\{v_{j1}, v_{j2}, \ldots, v_{jk}\}, \quad j1 < j2 < \ldots < jk,$$

we record the equivalence relationships

$$(v_{j1}, v_{ji}) \in e_V, \; i = 2, 3, \ldots, k.$$

This proceeds as follows. We assume that the vertices are numbered from 1 to n as $v_1, v_2, \ldots, v_n$ and use an array $rec[1..n]$ which is initialized with zeros. After finding a symmetry $Sym(\theta, R)$ we get a new set of pairs $(v_i, v_j) \in e_V, i < j$. For each of them, the array entry j is updated by

$$\text{if } (rec[j] = 0) \text{ or } (i < rec[j]) \text{ then } rec[j] := i$$

After all symmetries have been found, the array rec will look like

$$rec[l] = \begin{cases} 0, & \text{if } v_l = v_{j1} \\ j1, & \text{if } v_l = v_{ji}, \; i = 2, 3, \ldots, k \end{cases}$$

For v_{j1} this is true because there exists no $v_i, i < j1$, such that $(v_i, v_{j1}) \in e_V$. The entry $rec[j1]$ will never be updated and remains zero. For $v_{ji}, i = 2, 3, \ldots, k, rec[ji]$ will be updated to $j1$ after finding $(v_{j1}, v_{ji}) \in e_V$ and remains unchanged after that update. Finally, one scanning of the array rec will produce all equivalence classes of e_V.

Now we consider the complexity of our algorithm. For a polyhedron with m edges, there are $2m$ wedges. Thus, step 1 runs in time $O(m)$. The loop through steps 3 to 8 is repeated $2m$ times. In each loop, $O(m)$ operations are required. Totally, our algorithm has an $O(m^2)$ time complexity. Summing up the space requirement of the DCEL for representing a polyhdron and the array used in symmetry registration, our algorithm needs $O(m)$ space.

The time complexity analysis above corresponds to the worst case. If the object is not highly symmetrical it is most likely that, for many wedges (v'_i, v'_j, v'_k) of G, the coding process is stopped early because of either inequality of $C(v_i v_j)$ and $C(v'_i v'_j)$, or violation of the geometrical constraint. Thus it can be expected that the average time complexity is substantially better than the worst case.

5 An application to object recognition

Symmetry analysis has many applications, especially in computer vision and related fields. In this section we discuss a concrete application to object recognition. The recognition of 3-D objects is a central problem of computer vision and pattern recognition. Given an input image, the characteristic features of the objects actually contained in this input image, e.g. corners, lines, holes and surface patches, can be extracted. On the other hand, all objects that possibly occur in a scene are represented in terms of the same features and represented in a model base. Then the problem of model-based object recognition can be considered as one of searching for a consistent matching between the image features and the model features. That is, we need to assign each image feature to a corresponding model feature in a manner consistent with the assginment of the other image features.

For the sake of description clarity, we assume that there is only one object in the scene and the image features can be perfectly extracted from the sensory data. In our concrete case we use

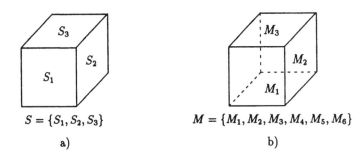

$$S = \{S_1, S_2, S_3\}$$

$$M = \{M_1, M_2, M_3, M_4, M_5, M_6\}$$

a)

b)

Figure 2: a) A scene. b) A cube.

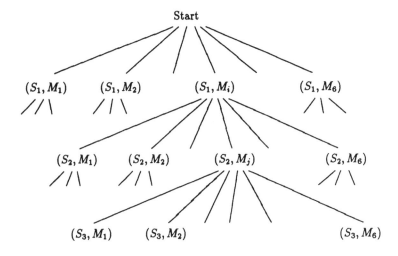

Figure 3: The search tree.

surface patches as features. In Fig. 2a) an input image containing a cube is shown in which three surface patches $S = \{S_1, S_2, S_3\}$ have been extracted. On the other hand, the cube is represented in the model base by six surface patches $M = \{M_1, M_2, M_3, M_4, M_5, M_6\}$. Then the recognition means assigning each $S_i \in S$ to some $M_j \in M$ such that two consistency constraints are satisfied. First, the surface patches S_i and M_j must be compatible. An incompatibility occurs, for instance, if S_i is a parallelogram and M_j is a triangle. Secondly, the assignment (S_i, M_j) must be consistent with all other assignments. For example, the assignment (S_{i1}, M_{j1}) is inconsistent with (S_{i2}, M_{j2}) if S_{i1} is a neighbor of S_{i2} but M_{j1} isn't a neighbor of M_{j2}. In Fig. 2a), (S_1, M_i) is consistent with (S_2, M_k) only if M_i is adjacent to M_k.

One way of actually searching for consistent matchings between the image features and the model features is tree search as shown in Fig. 3. Of the many variations of tree search, backtracking is the simplest one. Starting at a root node we construct a search tree in a depth-first fashion. At the first level of the tree, we consider assigning the first sensed surface patch to all possible model patches; at the next level, we assign the second sensed patch to all model patches, and so on. After each assignment (S_i, M_j) we test the consistency of S_i and M_j and the consistency of (S_i, M_j) and all assignments on the path from the root node to (S_i, M_j). As soon as an inconsistency is detected, we delete the assignment (S_i, M_j) and go to consider the next assignment for S_i. If all assignments for S_i have been checked and no candidate M_j for a consistent match (S_i, M_j) has been found, we go back to the father of (S_i, M_j) and consider the next assignment for S_{i-1}. A solution is found if some assignment (S_n, M_j) for the last sensed surface patch – S_3 in our example in Fig. 2 – is reached and all the consistency tests for (S_n, M_j) have been successfully passed through.

Since the search tree is potentially very large, it is crucial to apply constraints that efficiently restrict the portions of the search space that must be explored. In [7, 8] a set of effective constraints have been defined, one of which concerns the use of symmetry information.

Because of the high symmetry of the cube, there are many possible solutions in our example which are all equivalent. For the scene in Fig. 2a),

$$\{(S_1, M_1), (S_2, M_2), (S_3, M_3)\}$$

$$\{(S_1, M_2), (S_2, M_3), (S_3, M_1)\}$$

$$\{(S_1, M_3), (S_2, M_1), (S_3, M_2)\}$$

are three among them. As mentioned in section 2, it is possible to define, similar to e_V, a relation e_F and to build the equivalence classes of e_F. All faces of some equivalence class are equivalent in the sense that a symmetry rotation transforms one into another. In the case of the cube, we get a single equivalence class containing all the six faces. Note that the search space in Fig. 3 consists of six subspaces being the subtrees below $(S_1, M_i), i = 1, 2, \ldots, 6$. As all M_i's are equivalent, we conclude that a search in the subtree below (S_1, M_1) is enough for the purpose of object recognition, independent of whether the search in this subtree is successful or not. The search in the other five subtrees will only find equivalent solutions and thus not contribute to object recognition. The symmetry analysis enables us therefore to reduce the actual search space to one sixth of the potential search space.

The description above is based on some idealized assumptions. In a practical object recognition system, we have to consider a number of additional factors like the existence of multiple objects in the scene and the imperfect nature of feature detection. Furthermore, backtracking is usually not the best way of doing tree search. These questions are addressed in more details in [7, 8].

6   Discussion

In this paper we have proposed an algorithm for determining the rotational symmetries of polyhedral objects which is an extension of Weinberg's algorithm for determining isomorphisms of planar triply connected graphs. Our symmetry detection algorithm runs in $O(m^2)$ time and needs $O(m)$ space. As our algorithm has a a quadratic complexity, it is not optimal. The strength of our algorithm, however, lies in its simplicity and its inherent parallelism.

The algorithm described in section 4 is easy to understand and to code. As a matter of fact, our actual implementation has only about 120 lines of code. Although the worst case time complexity is quadratic, the average case occurring in practical applications can be expected substantially better than the worst case. The simplicity and the efficiency make our algorithm an attractive choice in applications where the polyhedra have a relatively small number of edges, or the symmetry detection takes place in an off-line phase as is the case in deductive model generation for industrial machine vision.

Our symmetry detection algorithm is of parallel nature. In fact, the coding process can be done independently for each of the $2m$ wedges. With m processors we get a parallel algorithm of the complexity $O(m)$.

References

[1] W. W. Rouse Ball, Mathematical recreations and problems, Macmillan, London, 1892. Numerous editions.

[2] C. Berge, Graphs, North-Holland, 1985.

[3] B. Bhanu and C. C. Chen, CAD-based 3D Object Representation for Robot Vision, *IEEE Computer* 20, No. 8, 1987, 19–35.

[4] R. C. Bolles and R. A. Cain, Recognising and locating partially visible objects, *Int. Journal of Robotics Research* 1, No. 3, 1982, 57–82.

[5] P. J. Flynn and A. K. Jain, CAD-based computer vision: From CAD models to relational graphs, *IEEE Trans. on Pattern Analysis and Machine Intelligence*, 13, No. 2, 1991, pp. 114–132.

[6] T. Glauser, E. Gmür, X. Y. Jiang and H. Bunke, Deductive generation of vision representations from CAD-models, in *Proc. of 6th Scand. Conf. on Image Analysis, Oulu, Finland, 1989*, pp. 645–651.

[7] X. Y. Jiang and H. Bunke, Recognition of overlapping convex objects using interpretation tree search and EGI matching, in *Proc. of SPIE Conf. on Applications of Digital Image Processing XII*, Vol. 1153, San Diego, 1989, pp. 611–620.

[8] X. Y. Jiang and H. Bunke, Recognizing 3-D objects in needle maps, in *Proc. of 10th Int. Conf. on Pattern Recognition, Atlantic City, New Jersey, 1990*, pp. 237–239.

[9] O. Ore, Theory of graphs, American Mathematical Society, 1962.

[10] R. J. Popplestone, Y. Liu and R. Weiss, A group theoretic approach to assembly planning, *AI magazine*, 11, No. 1, 1990, 82–97.

[11] F. R. Preparata, M. I. Shamos, Computational geometry, Springer-Verlag New York Inc., 1985.

[12] M. Shneier, A compact relational structure representation, in *Proc. of 6th IJCAI, Tokyo, Japan, 1979*, pp. 818–826.

[13] R. Waltzman, Geometric problem solving by machine visualization, CS-TR-2291, Center for Automation Research, University of Maryland, 1989.

[14] L. Weinberg, A simple and efficient algorithm for determining isomorphism of planar triply connected graphs, *IEEE Trans. on Circuit Theory*, 13, No. 2, 1966, pp. 142–148.

[15] J. D. Wolter, R. A. Volz and T. C. Woo, Automatic generation of gripping positions, University of Michigan, Ann Arbor, February, 1984.

[16] J. D. Wolter, T. C. Woo and R. A. Volz, Optimal algorithms for symmetry detection in two and three dimensions, *The Visual Computer*, 1, 1985, 37–48.

Moving Along a Street
(Extended Abstract)

Rolf Klein*

May 31, 1991

Abstract

A polygon with two distinguished vertices, s and g, is called a *street* iff the two boundary chains from s to g are mutually weakly visible. For a mobile robot with on-board vision system we describe a strategy for finding a short path from s to g in a street not known in advance, and prove that the length of the path created does not exceed $1 + \frac{3}{2}\pi$ times the length of the shortest path from s to g. Experiments suggest that our strategy is much better than this, as no ratio bigger than 1.8 has yet been observed. This is complemented by a lower bound of 1.41 for the relative detour each strategy can be forced to generate.

Key words: Shortest paths, simple polygons, path planning, uncertainty, robotics, navigation, computational geometry.

1 Introduction

How to quickly determine the *shortest path* between two points in a simple polygon P is a classical problem in computational geometry. An optimal solution was provided by Guibas and Hershberger [5] by proving that one can, in $O(n)$ preprocessing time, build up a search structure of size $O(n)$ that allows the shortest path between any two points in P to be computed within time $O(\log n + k)$, where n is the number of edges of P and k denotes the number of line segments the shortest path consists of. The preprocessing step requires a triangulation of P; due to Chazelle [2] or Seidel [12], this can be computed in $O(n)$ worst case or randomized time, too.

Such algorithms are based on the assumption that the whole polygon is known in advance. In real life, however, one often has to move through an environment without completely knowing it, but rather on the basis of *local* information provided by acoustical, visual, or tactile sensors. Given the importance of this problem it is quite surprising how few results exist; see e.g. [9, 8, 7, 3, 4], or [10, 11] for further references. Lumelsky and Stepanov [8] studied the case of a mobile robot equipped with a tactile sensor in an environment of obstacles. The robot is given the coordinates of the goal and of its own position in the plane; it starts heading straight to the goal until it hits an obstacle. Then it searches its contour for a point with minimum distance to the goal, and resumes from there. This simple strategy finds a path to the goal, if there is one, and the length of the path is bounded by 1.5 times the sum of the perimeters of all obstacles that are not farther away from the goal than the start point. Papadimitriou and Yanakakis [11] considered scenes of disjoint isothetic rectangles. They were able to bound the lenght of the generated path in terms of the length of the shortest path. Similar bounds were achieved by Eades, Lin, and Wormald [4] for barriers perpendicular to

---

*Praktische Informatik VI, FernUniversität–GH–Hagen, Elberfelder Straße 95, D–5800 Hagen, Germany. This work was partially supported by the Deutsche Forschungsgemeinschaft, grant KI-655, 2-1.

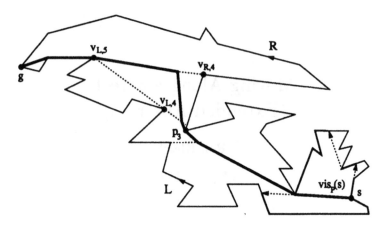

Figure 1: A street P, the visibility polygon of P at s, and the path found by our strategy.

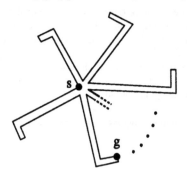

Figure 2: No strategy can guarantee a bounded relative detour in this case.

the line connecting start and goal, and by Blum, Raghavan, and Schieber [1] for more general convex obstacles; the latter paper also includes a randomized algorithm for non-convex obstacles.

In this paper we consider the following problem. Let P be a simple planar polygon with a start vertex, s, and a goal vertex, g. Assume that at vertex s a mobile robot is located that wants to get to g on as short a path as possible. The robot is equipped with a vision system that provides, for each point p in P, the *visibility polygon* $vis_P(p)$ of P at p; see Figure 1. The goal, g, is marked so that the robot can recognize it as soon as it sees it.

We do not discuss here the issues of image processing or the computational complexity involved. Rather, we are interested in a general strategy S such that the *relative detour*

$$D_S(P) := \frac{\text{length of the path created by strategy } S}{\text{length of the shortest path}}$$

becomes as small as possible. Note that just to find the goal represents no problem because the robot could simply follow the boundary until g is encountered (this is what the strategy of [8] mentioned above would in general do.) The difficulty is in keeping the detour small.

With general polygons one cannot hope for the relative detour to be bounded. In Figure 2, for example, there is no way of finding the goal other than by trying the streets leading away from the central "crossing" one by one.

In this paper we study a special class of polygons, based on the following observation. Racetracks and rivers like the Rhine contain many curves and bays, but (almost) no cul-de-sacs leading away from the main route. We formalize this property as follows.

Definition 1.1 Let P be a simple polygon with two distinguished vertices, s and g, and let L and R denote the oriented boundary chains leading from s to g. Then P is called a *street* iff L and R are mutually weakly visible, i.e. if each point of L can be seen from at least one point of R, and vice versa.

An example is shown in Figure 1. In a street, the situation depicted in Figure 2 cannot occur.

It follows from Definition 1.1 that each point of L can be connected to some point of R by a line segment contained in the polygon, and vice versa. On its way from s to g the robot is to cross all these line segments, so it sees the whole of L and R. Hence, a short path to the goal also represents a good solution to the *terrain acquisition problem* addressed e.g. in [7].

Our solution consists of two independent parts. In Section 2 we describe a *igh-level strategy* that finds a path from s to g subject to the following invariants. At each position p on this path, either the robot can see the goal (then it walks straight towards it), or the robot knows which of the corners visible ahead is visited by the shortest path from s to g (then it walks straight towards this vertex), or the robot can identify two corners ahead one of which must be visited by the shortest path from s to g, but it cannot tell which one; see Figure 5 where v, too, could be the goal. In this case, the robot chooses a point t on the line segment connecting the two vertices and walks straight in direction of this point. How to choose this point is left to a *low-level strategy*. However, it is crucial for the overall length of the generated path *how* this choice is made. There are some suggestive approaches that can result in an unbounded detour; see Section 3.

In Section 3 the low-level strategy *lad* is proposed that tries to minimize the *local absolute detour*, whenever an ambiguity arises in form of two candidate corners. Whereas promising low-level strategies are easily invented, it appears to be quite difficult to analyze them. We prove in Section 4 that for each street P the estimate

$$D_{lad}(P) < 1 + \frac{3}{2}\pi$$

holds. However, experiments show that our approach works much better than this bound suggests; we have not been able to construct a street P with a relative detour $D_{lad}(P) \geq 1.8$. On the other hand, we show in Section 3 that each strategy can be forced to produce a detour of at least $\sqrt{2} = 1.141...$, by choosing a suitably bad street.

Acknowledgement. The author wants to thank Joseph S. B. Mitchell and Günter Rote for helpful discussions during the seminar on computational geometry at Schloß Dagstuhl in October 1990, and Christian Icking for helpful discussions and for conducting the experiments at the UGH Essen.

2   A High Level Strategy

First we state the visibility properties of streets that will be used by the mobile robot on its way. Let P denote a street with start and goal vertices s and g. The polygonal chains L and R are ordered in direction from s to g. For simplicity, we assume that no three vertices of P are colinear.

Lemma 2.1 *The situation shown in Figure 3 cannot occur in a street. Neither can the prolongation of edge o beyond w hit a point of L ahead of w. The same holds for chain R.*

In fact, in Figure 3, vertex v cannot see any point of R, contradicting Definition 1.1. We have shown in [6] that the conditions stated in Lemma 2.1 are also sufficient for a polygon to be a street. They can be tested in time $O(n \log n)$. Also, from each interior point of P points of L and of R are visible.

The *visibility polygon* $vis_P(p)$ from a point p in P contains the circular list of all pieces of the boundary of P that can be seen from p, called the *umbrella* of p; see Figure 1. Where two pieces

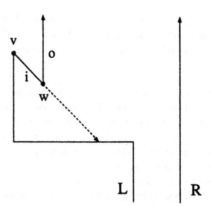

Figure 3: This situation cannot occur in a street.

meet, the one hit first by the ray from p is said to be *below* the other. Its endpoint is a reflex vertex of P, i.e. one whose internal angle is greater than π.

If the goal is not visible from p, the shortest path to g consists of a line segment leading to such a reflex vertex, followed by a polygonal chain that does not enter $vis_P(p)$ again.

The following observations are crucial.

Lemma 2.2 *As one scans the umbrella of $vis_P(p)$, all pieces belonging to L must appear consecutively and in clockwise order around p, whereas pieces of R appear consecutively in counterclockwise order, with respect to the orders on the chains.*

Lemma 2.3 *Suppose that from position p in P an initial piece of the outgoing edge of a reflex vertex v is visible. Then on each path from s to g in P there exists a position from where the incoming edge is visible.*

This fact does not imply that the robot needs to store all parts of P it has seen so far; see Corollary 2.10.

Next, the high-level strategy is listed.

```
PROCEDURE HighLevelStrategy;
CONST       s:    PointOfP;                    (* start *)
            g:    PointOfP;                    (* goal *)
VAR         p:    PointInP;                    (* current position *)
            p':   PointInP;                    (* here an event occurs *)
        vL, vR:   PointOfP;                    (* most advanced points on
                                                  L and R robot has so far
                                                  identified *)

BEGIN (* HighLevelStrategy *)
    p := s;
    determine vL and vR in visP(p);
    WHILE vL ≠ vR DO                           (* g not visible *)
        IF p, vL, vR are colinear
            THEN                               (* Case 1 *)
                p := the closer one of (vL, vR);
                walk straight to p;
                determine vL and vR in visP(p);
            ELSE                               (* Case 2 *)
                choose t in vLvR;              (* by low-level strategy *)
```

127

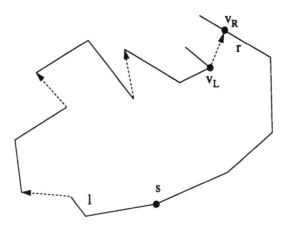

Figure 4: Here the shortest path must visit v_L.

```
        walk straight towards t UNTIL event occurs at some point p';
        p := p';
        update v_L and v_R in vis_P(p)
    END (* IF *)
  END (* WHILE *);
  walk straight to g
END HighLevelStrategy.
```

The path hereby created consists of a chain of line segments $l_1 \ldots l_m$ in P, where $l_i = \overline{p_{i-1}p_i}$, $p_0 = s$, and $p_m = g$. At the start point, p_{i-1}, of each new line segment the robot determines two points, $v_{L,i}$ and $v_{R,i}$, in the part of $vis_P(p_{i-1})$ ahead. These are the robot's orientation marks, enjoying the following properties.

Invariant 2.4 *Assume that the robot has arrived at a point $z \in l_i - \{p_i\}$ where $1 \leq i < m$. Then the following holds.*

1. *Exactly one of the cases shown in Figure 6 applies to l_i.*

2. *So far the robot has seen no part of P ahead of $v_{L,i}$ or $v_{R,i}$ except point c_i in Case 1.2 and segment $c_{i,z}$ in Case 2.*

3. *For all possible prolongations into a street of the part of P the robot has so far seen, $v_{L,i} \in L$ and $v_{R,i} \in R$ hold. The shortest path from s to the goal visits $v_{L,i}$ or $v_{R,i}$. In Case 1, the endpoint of l_i, p_i, lies on the shortest path. In Case 2, $\alpha_i < \pi$ holds.*

As Figure 6 shows, Case 1 of the above algorithm has two subcases. Though 1.2 is but a degenerate case of 2, we subsume it under Case 1 because the next vertex visited by the shortest path to s is known.

It remains to explain *how* the robot determines v_L and v_R. This process is intrinsically incremental, in that the robot would not be able to determine its orientation marks correctly if it were to start from some position in the middle of P (e.g. the i-dot in $vis_P(s)$ in Figure 1).

The construction is based on the following additional invariant. We put $v_{L,0} := v_{R,0} := s$.

Invariant 2.5 *Assume that $i = 0$ holds or that p_i is endpoint of a Case 1 type segment, and that p_i cannot see g. Then the following holds for the pieces in the umbrella of p_i that lie in clockwise order between the old orientation marks $v_{L,i}$ and $v_{R,i}$.*

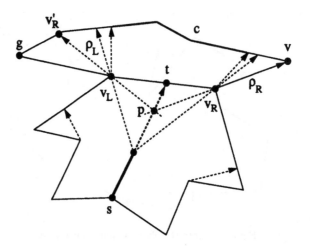

Figure 5: The shortest path visits v_L, but if the goal were at v then it would run through v_R.

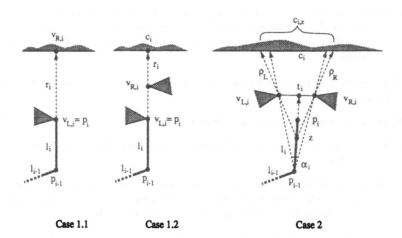

Case 1.1   Case 1.2   Case 2

Figure 6: The cases of Invariant 2.4.

- *No piece of L is below its left neighbor.*

- *No piece of R is below its right neighbor.*

Lemma 2.6 *In the situation described in Invariant 2.5 three orderings are possible among the pieces between $v_{L,i}$ and $v_{R,i}$ (including those containing $v_{L,i}$ and $v_{R,i}$.)*

1. *Each piece lies above its left neighbor. Only the rightmost piece, r, belongs to R.*

2. *Each piece lies above its right neighbor. Only the leftmost piece, l, belongs to L.*

3. *There is a unique piece, c, lying above both its neighbors. Its left neighbor belongs to L, its right neighbor to R.*

Proof: Assume there is a piece c that lies above both its neighbors. Its left neighbor is below c, so it must belong to L, due to Lemma 2.3. Similarly, the right neighbor of c belongs to R. Lemma 2.2 implies that there can be at most one piece like c.

If there is no such piece then 1) or 2) must apply because no piece can be below both its neighbors, due to Invariant 2.5. □

For $i = 0$, the first and the third alternative are depicted in Figure 4 and Figure 5, respectively. Here the leftmost and the rightmost piece are joined at $v_{L,0} = v_{R,0} = s$.

Next, we consider a Case 2 type segment l_i, see Figure 6. As the robot moves towards t_i, the rays ρ_L and ρ_R emanating from its current position, z, rotate about their pivots, $v_{L,i}$ and $v_{R,i}$, and segment $c_{i,z}$ grows longer. The following lemma shows that the robot will obtain more information before it arrives at t_i. This event marks the endpoint, p_i, of segment l_i. In the following we drop the index i, and denote $v_{L,i+1}$ by v'_L, etc. The events no. 2-4 are illustrated by the Figures 5, 8 ii), and 9, correspondingly.

Lemma 2.7 *Assume Case 2 applies to v_L and v_R; cf. Figure 6. Then one of the following events occurs before the robot arrives at $t \in \overline{v_L v_R}$ or hits the boundary of P.*

1. *The goal becomes visible.*

2. *The growing segment c reaches v_L or v_R.*

3. *An endpoint of c is encountered by one of the rays ρ_L, ρ_R.*

4. *One of the rays is blocked by a reflex vertex.*

Now we describe how the robot determines the new orientation marks v'_L and v'_R on arriving at p.

Determination of v'_L and v'_R

1. p sees g. Let $v'_L := v'_R := g$.

2. $p = s$ or p is endpoint of a Case 1 type segment.

   (a) Lemma 2.6 1) holds. Let v'_R be the left endpoint of piece r, and let v'_R be the vertex of r's left neighbor below v'_R; see Figure 4.

   (b) Lemma 2.6 2) holds. Symmetrically.

   (c) Lemma 2.6 3) holds. Let v'_L and v'_R be the reflex vertices of the neighbors of c; see Figure 5.

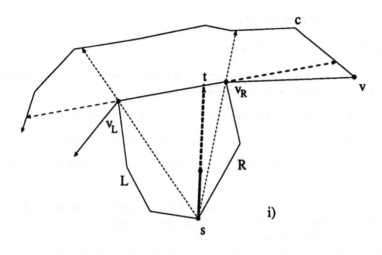

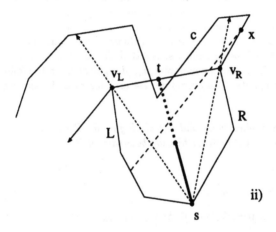

Figure 7: In i), vertex v cannot see any point of L. In ii), the robot crosses the line segment connecting x with L before hitting the boundary.

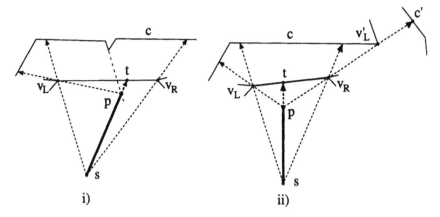

Figure 8: i) Part of c may become invisible. ii) Event no. 3.

3. p is endpoint of a Case 2 type segment. We distinguish between the different events that may have occurred at p.

   (a) Event no. 1. See 1) above.

   (b) Event no. 2 shown in Figure 5. Let v'_R be the hit point of $\rho_L = \overrightarrow{pv_L}$, and $v'_L := v_L$.

   (c) Event no. 3 shown in Figure 8 ii). Let v'_L be the reflex vertex where $\rho_R = \overrightarrow{pv_R}$ slips off c, and $v'_R := v_R$.

   (d) Event no. 4 shown in Figure 9. Let v'_R be the reflex vertex blocking ρ_R, and $v'_L := v_L$.

The symmetric versions of 3 (a), 3(b), and 3(c) are treated similarly.

The algorithm HighLevelStrategy is now completely specified. Its correctness is stated in the following theorem that can be shown by induction.

Theorem 2.8 *For each $i \geq 0$, Invariant 2.5 holds for p_i, and Invariant 2.4 holds for $l_{i+1} = \overline{p_i p_{i+1}}$. The sequences $(v_{L,i})_i$ and $(v_{R,i})_i$ are weakly increasing on L and R, correspondingly.*

Example. In Figure 1 the path from point p_3 on consists of two Case 2 segments with associated points $(v_{L,4}, v_{R,4})$ and $(v_{L,5}, v_{R,4})$. The endpoint of the second segment is determined by event no. 2. The main result of this section is stated in the following theorem.

Theorem 2.9 *Let P be a street consisting of n edges. Then algorithm HighLevelStrategy finds a path w from s to g in P that consists of $m = O(n)$ line segments. If $l_i l_{i+1} \ldots l_j$ is a sequence of Case 2 segments of w followed by a sequence $l_{j+1} l_{j+2} \ldots l_k$ of Case 1 segments then for each point z in $l_i l_{i+1} \ldots l_j$ the first vertex visited by the shortest path from z to g is p_{j+1}, and $l_{j+2} \ldots l_k$ is a piece of the shortest path from s to g.*

Corollary 2.10 *The memory size needed by the robot does not depend on the complexity of the street but only of the maximum complexity of the visibility polygons encountered.*

3 Minimizing the Local Absolute Detour

The problem not settled by the high-level strategy is *how* to choose the target point, t in $\overline{v_L v_R}$, in Case 2; see Figure 9.

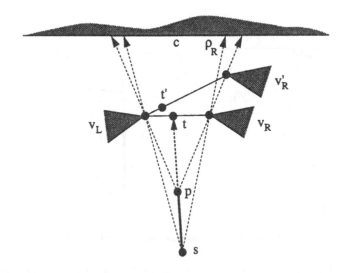

Figure 9: Event no. 4.

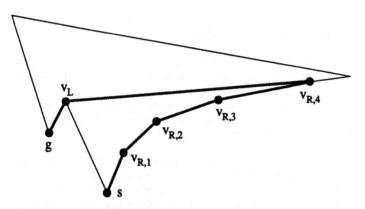

Figure 10: To walk towards the closer candidate corner can result in an unbounded detour.

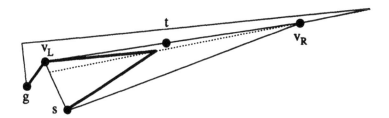

Figure 11: To head for the middle of $\overline{v_L v_R}$ can result in an unbounded detour, too.

An obvious idea is to choose the closer one of v_L and v_R. However, by this strategy the robot can be lured off the shortest path arbitrarily far, see Figure 10.

Another obvoius approach could be to head for the point in the middle of $\overline{v_L v_R}$. But in general, this strategy does not work either, as Figure 11 shows.

The strategy we propose tries to minimize the *local absolute detour*. Suppose the robot is for the first time in the situation of Case 2, as shown in Figure 9. At the latest upon arriving at $t \in \overline{v_L v_R}$ an event will occur. If v_L turns out to be the correct corner then the robot has to walk from t to v_L, causing the absolute detour $D_L(t) = st + tv_L - sv_L$, where vw denotes the distance between the points v and w. Otherwise, it must walk to v_R, resulting in the detour $D_R(t) = st + tv_R - sv_R$.

Lemma 3.1 *The maximum of $D_L(t)$ and $D_R(t)$ becomes minimal iff t is chosen such that*

$$v_L t = \frac{sv_L - sv_R + v_L v_R}{2}.$$

Proof: An application of the law of cosine shows that the function $D_L(t)$ is strictly increasing from 0 to a value greater than 0 as t moves from v_L to v_R; similarly, $D_R(t)$ is strictly decreasing from a positive value to 0. Thus, the maximum of both becomes minimal at the unique point t where the values are equal. □

In Figure 9 the robot chooses t by the above formula and starts walking towards t. On arriving at p', the robot sees vertex v_R' and chooses its next target point t' on $\overline{v_L, v_R'}$ by the same rationale. But this time the length of the shortest path from s to v_R', $sv_R + v_R v_R'$, is taken into account, so t' is determined by

$$v_L t' = \frac{sv_L - (sv_R + v_R v_R') + v_L v_R'}{2}.$$

Generally, the next target point t_{k+1} in $\overline{v_{L,k+1} v_{R,k+1}}$ is determined by low-level strategy *lad* according to

$$v_{L,k+1} t_{k+1} := \frac{A_{k+1} - B_{k+1} + v_{L,k+1} v_{R,k+1}}{2}.$$

Here A_{k+1} denotes the lenght of the left convex chain leading from the last point p_{i-1} that was safely on the shortest path, through the consecutive corners $v_{L,i}, v_{L,i+1}, \dots, v_{L,k+1}$.

Lemma 3.2 *The point t_{k+1} lies in $\overline{v_{L,k+1} v_{R,k+1}}$ and minimizes the maximum of the possible absolute detours*

$$D_L(t) := \sum_{j=i}^{k} p_{j-1} p_j + p_k t + t v_{L,k+1} - A_{k+1}$$

$$D_R(t) := \sum_{j=i}^{k} p_{j-1} p_j + p_k t + t v_{R,k+1} - B_{k+1}$$

where t ranges in $\overline{v_{L,k+1} v_{R,k+1}}$.

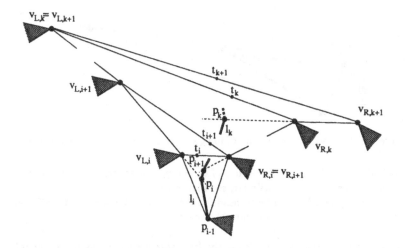

Figure 12: Constructing the next target point by strategy *lad*.

The performance of strategy *lad* depends only on its behavior in convex funnels.

Definition 3.3 A *funnel* (C_L, C_R) consists of a left convex chain, C_L, and a right convex chain, C_R leading from a common startpoint, p_0, to endpoints v_L and v_R that are mutually visible.

When started at p_0, strategy *lad* produces a path of line segments $\overline{p_{j-1}p_j}$ and associated verte sequences $(v_{L,j})_j$ and $(v_{R,j})_j$. Assume that from position p_k the robot sees for the first time bot endpoints, $v_L = v_{L,k+1}$ and $v_R = v_{R,k+1}$.

Lemma 3.4 *For each constant $D > 0$ the following assertions are equivalent.*

1. *For each funnel (C_L, C_R)*

$$\sum_{j=i}^{k} p_{j-1}p_j + p_k t_{k+1} + t_{k+1} v_L \leq D \text{ length of } C_L$$

holds, where $k+1$ is the length of the funnel.

2. *For each street P we have $D_{lad}(P) \leq D$.*

In practice, strategy *lad* works very well. Though we have deliberately tried to create bad fun nels, we have not been able to construct a funnel whose relative detour exceeds $D = 1.8$. This i complemented by the following lower bound.

Theorem 3.5 *For each possible strategy S*

$$\inf D_S(P) \geq \sqrt{2} = 1.414\ldots$$

holds, the infimum being taken over all streets P.

Proof: Let P denote the polygon depicted in Figure 13. P is not a complete street since th goal has not yet been specified. The robot cannot look into the caves before it reaches the dotte line, b. Suppose that its first point of contact lies to the right of h, depending on strategy S. In thi case, the goal is put into the left cave. Then the total length of the robot's path from s to g is a least as large as

$$\text{length of } h + \frac{1}{2} \text{ length of } b = \frac{1}{\sqrt{2}} + \frac{1}{\sqrt{2}} = \sqrt{2}$$

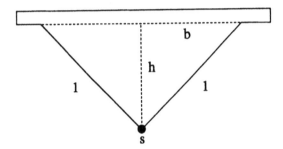

Figure 13: Establishing a lower bound for the relative detour.

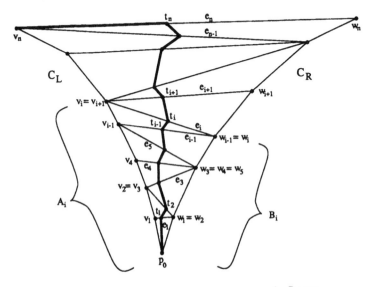

Figure 14: Point t_i is determined by $v_i t_i := \frac{A_i - B_i + v_i w_i}{2}$.

whereas the shortest path from s to g is of length 1. □

One might object that a street whose "breadth" exceeds its "length" makes a poor example. But rather than placing the goal in one of the caves, we could as well glue on another copy of P, and iterate the construction. This would lead to a street of unbounded length and bounded breadth.

4 An Upper Bound for the Global Relative Detour

In this section we show that strategy *lad* presented in Section 3, plugged into the frame of the high-level strategy discussed in Section 2, does guarantee an upper bound for the relative detour in arbitrary streets. In view of Lemma 3.4 it is sufficient to study funnels, as defined in Definition 3.3.

Since it is analytically very difficult to estimate the length of the path created in a funnel, consider the *longer* path depicted in Figure 14 that would result if the robot did not react to the new visibility information obtained at point p_i but always continues walking to its target point, t_i, for each i.

The next theorem provides an upper bound for the length of the path shown in Figure 14.

Theorem 4.1 *With the above notations the following holds.*

$$p_0 t_1 + \sum_{j=2}^{n} t_{j-1} t_j + t_n v_n \le (1 + \frac{3}{2}\pi) A_n.$$

A fortiori, this yields an upper bound for the relative detour caused by strategy *lad*.

Theorem 4.2 *For each street P*

$$D_{lad}(P) \le 1 + \frac{3}{2}\pi = 5.71\ldots.$$

In order to prove Theorem 4.1 we have to estimate the length of the path depicted in Figure 14 against the length of the left convex chain, C_L. W.l.o.g we may assume that each vertex of the funnel is the endpoint of an edge e_i. Then the edges e_i split the funnel into two types of triangles. If $w_{i-1} = w_i$ holds for two consecutive edges e_i, e_{i-1}, then the included triangle T_{i-1} shares its third side with C_L. If $v_i = v_{i+1}$ then the included triangle T_i has its third side on C_R. The bottommost triangle of the funnel is special; we define it to be of the former type by putting $v_0 := p_0$, $w_0 := w_1$, and $t_0 := p_0$.

The following lemma shows that the target points can be computed incrementally.

Lemma 4.3 *Assume that $v_i = v_{i+1}$ holds. Then*

1. $v_i t_{i+1} = \frac{1}{2}(v_i t_i - t_i w_i - w_i w_{i+1} + v_i w_{i+1})$,

2. $t_{i+1} w_{i+1} = \frac{1}{2}(t_i w_i + w_i w_{i+1} - v_i t_i + v_i w_{i+1})$,

3. $v_i t_{i+1} \le v_i t_i$.

Symmetric formulae hold in the case $w_{i-1} = w_i$.

Next, we distribute the length of the path to be estimated among the triangles T_i.

Definition 4.4 *Let e_i, $i < n$, be an edge in the funnel. Then the cost of the triangle T_i above e_i is defined by*
$$cost\ (T_i) := t_i t_{i+1} + v_{i+1} t_{i+1} - v_i t_i.$$

Note that for the special triangle T_0 we obtain $cost\ (T_0) = p_0 t_1 + v_1 t_1$, since $t_0 = p_0 = v_0$. Clearly, the sum of these quantities telescopes into the length of the path,

$$\sum_{i=0}^{n-1} cost\ (T_i) = p_0 t_1 + \sum_{j=2}^{n} t_{j-1} t_j + t_n v_n.$$

The following lemma provides the main tool for estimating the cost contribution of a single triangle.

Lemma 4.5 *Let α_i denote the angle between e_i and e_{i+1}.*

1. If $v_i = v_{i+1}$ then $cost\ (T_i) \le v_i t_i \sin \alpha_i$

2. If $w_i = w_{i+1}$ then $cost\ (T_i) \le t_i w_i \sin \alpha_i + v_i v_{i+1}$

The triangles in the funnel can be grouped into *left fans*, denoting maximal sequences of triangles that share but a vertex with C_L, and *right fans*. By convention, the bottommost triangle, T_0, is (part of) a right fan. We define the *cost of a fan* to be the sum of the costs of its constituing triangles. From Lemma 4.5 follows

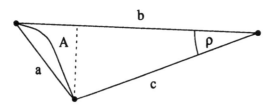

Figure 15: Notations of Lemma 4.7.

Lemma 4.6 *Let* $F = T_i T_{i+1} \ldots T_k$ *be a fan in the funnel.*

1. If F is a left fan then

$$\text{cost}\,(F) \leq v_i t_i \sum_{j=i}^{k} \alpha_j =: Bcost\,(F)$$

2. If F is a right fan then

$$\text{cost}\,(F) \leq t_i w_i \sum_{j=i}^{k} \alpha_j + \sum_{j=i}^{k} v_j v_{j+1} =: Bcost\,(F)$$

Finally, we need the following technical result.

Lemma 4.7 *Let* a, b, *and* c *be the sides of a triangle, let* ρ *denote the angle opposite to* a, *and let* A *be a curve connecting the endpoints of* a; *see Figure 15. Then the following holds for the lengths of these pieces.*

$$c\rho + \frac{1}{2}(A - c + b)\pi \leq \frac{3}{2}A\pi.$$

Now we can prove Theorem 4.1.
Proof: We show by induction on the number of left fans in the funnel that

$$\sum_{F \text{ fan}} Bcost\,(F) \leq \left(1 + \frac{3}{2}\right)A_n$$

holds, where $Bcost$ is as defined in Lemma 4.6.

A funnel without a left fan consists of a single right fan F that includes the bottommost triangle; see Figure 16 i).
Thus,

$$Bcost\,(F) = t_0 w_0 \sum_{j=0}^{n-1} \alpha_j + A_n.$$

An application of Lemma 4.7 to the triangle (v_0, w_0, v_n) yields

$$t_0 w_0 \sum_{j=0}^{n-1} \alpha_j \leq \frac{3}{2}A_n\pi.$$

Therefore, the claim holds for funnels without left fans.

Now assume that the funnel does contain left fans. If it ends with a right fan, F, then the same argument shows that $Bcost\,(F)$ cannot exceed $1 + 3\pi/2$ times the length of the last segment of C_L bordering F. Therefore, we can assume that the funnel ends with a left fan, F, as shown in Figure 16 ii). Let e_i and e_j denote the lowest and the highest edges of the funnel F' below F. Let α and ρ denote the angles at the apices of F and F', and let δ be the angle between e_i and the tangent ray, r, from v_i that touches C_R at some point w_l.

138

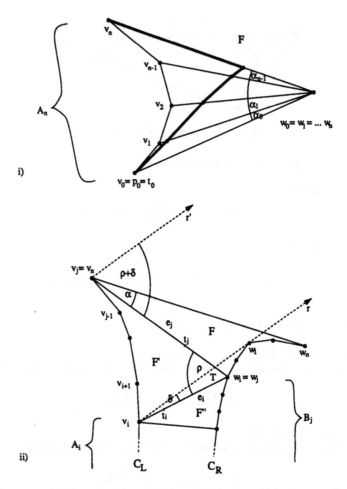

Figure 16: i) A funnel without left fans. ii) Cutting off the last two fans.

The ray r' from v_n parallel to r cannot hit C_R. Since $\overline{v_n w_n}$ does meet C_R we have $\alpha < \rho + \delta$. On the other hand, r crosses e_j, giving $\alpha < \rho + \delta \leq \pi$.

Let $A_{i,j}$ denote the length of the left chain between v_i and v_j; from the defining equations

$$v_j t_j = \frac{1}{2}(A_i + A_{i,j} - B_j + v_j w_j)$$

$$v_i t_i = \frac{1}{2}(A_i - B_j + v_i w_j)$$

one obtains by subtraction

$$v_j t_j = v_i t_i + \frac{1}{2}(A_{i,j} - v_i w_j + v_j w_j).$$

Hence, by Lemma 4.6,

$$
\begin{aligned}
Bcost\ (F) + Bcost\ (F') &= v_j t_j\, \alpha + t_i w_j\, \rho + A_{i,j} \\
&< v_j t_j (\rho + \delta) + t_i w_j\, \rho + A_{i,j} \\
&= v_i t_i\, \delta + v_i w_j\, \rho + \frac{1}{2}(A_{i,j} - v_i w_j + v_j w_j)(\rho + \delta) + A_{i,j} \\
&\leq v_i t_i\, \delta + v_i w_j\, \rho + \frac{1}{2}(A_{i,j} - v_i w_j + v_j w_j)\pi + A_{i,j}.
\end{aligned}
$$

According to Lemma 4.7 we have

$$v_i w_j\, \rho + \frac{1}{2}(A_{i,j} - v_i w_j + v_j w_j)\pi \leq \frac{3}{2} A_{i,j} \pi.$$

Thus,

$$Bcost\ (F) + Bcost\ (F') < v_i t_i\, \delta + (1 + \frac{3}{2}\pi) A_{i,j}.$$

Let Q' denote the funnel that results from the original funnel, Q, by removing the fans F and F', and adding the left fan T defined by (v_i, w_i, w_l) to the left fan F'' on top of the rest. By Lemma 4.3 3), we have $Bcost\ (F'') + v_i t_i\, \delta \leq Bcost\ (F''T)$ for the resulting left fan $F''T$. Hence,

$$
\begin{aligned}
\sum_{G \in Q} Bcost\ (G) &< \sum_{G \in Q - \{F'', F', F\}} Bcost\ (G) + Bcost\ (F'') + v_i t_i\, \delta + (1 + \frac{3}{2}\pi) A_{i,j} \\
&\leq \sum_{G \in Q - \{F'', F', F\}} Bcost\ (G) + Bcost\ (F''T) + (1 + \frac{3}{2}\pi) A_{i,j} \\
&= \sum_{G \in Q'} Bcost\ (G) + (1 + \frac{3}{2}\pi) A_{i,j} \\
&\leq (1 + \frac{3}{2}\pi)(A_i + A_{i,j})
\end{aligned}
$$

by induction hypothesis, Q' having one left fan less than Q. This completes the proof of Theorem 4.1. $\square$

5 Concluding Remarks

We have introduced a class of simple polygons in order to describe streets of varying breadth that may contain many curves but no crossings. Without knowing the street in advance, a mobile robot equipped with an on-board vision system can find a path from the start to the goal large portions of which are part of the shortest path. There are, however, situations where the robot cannot know if the street ahead is turning left or right; then a deviation from the shortest path is unavoidable.

In order to keep the deviation short, we have designed a strategy that tries to minimize the local absolute detour and thus guarantees the overall relative detour to be bounded.

One challange is to close the gaps between the proven upper bound of 5.72, the empirical upper bound of 1.8, and the lower bound of 1.4. Though it seems reasonable to study a continuous model (with curves instead of polygonal chains) it is not clear if the theory of differential equations can help.

Another question addresses different low-level strategies. One alternative is strategy *spl* that always follow the shortest path to the line segment $\overline{v_L v_R}$. But despite being simpler than *lad*, strategy *spl* is still difficult to analyze. Also, one can construct examples where the detour caused by *spl* exceeds 1.8, our empirical bound for *lad*. This approach is currently under investigation.

A further problem concerns the generalization to a kinodynamic model, where the robot has a unit mass and is, in each direction, capable of a maximal velocity and acceleration. Here no longer a short path is asked for, but a fast trajectory that has to be safe in the sense that the robot can always stay on the street no matter how the next curve looks like. An additional challenge arises if the robot's speed is so large that the time needed for image processing and for deciding about the next action must be taken into account—a situation well known from real life.

References

[1] A. Blum, P. Raghavan, and B. Schieber. Navigating in unfamiliar geometric terrain. Tech. Rep. No. RC 16452 (#73101) 1/17/91, Computer Science, IBM Research Division, T. J. Watson Research Center, Yorktown Heights, USA, 1991.

[2] B. Chazelle. Efficient polygon triangulation. Preprint, 1990.

[3] A. Datta and K. Krithivasan. Path planning with local information. In *Proc. Foundations of Software Technology and Theoretical Computer Science*, New Delhi, India, 1988.

[4] P. Eades, X. Lin, and N. C. Wormald. Performance guarantees for motion planning with temporal uncertainty. Tech. Rep. No. 173, Key Center for Software Technology, Dept. of Computer Sc., The University of Queensland, Australia, 1990.

[5] L. J. Guibas and J. Hershberger. Optimal shortest path queries in a simple polygon. In *Proc. 3rd ACM Symposium on Computational Geometry*, pages 50–63, Waterloo, 1987.

[6] Ch. Icking and R. Klein. The two guards problem. In *Proc. 7th ACM Symposium on Computational Geometry*, North Conway, 1991.

[7] V. J. Lumelsky, S. Mukhopadhyay, and K. Sun. Dynamic path planning in sensor-based terrain acquisition. *IEEE Transactions on Robotics and Automation*, 6(4):462–472, 1990.

[8] V. J. Lumelsky and A. A. Stepanov. Path-planning strategies for a point mobile automaton moving amidst unknown obstacles of arbitrary shape. *Algorithmica*, 2:403–430, 1987.

[9] J. S. B. Mitchell. An autonomous vehicle navigation algorithm. In *Proc. SPIE Applications of Artificial Intelligence*, 485:153–158, 1984.

[10] J. S. B. Mitchell. Algorithmic approaches to optimal route planning. In *Proc. SPIE Conference on Mobile Robots*, 1990.

[11] C. H. Papadimitriou and M. Yanakakis. Shortest paths without a map. In *Proc. 16th ICALP*, 610–620, 1989.

[12] R. Seidel. A simple and fast incremental randomized algorithm for computing trapezoidal decompositions and for triangulating polygons. Tech. Rep. B 90-07, FU Berlin, FB Mathematik, Serie B Informatik, 1990.

Planar Geometric Reasoning With the Theory of Hints

Paul-André Monney

Institute for Automation and Operations Research
University of Fribourg
CH-1700 Fribourg (Switzerland)
E-mail Monney@cfruni51

Abstract

Consider a certain number of objects in the plane. It is supposed that we dispose of vague information concerning the location of some objects with respect to some other objects. In view of this information, the problem to be addressed is how to draw conclusions about the position of a particular object s with respect to another object r. An adaptation of the well known mathematical theory of evidence called theory of hints will be used for the representation of uncertain and imprecise information. A hint is a body of evidence about the location of an object. Once this formalism is adopted, the process of combining several hints follows logically. But there remains the important task of the explicit computation of the combined hint from those which are available. The major part of the paper is dedicated to this problem and several efficient computational methods are presented.

1. Introduction to the theory of hints.

The subject of this paper is the location of objects in the Euclidian plane under partial information. So there is a need to represent uncertain and imprecise information about relative position of objects. For this purpose, the well founded and persuasive *theory of hints* will be used (Kohlas, Monney, 1990 a). This general theory originates o Glenn Shafer's mathematical theory of evidence (Shafer, 1976). Its most essential eatures will be presented in this section.

Suppose that we must find the unique correct answer $\theta^* \in \Theta$ to a given question. For example, one can be in charge to find the actual murder in a given list of suspects. n order to answer the question, we dispose of a piece of information which is usually imprecise and uncertain. Typically, the uncertainty is about the correct interpretation to give to the information. This set of exhaustive and mutually exclusive interpretations is denoted by $\Omega = \{\omega_1, \omega_2, \ldots, \omega_n\}$ and the probability $p(\omega_i)$ is the likelyhood that interpretation ω_i is correct. If ω_i is correct, then θ^* necessarily lies in a subset $\Gamma(\omega_i) \subseteq \Theta$ which is called a *focal set*. For example, suppose that some blond hair has been found near the corpse. This is an information which can be interpreted in two different ways ω_1 and ω_2. The first interpretation ω_1 asserts that these hairs really belong to the murder, in which case $\Gamma(\omega_1)$ is composed of all blond suspects. In the second interpretation ω_2, the hairs do not belong to the murder and thus $\Gamma(\omega_2) = \Theta$. Note that $\Gamma(\omega_2)$ is different from the set of non blond suspects because it is not excluded that ω_2 is true and the murder is blond. Accordingly, a *hint* is a quadruple $\mathcal{H} = (\Omega, p, \Gamma, \Theta)$ where Ω is a (finite) set of interpretations, p a probability measure on Ω and Γ a multivalued mapping from Ω to Θ.

Given some subset $H \subseteq \Theta$, one may wonder whether the unknown answer θ^* lies in H or not. This question cannot be answered with certainty in the light of the hint. But since each interpretation ω for which $\Gamma(\omega) \subseteq H$ assures the hypothesis, it is quite natural to define the *degree of support* of H by

$$sp(H) = p(\{\omega \in \Omega : \Gamma(\omega) \subseteq H\}). \tag{1}$$

On the other hand, H appears as *possible* under the interpretation ω if $\Gamma(\omega) \cap H \neq \emptyset$. Hence the *degree of plausibility* of H is given by

$$pl(H) = p(\{\omega \in \Omega : \Gamma(\omega) \cap H \neq \emptyset\}). \tag{2}$$

Some elementary properties of these definitions are $sp(\emptyset) = pl(\emptyset) = 0$, $sp(\Theta) = pl(\Theta) = 1$, $sp \leq pl$ and $sp(H) = 1 - pl(H^c)$ for all $H \subseteq \Theta$.

Now the question of combining several hints is addressed. Suppose that two independent hints concerning the same question are available. How should a hypothesis H be evaluated in the light of both hints simultaneously ? To answer this question, a combined hint is defined and H is evaluated according to this new hint.

Let $\mathcal{H}_1 = (\Omega_1, p_1, \Gamma_1, \Theta)$ and $\mathcal{H}_2 = (\Omega_2, p_2, \Gamma_2, \Theta)$ be two hints on Θ. If $\omega = (\omega_1, \omega_2) \in \Omega' = \Omega_1 \times \Omega_2$ is the correct combined interpretation, then $\Gamma(\omega) = \Gamma_1(\omega_1) \cap \Gamma_2(\omega_2)$ is the smallest subset of Θ which necessarily contains θ^*. In case of independence, the likelyhood that ω is correct equals $p'(\omega) = p_1(\omega_1)p_2(\omega_2)$. It is not excluded that $\Gamma_1(\omega_1) \cap \Gamma_2(\omega_2)$ is empty for some pairs of interpretations and these are certainly not correct. This is a new information which indicates that (Ω', p') must be conditionned on the event $\Omega = \Omega' - K$ where $K = \{(\omega_1, \omega_2) : \Gamma_1(\omega_1) \cap \Gamma_2(\omega_2) = \emptyset\}$. The new combined hint is then given by

$$\mathcal{H}_1 \oplus \mathcal{H}_2 = (\Omega, p, \Gamma, \Theta) \tag{3}$$

where $p(\omega) = p'(\omega)/p'(\Omega)$. This way to combine hints is called *Dempster's rule* of combination (Dempster, 1967). Two hints are called equivalent if they have equal support functions. In this sense, the combination of hints is associative and commutative, so that the combination of more than two hints is well defined.

Now the notion of representation of a hint with respect to a product space will be introduced. Suppose that $\Theta = \prod\{\Theta_i : i \in R\}$ and for $I \subseteq R$ let $\Theta_I = \prod\{\Theta_i : i \in I\}$. Given $J \subseteq I \subseteq R$, the *representation* of a hint $\mathcal{H} = (\Omega, p, \Gamma, \Theta_J)$ on Θ_I is $\mathcal{H}/I = (\Omega, p, \Gamma \times \Theta_{I-J}, \Theta_I)$. Conversely, the representation of $\mathcal{H} = (\Omega, p, \Gamma, \Theta_I)$ on Θ_J is $\mathcal{H}/J = (\Omega, p, \Gamma_J, \Theta_J)$ where $\Gamma_J(\omega)$ is the projection of $\Gamma(\omega)$ on Θ_J. Note that if $\mathcal{H}$ is defined on Θ_J, then $\mathcal{H}/I/J = \mathcal{H}$, but if $\mathcal{H}$ is defined on Θ_I, then in general $\mathcal{H}/J/I \neq \mathcal{H}$ unless $\Gamma(\omega)$ is a cylinder set with base in Θ_J for all $\omega \in \Omega$.

2. Describing geometric information by hints.

This section contains the presentation of the problem to be treated together with some results. Suppose that a certain number n of objects are embedded in the Euclidian plane $\mathcal{E}$. These objects are idealized in the sense that they are assimilated with points in $\mathcal{E}$. Let $V = \{i, j, k, \ldots\}$ denote the set of all objects. For each object $i \in V$, a positively oriented orthonormal coordinate system f_i is placed on i so that all $x-$axis are parallel. According to Davis (1987), the coordinate systems $f_i, i \in V$ are called *frobs*. Then let $d_{ij} = (d_{ij}(1), d_{ij}(2))$ represent the position of the object j with respect to the frob f_i (see Figure 1).

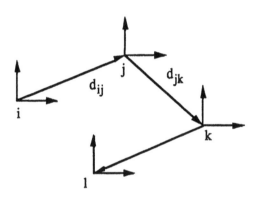

Figure 1

This type of model has already been considered by several authors like e.g. Brooks (1981) and McDermott, Davis (1984).

Of course $d_{ii} = 0$, $d_{ij} = -d_{ji}$ and more generally the Chasles type relation

$$d_{ik} = d_{ij} + d_{jk} \tag{4}$$

is satisfied for all triples of objects $(i, j, k) \in V^3$. If $m = n^2$, then a vector $d = (d(1), d(2))$ in the linear space $\mathcal{V} = \mathcal{R}^m \times \mathcal{R}^m$ is called *structural* if (4) is satisfied for all $(i, j, k) \in V^3$. This term is motivated by the fact that relation (4) is imposed by the structure of the Euclidian plane. Now some important results on structural vectors are presented.

A vector $t = (t_{ij} : (i,j) \in V^2) \in \mathcal{R}^m$ is called a *tension* if there exists a mapping $\pi : V \to \mathcal{R}$ such that

$$t_{ij} = \pi(j) - \pi(i) \qquad (5)$$

for all $(i,j) \in V^2$. This definition corresponds to the usual definition of a tension for the special case of the complete graph over V. (see for example Sakarovitch, 1984). The mapping π is called a *potential*.

Theorem 1.

A vector $t \in \mathcal{R}^m$ is a tension if and only if $t_{ik} = t_{ij} + t_{jk}$ for all $(i,j,k) \in V^3$.

Proof Suppose that t is a tension and let π be an associated potential. Then $t_{ik} = t_{ij} + t_{jk}$ becomes $\pi(k) - \pi(i) = \pi(j) - \pi(i) + \pi(k) - \pi(j)$ which is always satisfied.

Conversely, suppose that $t_{ik} = t_{ij} + t_{jk}$ is satisfied for all $(i,j,k) \in V^3$. We shall prove that t is a tension by exhibiting a potential π. First, define $\pi(i') = 0$ for an arbitrary object $i' \in V$, and let $\pi(i) = t_{i'i}$ for each other object $i \in V$. Then (5) becomes $t_{ij} = t_{i'j} - t_{i'i}$, which is satisfied by hypothesis. $\triangle$

The following immediate corollary makes the connection between structural vectors and tensions.

Corollary 1.

A vector $d = (d(1), d(2)) \in \mathcal{V}$ is structural if and only if both $d(1)$ and $d(2)$ are tensions.

If K denotes the complete graph over V, let $T = (V, U)$ represent an arbitrary partial graph of K which is a tree. If an arc $u \in U$ is removed from T, then the remaining graph will decompose into two connected components $T_1 = (V_1, U_1)$ and $T_2 = (V_2, U_2)$. For any arc $u \in U$, a vector $\tau(u) \in \mathcal{R}^m$ is defined as follows:

$$\tau(u)_{ij} = \begin{cases} 1 & \text{if } i \in V_1, j \in V_2 \\ -1 & \text{if } i \in V_2, j \in V_1 \\ 0 & \text{otherwise} \end{cases} \qquad (6)$$

for all $(i,j) \in V^2$. In fact, this vector is a tension because the mapping $\mu : V \to \mathcal{R}$ given by

$$\mu(i) = \begin{cases} 1 & \text{if } i \in V_2 \\ 0 & \text{if } i \in V_1 \end{cases} \qquad (7)$$

is an associated potentioal. If $\mathcal{T}$ denotes the set of all tensions in $\mathcal{R}^m$ relative to K, then the following theorem can be formulated.

Theorem 2.

$\mathcal{T}$ is a linear subspace of $\mathcal{R}^m$ and $dim(\mathcal{T}) = n - 1$.

This is a well known result from the theory of flows and tensions over a graph (Sakarovitch, 1984). The next theorem shows that a basis of $\mathcal{T}$ can be deduced from the tree $T = (V, U)$. Its proof can be found in the book of Berge et al. (1962).

Theorem 3.

The $(n-1)$ tensions in $\tau(U) = \{\tau(u) : u \in U\}$ form a basis of $\mathcal{T}$. Moreover, if $t \in \mathcal{T}$ and if $t(u)$ denotes the value of the tension t on the arc $u \in U$, then

$$t = \sum \{t(u)\tau(u) : u \in U\} \tag{8}$$

The next corollary is an immediate consequence of this theorem.

Corollary 2. A structural vector $d = (d_{ij} : (i,j) \in V^2)$ is uniquely defined by its values on the arcs of a spanning tree over V.

This corollary shows that strong linear dependencies exist among relations (4).

Besides the so-called *structural information* about positions of objects expressed by relations (4), suppose that we dispose of another type of information called *external information*. This information is given in the form of a certain number of independent hints $\mathcal{H}_{ij}$ for all $(i,j) \in E \subseteq V^2$. More precisely, $\mathcal{H}_{ij}$ is a hint on the d_{ij}–plane expressing our knowledge about the position of object j with respect to frob f_i.

It is convenient to represent the relation (4) as a hint $\mathcal{H}(i,j,k)$ whose unique interpretation is mapped to the focal set

$$F(i,j,k) = \{(d_{ik}, d_{ij}, d_{jk}) \in (\mathcal{R}^2)^3 : d_{ik} = d_{ij} + d_{jk}\}. \tag{9}$$

Now the problem is to combine both structural and external information in order to deduce information about positions of objects. After all hints have been represented with respect to the product set $\mathcal{V}$ as described in the previous section, this operation is performed using Dempster's rule of combination.

All available information is summarized in the graph $G = (V, E)$ together with the hints $\mathcal{H}_{ij}$ associated to the arcs $(i,j) \in E$. If we are interested in the location of a particular object s with respect to a certain frob f_r, then the problem is to compute the representation of the hint

$$\mathcal{H}(G) = \oplus\{\mathcal{H}_{ij} : (i,j) \in E\} \oplus \{\mathcal{H}(i,j,k) : (i,j,k) \in V^3\} \tag{10}$$

with respect to the factor corresponding to d_{rs}. This hint will be denoted by $\mathcal{H}(G)/(r,s)$ and it can be used to judge the hypothesis that the object s is in a certain domain $H \subseteq \mathcal{E}$ with respect to f_r (see section 1).

If $\mathcal{H}_{ij} = (\Omega_{ij}, p_{ij}, \Gamma_{ij}, \mathcal{R}^2)$, then an element

$$\omega = (\omega_{ij} : (i,j) \in E) \tag{11}$$

in

$$\Pi\{\Omega_{ij} : (i,j) \in E\} = \Pi \tag{12}$$

is a sample of interpretations. The focal sets $\Gamma_{ij}(\omega_{ij})$ will be written $\Gamma_{ij}(\omega)$ for a fixed sample $\omega \in \Pi$. The application of Dempster's rule amounts to compute the corresponding focal sets

$$\Lambda(\omega) = \{d \in \mathcal{V} : d(1), d(2) \in \mathcal{T} \text{ and } d_{ij} \in \Gamma_{ij}(\omega) \text{ for all } (i,j) \in E\} \qquad (13)$$

for all $\omega \in \Pi$. The elements of $\Lambda(\omega)$ are called *compatible* vectors. Thus the effective question to be treated is how to determine projections of sets of compatible vectors. This is precisely the subject of the next section.

3. Computing focal sets.

Consider a graph $G = (V, E)$ and suppose that $\Gamma_{ij}(\omega)$ is always a *convex polygon*. In this case, how can we compute the projection of $\Lambda(\omega)$ to the variable d_{rs} ? Corollary 1 implies that the determination of $\Lambda(\omega)$ amounts to find

$$\Psi(\omega) = \{\pi \in (\mathcal{R}^2)^n : d_{ij} = \pi_j - \pi_i \in \Gamma_{ij}(\omega) \text{ for all } (i,j) \in E\}. \qquad (14)$$

Because $\Gamma_{ij}(\omega)$ is a convex polygon, the constraint $d_{ij} \in \Gamma_{ij}(\omega)$ is equivalent to a set of linear inequalities of the form

$$\alpha(\omega)d_{ij}(1) + \beta(\omega)d_{ij}(2) \leq \gamma(\omega). \qquad (15)$$

(see Figure 2 where the equation of the line (D) is given by (15) with the equality sign).

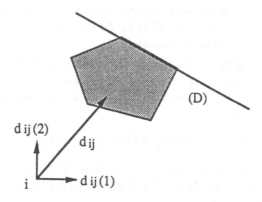

Figure 2

In terms of potentials, the inequality (15) reads

$$\alpha(\omega)\pi_j(1) - \alpha(\omega)\pi_i(1) + \beta(\omega)\pi_j(2) - \beta(\omega)\pi_i(2) \leq \gamma(\omega). \qquad (16)$$

If k_{ij} denotes the number of edges of the polygon $\Gamma_{ij}(\omega)$, then there exists a $p \times q$ matrix $A(\omega)$ and a p–vector $b(\omega)$ such that

$$\Psi(\omega) = \{\pi \in (\mathcal{R}^2)^n : A(\omega)\pi \le b(\omega)\} \tag{17}$$

where $p = \sum_{(i,j)\in E} k_{ij}$ and $q = 2n$. The previous section shows that we can fix $\pi_r = 0$ and from now on it will be supposed that the system $A(\omega)\pi \le b(\omega)$ incorporates this additional condition. Let $\Psi(\omega)/(r,s)$ denote the projection of $\Psi(\omega)$ to $\pi_s = (\pi_s(1), \pi_s(2)) = d_{rs}$.

The set $\Psi(\omega)$ is a convex polytope and its projection is a convex polygon in the d_{rs}–plane. Its vertices can be found by performing a parametric analysis on the objective function of the associated linear program

$$\begin{aligned} A(\omega)\pi &\le b(\omega) \\ \pi_s(1) + \mu\pi_s(2) &= z \; (Max) \end{aligned} \tag{18}$$

In fact, phase I of the simplex method indicates whether $\Psi(\omega)$ is empty, and if this is not the case, then an initial basic feasible solution is available. Starting from this solution, a sensitivity analysis on the parameter μ provides a set of critical values corresponding to the vertices of $\Psi(\omega)/(r,s)$. For details on the parametric analysis of the objective function of a linear program, see for example Simmonard, 1972. Commercial linear programming packages usually perform this kind of analysis without difficulty. On the other hand, since $p(\omega)$ is known, it is interesting to note that (18) can be viewed as a stochastic linear program (Kall, 1976).

Unfortunately, the method just described suffers from a heavy drawback because the parametric analysis must in principle be repeated for each sample $\omega \in \Pi$. So we are confronted with the problem of combinatorial growth of the cardinality of Π with respect to the number of hints $\mathcal{H}_{ij}$ involved in the model. This shows that simplifying assumptions must be looked for. A rather realistic one is to suppose that the focal sets of any hint $\mathcal{H}_{ij}$ are polygons with *parallel* edges (see Figure 3).

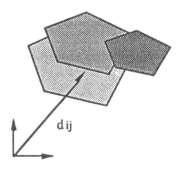

dij

Figure 3

Polygons of two different hints are of course not necessarily parallel. This hypothesis implies that the matrices $A(\omega)$ no longer depend on ω and the linear program (18) becomes much easier to analyse.

An even more simple situation arises when the polygons are axis-parallel rectangles, i.e.

$$\Gamma_{ij}(\omega) = [l_{ij}^{(1)}(\omega), u_{ij}^{(1)}(\omega)] \times [l_{ij}^{(2)}(\omega), u_{ij}^{(2)}(\omega)] \tag{19}$$

for all $(i,j) \in E$ and for all $\omega \in \Pi$. In this case $\Lambda(\omega)$ is the set of all vectors $d = (d(1), d(2))$ such that $d(1)$ and $d(2)$ are tensions and

$$l_{ij}^{(1)}(\omega) \le d_{ij}(1) \le u_{ij}^{(1)}(\omega)$$
$$l_{ij}^{(2)}(\omega) \le d_{ij}(2) \le u_{ij}^{(2)}(\omega) \tag{20}$$

for all $(i,j) \in E$. But this shows that $\Lambda(\omega) = \Lambda^{(1)}(\omega) \times \Lambda^{(2)}(\omega)$ where

$$\Lambda^{(k)}(\omega) = \{d(k) \in \mathcal{T} : d_{ij}(k) \in [l_{ij}^{(k)}(\omega), u_{ij}^{(k)}(\omega)]\} \tag{21}$$

for $k = 1, 2$. Hence the question whether $\Lambda(\omega)$ is empty is answered by the following compatible tension theorem (Ghouila-Houri, 1960). Given a cycle μ in a graph, we denote by μ^+ the set of all arcs in μ that are the direction that the cycle is traversed, and we denote by μ^- the set of all the other arcs in μ.

Theorem 4.

The set $\Lambda^{(k)}(\omega)$ is not empty if and only if the following condition is satisfied for all cycles μ in $G = (V, E)$:

$$\sum_{a \in \mu^+} u_a^{(k)}(\omega) \ge \sum_{a \in \mu^-} l_a^{(k)}(\omega). \tag{22}$$

The projection of $\Lambda(\omega)$ to d_{rs} is also a rectangle

$$[l^{(1)}(\omega), u^{(1)}(\omega)] \times [l^{(2)}(\omega), u^{(2)}(\omega)] \tag{23}$$

where $[l^{(k)}(\omega), u^{(k)}(\omega)]$ is the interval corresponding to the projection of $\Lambda^{(k)}(\omega)$ to $d_{rs}(k)$. The determination of $u^{(k)}(\omega)$ is the classical maximum tension problem. The following theorem is a consequence of a labelling algorithm due to J.C. Herz (for details, see Berge, 1985).

Theorem 5.

If $\tilde{a}$ denotes the arc (r, s) which is not necessarily in E and if $\mathcal{C}$ is the set of all cycles μ in $(V, E \cup \{\tilde{a}\})$ such that $\tilde{a} \in \mu^+$, then

$$u^{(k)}(\omega) = min \left\{ \sum_{a \in \mu^-} u_a^{(k)}(\omega) - \sum_{a \in \mu^+ - \{\tilde{a}\}} l_a^{(k)}(\omega) : \mu \in \mathcal{C} \right\}. \tag{24}$$

In this formula, note that if the minimum were to be taken over all cycles μ such that $\mu^+ = \{\tilde{a}\}$, then $u^{(k)}(\omega)$ would be equal to the length of the shortest path from r to s in $G = (V, E)$ with arc costs $u_{ij}^{(k)}(\omega)$. Since this is only a subset of $\mathcal{C}$, the correct value of $u^{(k)}(\omega)$ is in general smaller than this length. Using theorem 5, a similar formula for $l^{(k)}(\omega)$ can easily be deduced:

$$l^{(k)}(\omega) = max \; \{ \sum_{a \in \mu^-} l_a^{(k)}(\omega) - \sum_{a \in \mu^+ - \{\tilde{a}\}} u_a^{(k)}(\omega) : \mu \in \mathcal{C}\}. \tag{25}$$

Again, $l^{(k)}(\omega)$ is in general greater than the length of the longest path from r to s in G with arc costs $l_{ij}^{(k)}(\omega)$.

Without loss of generality, we can suppose that $G = (V, E)$ is connected. In fact, if this is not the case and r and s belong to different components, then $\Lambda(\omega) \neq \emptyset$ implies that $\Lambda(\omega)/(r, s) = \mathcal{R}^2$. Moreover, if G is not connected and r and s belong to the same connected component, then $\Lambda(\omega) \neq \emptyset$ implies that the other components can be discarded for the computation of $\Lambda(\omega)/(r, s)$ (see section 4 below).

Since the constraint $d_{ij}(k) \geq l_{ij}^{(k)}(\omega)$ is equivalent to

$$-d_{ij}(k) \leq -l_{ij}^{(k)}(\omega), \tag{26}$$

consider the network $R = (V, \bar{E}, c^{(k)}(\omega))$ where

$$\bar{E} = E \cup \{a = (i, j) \in V^2 : \bar{a} = (j, i) \in E\} \tag{27}$$

and $c^{(k)}(\omega)$ is the mapping from $\bar{E}$ to $\mathcal{R}$ given by

$$c_a^{(k)}(\omega) = \begin{cases} u_a^{(k)}(\omega) & \text{if } a \in E \\ -l_{\bar{a}}^{(k)}(\omega) & \text{if } a \in \bar{E} - E. \end{cases} \tag{28}$$

for all $a \in \bar{E}$.

If A denotes the node-arc adjacency matrix of R, then $\Lambda^{(k)}(\omega)$ is expressed in terms of potentials by

$$\Psi_k(\omega) = \{\pi(k) \in \mathcal{R}^n : -A^T \pi(k) \leq c^{(k)}\}. \tag{29}$$

If we fix $\pi_r(k) = 0$, then $d_{rs}(k) = \pi_s(k)$ and the upper limit $u^{(k)}(\omega)$ of the interval $\Lambda_{rs}^{(k)}(\omega)$ is given by the optimal value of z in the linear program

$$\begin{aligned} \pi_r(k) &= 0 \\ -A^T \pi(k) &\leq c^{(k)}(\omega) \qquad (P) \\ \pi_s(k) &= z \; (Max) \end{aligned}$$

Theorem 6.

The linear program (P) has a feasible solution if and only if R contains no circuit of negative length. If this condition is satisfied, then the optimal value of $\pi_s(k)$ (in other words $u^{(k)}(\omega)$) is equal to the shortest distance from r to s in R.

This is the maximum tension theorem (Sakarovitch, 1984) and every shortest path algorithm which can deal with circuits and negative arc costs can be used to compute $u^{(k)}(\omega)$.

A similar development can be done for $l^{(k)}(\omega)$. This time, consider the network $R' = (V, \bar{E}, c'^{(k)}(\omega))$ where

$$c_a'^{(k)}(\omega) = \begin{cases} l_a^{(k)}(\omega) & \text{if } a \in E \\ -u_{\bar{a}}^{(k)}(\omega) & \text{if } a \in \bar{E} - E. \end{cases} \tag{30}$$

In this case, $l^{(k)}(\omega)$ is given by the optimal value of y in the linear program

$$\pi_r(k) = 0$$
$$-A^T \pi(k) \geq c'^{(k)}(\omega) \qquad (P')$$
$$\pi_s(k) = y \ (Min)$$

Theorem 7.

The linear program (P') has a feasible solution if and only if R' contains no circuit of positive length. If this condition is satisfied, then the optimal value of $\pi_s(k)$ (in other words $l^{(k)}(\omega)$) is equal to the longest distance from r to s in R'.

The hypothesis that focal sets are axis-parallel rectangles might not be sufficient to make computations feasible. In fact, for each combined interpretation ω, there are four optimal distances to be computed in different networks. A further simplifying assumption is to consider that all hints $\mathcal{H}_{ij}$ come from the same source (expert or estimation procedure). This source is working with some unknown degree of precision δ whose value is an element of the finite list $\Delta = \{\delta_1, \ldots, \delta_m\}$. The likelyhood that δ_i is the correct degree is given by the probability $p(\delta_i)$. The information given by the source concerning the position of the object j with respect to f_i is interpreted in different ways according to its degree of precision. Thus this information can be represented in the form of a hint

$$\mathcal{H}_{ij} = (\Delta, p, \Gamma_{ij}, \mathcal{R}^2). \tag{31}$$

Since the focal sets of $\mathcal{H}_{ij}$ depend on the degree of precision of the source, it is reasonable to assume that they are nested rectangles

$$\Gamma_{ij}(\delta_1) \subseteq \ldots \subseteq \Gamma_{ij}(\delta_m). \tag{32}$$

This type of hints is called *consonant* by Shafer, 1976. Moreover, all hints $\mathcal{H}_{ij}$ are dependent because they come from the same source. This model is referred as the *consonant dependent* model. Of course, the Dempster's rule given in section 1 must be adapted to cope with these dependencies (Kohlas, Monney, 1990 a). This requires to compute the focal sets

$$\Psi(\delta) = \{\pi \in (\mathcal{R}^2)^n : d_{ij} = \pi_j - \pi_i \in \Gamma_{ij}(\delta)\} \tag{33}$$

for all $\delta \in \Delta$. This might considerably reduce the amount of optimal path problems to solve because the number of elements in Δ is usually much smaller than the number of elements in Π. Moreover, it is easily verified that the resulting hint about d_{rs} is also consonant. In consequence, if pl denotes its plausibility function, then

$$pl(H) = max \; \{pl(x) : x \in H\} \tag{34}$$

for any hypothesis $H \subseteq \mathcal{R}^2$ about the location of the object r with respect to f_s (Kohlas, Monney, 1990 a). The function $pl : \mathcal{R}^2 \to [0,1]$ can thus be considered as a kind of density function. This becomes important when representation questions are addressed (see section 5 below).

It is interesting to note that this model has tight connections with possibility theory and hence with fuzzy set theory. In fact, if $\Gamma_a(\delta) = \Gamma_a^{(1)}(\delta) \times \Gamma_a^{(2)}(\delta)$ for all $a \in E$, then

$$\Gamma_a^{(k)}(\delta_1) \subseteq \ldots \subseteq \Gamma_a^{(k)}(\delta_m) \tag{35}$$

for $k = 1,2$ and the plausibility function of the hint

$$\mathcal{H}_a^{(k)} = (\Delta, p, \Gamma_a^{(k)}, \mathcal{R}) \tag{36}$$

is called a *possibility measure* by Dubois, Prade (1988). The combination of consonant dependent hints can be nicely expressed in terms of plausibility functions (Kohlas, Monney, 1990 a). The result corresponds essentially to the formula given by Dubois and Prade to combine possibility functions, or equivalently to intersect fuzzy sets.

4. Reduction and decomposition methods.

Before the results presented in the previous section are applied, it is sometimes possible to reduce or to decompose the original problem to obtain a smaller and hence simpler problem. These procedures are general and do not depend on the particular form of the focal sets $\Gamma_{ij}(\omega)$.

We begin with the *reduction* methods. Suppose that an object $j \in V$ different from r and s has exactly two incident arcs $(i,j) \in E$ and $(j,k) \in E$. The original graph $G = (V,E)$ is then replaced by a new graph $G' = (V',E')$ where $V' = V - \{j\}$ and $E' = E - \{(i,j),(j,k)\} + \{(i,k)\}$. Note that if $(i,k) \in E$, then G' contains two parallel arcs from i to k. The set affected to the new arc (i,k) in G' is $\Gamma_{ij}(\omega) + \Gamma_{jk}(\omega)$ (see Figure 4).

As far as the computation of the projection to d_{rs} is concerned, the two graphs G and G' are equivalent as the following theorem says.

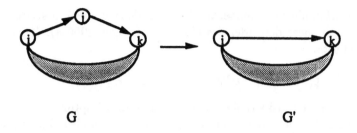

G   G'

Figure 4

Theorem 8.

If $\Lambda'(\omega)$ is the set of compatible vectors with respect to G', then $\Lambda(\omega)/(r,s) = \Lambda'(\omega)/(r,s)$.

Proof. If $d''_{rs} \in \Lambda'(\omega)/(r,s)$, then there exists $d' \in \mathcal{V}'$ such that $d'(1) \in \mathcal{T}', d'(2) \in \mathcal{T}'$ and $d''_{hl} \in \Gamma'_{hl}(\omega)$ ($\mathcal{V}', \mathcal{T}'$ and Γ' are defined like $\mathcal{V}, \mathcal{T}$ and Γ, but for V'). Let $\pi' = (\pi'(1), \pi'(2))$ represent potentials associated to these tensions. But there exist $d_{ij} \in \Gamma_{ij}(\omega)$ and $d_{jk} \in \Gamma_{jk}(\omega)$ such that $d'_{ik} = d_{ij} + d_{jk}$. This permits to define an extension π of π' from V' to V by setting $\pi_j = \pi_i + d_{ij}$.

The tension d associated to this potential π is in $\Lambda(\omega)$ by construction, the only point to verify is that $\pi_k - \pi_j \in \Gamma_{jk}(\omega)$. But $\pi_k - \pi_j = \pi_k - \pi_i - d_{ij} = d'_{ik} - d_{ij} = d_{jk} \in \Gamma_{jk}(\omega)$. Since $d_{rs} = d''_{rs}$, this implies that $d''_{rs} \in \Lambda(\omega)/(r,s)$ and so $\Lambda'(\omega)/(r,s) \subseteq \Lambda(\omega)/(r,s)$.

On the other hand, if $d \in \Lambda(\omega)$, then its restriction to $\mathcal{V}'$ is an element of $\Lambda'(\omega)$ because $d_{ik} = d_{ij} + d_{jk} \in \Gamma_{ij}(\omega) + \Gamma_{jk}(\omega)$. This implies that $\Lambda(\omega)/(r,s) \subseteq \Lambda'(\omega)/(r,s)$ because $j \notin \{r,s\}$ and hence $\Lambda(\omega)/(r,s) = \Lambda'(\omega)/(r,s)$. $\triangle$

This way to reduce a graph G is called a *series reduction*. It can easily be verified that $\mathcal{H}(G)$ remains unchanged if an arc $(i,j) \in E$ is replaced by its opposite arc (j,i) and $\Gamma_{ji}(\omega) = -\Gamma_{ij}(\omega)$. So the orientation of the arcs in E is not essential. This permits to perform reductions which would otherwise not be possible, like e.g. when (i,j) and (k,j) are the only arcs in E which are incident to j.

Another type of reduction called *parallel reduction* is a natural complement to the series reduction. Suppose that two parallel arcs from i to j in E have $\Gamma'_{ij}(\omega)$ and $\Gamma''_{ij}(\omega)$ as associated focal sets. The new graph G' obtained by replacing them by a single new arc with $\Gamma'_{ij}(\omega) \cap \Gamma''_{ij}(\omega)$ is equivalent to G as the next theorem asserts (see Figure 5).

Theorem 9.

If $\Lambda'(\omega)$ is the set of compatible vectors with respect to G', then $\Lambda(\omega) = \Lambda'(\omega)$. Hence $\Lambda(\omega)/(r,s) = \Lambda'(\omega)/(r,s)$.

Proof. This is a direct consequence of the definition of a compatible set. $\triangle$

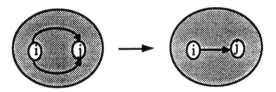

Figure 5

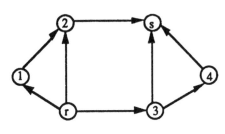

Figure 6

Series and parallel reductions permit to simplify the computation of $\Lambda(\omega)/(r,s)$ because series reduction reduces the number of variables to consider. It is even sometimes possible to reduce a graph to a single arc from r to s (see Figure 6).

In this case its associated set is precisely $\Lambda(\omega)/(r,s)$. Note that in some occasions it may prove useful to reverse the orientation of some arcs in E to perform parallel reductions.

There still remains the question how to perform explicitly series and parallel reductions for the special case of convex polygons and rectangles. In other words, how can we compute the intersection and the addition of two convex polygons P_1 and P_2 (the case of rectangles is quite clear because this corresponds to the intersection and the addition of intervals). As far as they are given by their vertices, the first task is a well known problem in computational geometry and good algorithms can be used (Preparata et al. 1985). To answer the second question, proceed as follows. A point x in a convex polygon P is a vertex if and only if there does not exist two different points x_1 and x_2 in P such that $x = (x_1 + x_2)/2$. If P^* denotes the vertices of P, then the following inclusion holds.

Theorem 10.
$$(P_1 + P_2)^* \subseteq P_1^* + P_2^*.$$

Proof. Let $x \in (P_1 + P_2)^*$ and select any $x_1 \in P_1$ and $x_2 \in P_2$ such that $x = x_1 + x_2$. We prove that both $x_1 \in P_1^*$ and $x_2 \in P_2^*$. If this were not the case, then $x_1 \notin P_1^*$ or

$x_2 \notin P_2^*$. Without loss of generality, let's assume that at least $x_1 \notin P_1^*$. Then there exist $y_1 \neq y_2 \in P_1$ such that $x_1 = (y_1 + y_2)/2$. Thus

$$
\begin{aligned}
x = x_1 + x_2 &= (y_1 + y_2)/2 + (x_2 + x_2)/2 \\
&= (y_1 + x_2)/2 + (y_2 + x_2)/2
\end{aligned}
\tag{37}
$$

and x is not a vertex because $y_i + x_2 \in P_1 + P_2$ for $i = 1, 2$. But this is a contradiction. $\triangle$

For a set $X \subseteq \mathcal{E}$, let $conv\,(X)$ denote the convex hull of X. Since there exist good algorithms to compute the convex hull of a finite set of points (Preparata et al., 1985), the following corollary provides a tractable method to compute $P_1 + P_2$.

Corollary 3.
$P_1 + P_2 = conv\,(P_1^* + P_2^*)$.

Proof. Since the convex hull operation preserves inclusion, theorem 10 implies that

$$
P_1 + P_2 = conv\,[(P_1 + P_2)^*] \subseteq conv\,(P_1^* + P_2^*) \subseteq conv\,(P_1 + P_2) = P_1 + P_2
$$

because $P_1 + P_2$ is a convex polygon and $conv\,(P^*) = P$ for all convex polygons. $\triangle$

Beside these reduction techniques, there exist two possibilities to *decompose* a graph into smaller and hence simpler parts. A vertex $k \in V$ is called a *cut-vertex* if its elimination from V disconnects the graph G into two or more connected components. If V_j denotes the vertices in the j-th component together with vertex k, then let $G_j = (V_j, E_j)$ denote the subgraph of G over V_j (see Figure 7).

Figure 7

The various graphs G_j form what is called a *series decomposition* with respect to k. The next theorem asserts that under certain conditions the computations can be limited to a single component.

Theorem 11.
Suppose that r and s belong to the same component G_i and let $\Lambda_i(\omega)$ denote the set of compatible vectors with respect to G_i. If $\Lambda(\omega)$ is not empty, then

$$
\Lambda(\omega)/(r, s) = \Lambda_i(\omega)/(r, s).
\tag{38}
$$

Proof. Let $d \in \Lambda(\omega)$ and π an associated potential. Then the tension associated to the restriction of π to V_i shows that $d_{rs} \in \Lambda_i(\omega)/(r,s)$.

Conversely, since $\Lambda(\omega) \neq \emptyset$, there exists a potential π'' on $V - V_i$ such that $\pi''_l - \pi''_h \in \Gamma_{hl}(\omega)$ for all $(h,l) \in E - E_i$. Let $d' \in \Lambda_i(\omega)$ and let π' be an associated potential on V_i. Then the tension d associated to the potential π on V given by

$$\pi_h = \begin{cases} \pi'_h & \text{if } h \in V_i \\ \pi''_h & \text{otherwise} \end{cases} \tag{39}$$

is in $\Lambda(\omega)$ and hence $d'_{rs} = d_{rs} \in \Lambda(\omega)/(r,s)$ which demonstrates the theorem. $\triangle$

Theorem 12.

Suppose that r belongs to G_i and s to G_j with $i \neq j$. If $\Lambda(\omega)$ is not empty, then

$$\Lambda(\omega)/(r,s) = \Lambda_i(\omega)/(r,k) + \Lambda_j(\omega)/(k,s). \tag{40}$$

Proof. Let $d \in \Lambda(\omega)$ and let π be an associated potential. The tensions d' and d'' associated to the restriction π' and π'' of π to V_i and V_j are in $\Lambda_i(\omega)$ and $\Lambda_j(\omega)$ respectively. But

$$\begin{aligned} d_{rs} = \pi_s - \pi_r &= (\pi_k - \pi_r) + (\pi_s - \pi_k) \\ &= d'_{rk} + d''_{ks} \in \Lambda_i(\omega)/(r,k) + \Lambda_j(\omega)/(k,s). \end{aligned} \tag{41}$$

Conversely, since $\Lambda(\omega) \neq \emptyset$, there exists a potential $\tilde{\pi}$ on $V - (V_i \cup V_j)$ such that $\tilde{\pi}_l - \tilde{\pi}_h \in \Gamma_{hl}(\omega)$ for all $(h,l) \in E - (E_i \cup E_j)$. Let $d' \in \Lambda_i(\omega)$, $d'' \in \Lambda_j(\omega)$ and let π', π'' be associated potentials on V_i and V_j respectively. Since potentials are defined up to an additive constant, we can suppose that $\pi'_k = \pi''_k$. Then the tension d associated to the potential π on V given by

$$\pi_h = \begin{cases} \pi'_h & \text{if } h \in V_i \\ \pi''_h & \text{if } h \in V_j \\ \tilde{\pi}_h & \text{otherwise} \end{cases} \tag{42}$$

is in $\Lambda(\omega)$. Hence

$$\begin{aligned} d'_{rk} + d''_{ks} &= (\pi_k - \pi_r) + (\pi_s - \pi_k) \\ &= \pi_s - \pi_r = d_{rs} \in \Lambda(\omega)/(r,s) \end{aligned} \tag{43}$$

and the theorem is proved. $\triangle$

Another kind of decomposition arises when there exist two subsets V_1 and V_2 of V such that

(1) $V = V_1 \cup V_2$, $V_1 \cap V_2 = \{r,s\}$
(2) If E_i denotes the arcs in the subgraph of G over $V_i, i = 1,2$, then

$$E_1 \cap E_2 = \begin{cases} (r,s) & \text{if } (r,s) \in E \\ \emptyset & \text{if } (r,s) \notin E. \end{cases} \tag{44}$$

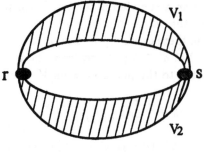

Figure 8

(see Figure 8).

If (r, s) is in E, then it will arbitrarily be removed from E_2 (but not from E_1) so that always $E_1 \cap E_2 = \emptyset$ and $E_1 \cup E_2 = E$. This kind of partition is called a *parallel decomposition* with respect to $\{r, s\}$. If $G = (V_1, E_1)$ and $G_2 = (V_2, E_2)$, then we can compute on graphs G_1 and G_2 indepently as the following theorem asserts. Let $\Lambda_1(\omega)$ and $\Lambda_2(\omega)$ denote the set of compatible vectors in G_1 and G_2 respectively.

Theorem 13

If G allows for a parallel decomposition with respect to $\{r, s\}$, then

$$\Lambda(\omega)/(r, s) = \Lambda_1(\omega)/(r, s) \cap \Lambda_2(\omega)/(r, s). \tag{45}$$

Proof. Let $d \in \Lambda(\omega)$ and let π be an associated potential. Then the tensions d' and d'' associated to the restrictions of π to V_1 and V_2 are in $\Lambda_1(\omega)$ and $\Lambda_2(\omega)$ respectively. Since $V_1 \cap V_2 = \{r, s\}$, this shows that $d_{rs} = d'_{rs} = d''_{rs} \in \Lambda_1(\omega)/(r, s) \cap \Lambda_2(\omega)/(r, s)$.

Conversely, let $\theta \in \Lambda_1(\omega)/(r, s) \cap \Lambda_2(\omega)/(r, s)$. Then there exist $d' \in \Lambda_1(\omega)$ and $d'' \in \Lambda_2(\omega)$ such that $d'_{rs} = d''_{rs} = \theta$. Let π' and π'' be potentials associated to d' and d'' respectively such that $\pi'_r = \pi''_r$. Since $d'_{rs} = d''_{rs}$ we must have $\pi'_s = \pi''_s$ and this implies that the potential π on V given by

$$\pi_h = \begin{cases} \pi'_h & \text{if } h \in V_1 \\ \pi''_h & \text{if } h \in V_2 \end{cases} \tag{46}$$

is well defined. But the tension d associated to π is in $\Lambda(\omega)$ and $d_{rs} = \pi_s - \pi_r = \theta$ shows that $\theta \in \Lambda(\omega)/(r, s)$. $\triangle$

The three last theorems about decompositions together with those concerning the reductions can be formulated in terms of hints $\mathcal{H}(G)$ instead of focal sets (Kohlas, Monney, 1990 b,c,d).

When these simplification methods cannot be applied and because the analysis has to be carried out in limited time and space, a possibility to cope with complexity is to use only part of the available information represented by the hints $\mathcal{H}_{ij}$. This restriction may possibly allow for further reduction or decomposition, but in any case computations will be simpler. But this raises the question of which hints should be dropped. To that respect, very uncertain and imprecise information can be discarded. This requires that a precise definition of the *content of information* of a hint should be given. More research is needed in this domain.

Another possibility is to consider approximations obtained by grid models. Moreover, an approximation procedure using all information and based on the Floyd algorithm can be applied (Kohlas, Monney, 1990 b).

5. Approaches to represent the resulting hint.

The knowledge of $\mathcal{H}(G)/(r,s)$ permits to evaluate, to judge the hypothesis that the object s is in a certain region H with respect to frob f_r. Since there are of course infinitely many possible regions to inspect, an explicit list of all support-plausibility couples is impossible. Moreover, even if this were possible, this would not give a good overall idea of the available information encoded in the hint $\mathcal{H}(G)/(r,s)$. Thus it is desirable to look for methods providing good insight about the location of s with respect to f_r in the light of the hint. One might think of the existence of a density $h : \mathcal{E} \to \mathcal{R}$ of the hint from which support and plausibility could be easily derived. Unfortunately, this is in general not possible, unless the dependent consonant model is used. This is not even the case when all $\mathcal{H}_{ij}$ are only consonant because the combination of consonant hints is no longer consonant. For details, see Shafer, 1976 or Kohlas, Monney, 1990 a.

However, the following three approaches can be applied.

(1) *Overall idea of plausibility.*
   The first idea is to consider a grid over the d_{rs}−plane. Then a plausibility value is assigned to each rectangle induced by the grid. These rectangles are drawn in a certain grey level according to their plausibility (the higher the plausibility, the darker the rectangle). This procedure yields an overall idea of the plausibility function of $\mathcal{H}(G)/(r,s)$. If this hint is consonant, like in the dependant consonant model, then drawing points according to their plausibilities gives a good picture of the result (see also (34)).

(2) *Confidence regions.*
   If sp and pl denote the support and plausibility functions of $\mathcal{H}(G)/(r,s)$ respectively, then the goal is to find a region H such that $sp(H) \geq a$ and $pl(H) \geq b$ for certain values a, b in $\mathcal{R}$. These conditions require that the region H has a sufficient support whereas its doubt is negligeable. Of course, this region is in general not unique. To find one, proceed as follows. Let $F_1, F_2, \ldots, F_n$ denote the focal sets of $\mathcal{H}(G)/(r,s)$ and their respective cummulated probabilities $m_1, m_2, \ldots, m_n$. If $I = \{1, 2, \ldots, n\}$, then the family

$$\mathcal{A}(a) = \{J \subseteq I : \sum_{i \in J} m_i \geq a\} \tag{47}$$

158

is a monotone family of subsets (i.e. $J \in \mathcal{A}(a)$ and $J' \supseteq J$ imply $J' \in \mathcal{A}(a)$) whose minimal elements are written $J_1, J_2, \ldots, J_r$. Let

$$H(J_i) = \cup\{F_k : k \in J_i\}. \tag{48}$$

If there exists at least one J_i for which $pl(H(J_i)) \geq b$, then take for H the domain $H(J_i)$ which satisfies this condition with less surface (the most precise). Otherwise take a domain of small surface

$$H = \cup\{F_k : k \in J\} \tag{49}$$

such that $pl(H) \geq b$ and $J \supseteq J_i$ for at least one index $i \in \{1, \ldots, r\}$.

(3) *Moving fixed shapes.*

Imagine a shape which can be moved in the plane and representing a hypothesis H. Its form can be selected among a predefined palette of possibilities. The color of the shape changes according to the support and plausibility while it is moved. The shape can thus be considered as a kind of visualizing window for the plausibility function.

Bibliography

[1] Berge C. (1985): *Graphs*. Second Revised Edition. North-Holland.
[2] Berge C., Ghouila-Houri A. (1962): *Programmes, jeux et réseaux de transports*. Dunod, Paris.
[3] Brooks R.A. (1981): Symbolic Reasoning Among 3-D Models and 2-D Images. *Artificial Intelligence* 17 285–348.
[4] Davis E. (1987): Constraint Propagation With Interval Labels. *Artificial Intelligence*, 32, 281–331.
[5] Dempster A.P. (1967): Upper and Lower Probabilities Induced by a Multivalued Mapping. *Annals of Mathematical Statistics*, 38, 325–339.
[6] Dubois D.; Prade H. (1988): *Théorie des possibilités*. Deuxième édition, Masson, Paris.
[7] Ghouila-Houri A. (1960): Sur l'existence d'un flot ou d'une tension prenant ses valeurs dans un groupe abélien. *Comptes Rendus de l'Académie des Sciences*, 250, 3931–3932.
[8] Kall P. (1976): *Stochastic Linear Programming*. Springer.
[9] Kohlas, J.; Monney P.A. (1990 a): Modeling and Reasoning With Hints. *Institute for Automation and Operations Research, University of Fribourg*, working paper 174.
[10] Kohlas, J.; Monney P.A. (1990 b): Propagating Belief Functions Through Constraint Systems. *Institute for Automation and Operations Research, University of Fribourg*, No. 171. To appear in *International Journal of Approximate Reasoning*.
[11] Kohlas, J.; Monney P.A. (1990 c): Les fonctions de crédibilité dans le raisonnement temporel sous incertitude. Recueil d'Articles de la Troisième Conférence Int. sur le Traitement d'Information et la Gestion d'Incertitudes dans les Systèmes à Base de Connaissances, Paris, 67-73.

[12] Kohlas, J.; Monney P.A. (1990 d): Temporal Reasoning Under Uncertainty With Belief Functions. To appear in the *Lecture Notes in Computer Science*.

[13] McDermott, D.; Davis E. (1984): Planning Routes Through Uncertain Teritory. *Artificial Intelligence*, 22, 107-156.

[14] Preparata F.P.; Shamos M.I. (1985): *Computational Geometry. An introduction.* Springer Verlag, New York.

[15] Sakarovitch M. (1984): *Optimisation combinatoire.* Tomes 1 et 2. Hermann, Paris.

[16] Shafer G. (1976): *A mathematical theory of evidence.* Princeton University Press.

[17] Simmonard M. (1972): *Programmation linéaire.* Tomes 1 et 2. Dunod, Paris.

Solving Algebraic Systems in Bernstein-Bézier Representation

Heinrich Müller, Matthias Otte
Institut für Informatik
Universität Freiburg

Abstract

The representation of polynomials in the Bernstein basis has advantages over the usual monomial basis since it allows a simple geometric interpretation of the coefficients. It is shown how this so-called Bézier representation can be used for the calculation of the solution manifold of algebraic systems. In this contribution, the manifold is represented by a hierarchy of cuts describing its complete topology. The location of the cuts is calculated by iterated subdivision excluding non-relevant partition elements by using the convex hull property of the Bézier representation.

1. Algebraic systems

A system of equations

$$f(x_1,\ldots,x_d) = 0, \ f : R^d \to R^m, \ x_i \in [0,1], \ i = 1,\ldots,d,$$

with

$$f(x_1,\ldots,x_d) := \sum_{i_1=0}^{n_1} \ldots \sum_{i_d=0}^{n_d} a_{i_1,i_2,\ldots,i_d} x_1^{i_1} x_2^{i_2} \cdots x_d^{i_d}, \ a_{i_1,i_2,\ldots,i_d} \in R^m,$$

is called an *algebraic system*. Algebraic systems have applications in computer graphics and geometric modeling [e.g. Arnon, 1983, Sederberg, 1984, 1985, Sederberg et al., 1984, 1988], in robotics [e.g. Schwartz, Sharir, Hopcroft, 1987, Schwartz, Yap, 1987], and many other areas.

The solutions of an algebraic system form a manifold which may be of considerable complexity. One common approach is to calculate a triangulated approximation of this manifold [Algower, Georg, 1982, Bloomenthal, 1988]. This may be done by tracking along the manifold. A different strategy is *subdivision* [e.g. Morgan, Shapiro, 1987]. The interval of interest, e.g. $[0,1]^d$, is subdivided according to some strategy in partition elements. Subdivision is iterated until an element contains a sufficiently simple part of the manifold. In figure 1, subdivision is applied on

$$x^4 - 2x^3 + x^2 - 3y^2x + 2y^4 = 0, \ x,y \in [0,1]^2.$$

The strategy of subdivision in this example is to replace a square by four congruent sub-squares.

The problem with these approaches is that the description of the manifold may require considerable space and time for its calculation. An alternative to this kind of "exhaustive enumeration" is to partition the solution set in a more adaptive way into regions of equivalent topology. An example is the cylindrical algebraic decomposition (cad) [Collins, 1975]. The idea of a cad is to find a partition of $[0,1]^{d-1}$ into cells so that the different branches of the solution set lie disjoint in the "cylinder" over a cell. This approach is iterated over the dimensions, i.e. the cells of $[0,1]^{d-1}$ are cylinders over $[0,1]^{d-2}$, and so on. In the 2-dimensional case of figure 2 the cylinders are stripes. Neighbored stripes differ in the number of traversing branches of the solution set.

The calculation of a cad is done by iterated projection. Projections are calculated by elimination of variables, using resultants. This involves a great deal of algebra and may lead to algebraic expressions of considerable size. The advantage of the cad is that the decomposition is

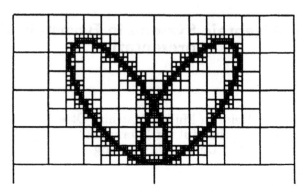

Figure 1: An approximation of $x^4 - 2x^3 + x^2 - 3y^2x + 2y^4 = 0$, $x, y \in [0,1]^2$, by subdivision. The depth of iteration is 8, the number of cells is 1397.

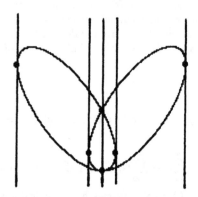

Figure 2: A cylindrical decomposition of $x^4 - 2x^3 + x^2 - 3y^2x + 2y^4 = 0$, $x, y \in [0,1]^2$.

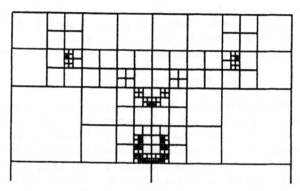

Figure 3: Restriction of the subdivision of $x^4 - 2x^3 + x^2 - 3y^2x + 2y^4 = 0$, $x, y \in [0,1]^2$, to points with vertical tangents and crossing points. The iteration depth is 8, the number of cells is 277.

more adapted to the requirements of the topology of the solution space than then straightforward subdivision approach. Thus the size of the resulting description can be expected smaller when compared on the same level of accuracy.

In this paper a combination of iterated subdivision and the idea of identifying topological interesting regions of the manifold is presented. The idea of our *approximate extremal subdivision* (aes) is to carry out an iterated subdivision of $[0,1]^d$. The criteria when a cell has to be divided further are oriented at the topology of the solution manifold. The criteria are so that under certain conditions they may be only satisfied by isolated points so that the subdivision converges towards these points. The convergency of the subdivision to isolated points instead towards the whole manifold leads to a considerably smaller number of elements of subdivision to be inspected. In figure 3 the criteria are chosen so that only those points with vertical tangents and crossing points are approximated. For 8 iterations, the number of cells to be inspected reduces to 277. The aes is defined in section 2. Algorithms for its calculation are sketched in section 3.

The criteria of subdivision consist in detecting whether a (usually multivariate) polynomial or a determinant of polynomials are completely positive or completely negative over a cell of subdivision. A sufficient condition is immediate from the convex hull property of the Bézier representation of polynomials. Bézier polynomials also harmonize quite naturally with iso-oriented subdivision strategies, due to the algorithm of de Casteljau. Further, derivatives required in the criteria can be calculated extremely simple in Bézier representation. Finally, the usual polynomial arithmetic carries over almost immediately on this representation. The criteria are presented in section 4. Their calculation by Bézier techniques is discussed in section 5.

2. Extremal subdivisions

An extremal subdivision of an algebraic system $\mathbf{f} = \mathbf{0}$, $\mathbf{f} : [0,1]^d \rightarrow I\!\!R^m$, consists of a subdivision of the interval $[0,1]$ on the d-th axis by critical values $c_0, \ldots, c_p$, $c_0 = 0 < c_1 < \ldots < c_{p-1} < c_p = 1$. The subdivision has the property that the number of connected components of the solution set of the intersection $\mathbf{f}(x_1, \ldots, x_{d-1}, c)$, $c_k < c < c_{k-1}$ is invariant. This means that during the sweep of the hyperplane $x_d = c$ from c_k to c_{k+1} the connected components of the intersection of the solution set of $\mathbf{f} = \mathbf{0}$ with the hyperplane remain the same. In this sweep interpretation, the number of connected components only may change if a local extremum w.r.t. to the direction of the d-axis occurs. Each of these extrema corresponds to one of the c_k.

Besides the c_k's corresponding to the extrema of the sweep, between any two of them at least one more c_k has to occur in the sequence. It represents the topology between the critical cuts.

Finally, since in practice necessary criteria for an extremum can more easily be checked than sufficient and necessary criteria, we allow more c_k than just those of the local extrema.

In order to obtain a complete extremal subdivision, each of the intersections $\mathbf{f}(x_1, \ldots, x_{d-1}, c_k) = \mathbf{0}$, $k = 0, \ldots, p$ is recursively treated the same way with respect to the $(d-1)$-axis.

The connected components of each subdivision are labeled so that those of them belonging to the same global connected component of the solution set get the same label. Local components belonging together are identified by curve tracking along the sweeping direction between two neighbored cuts, starting with those of smallest dimension.

In practice it is impossible to calculate an extremal subdivision exactly. An *approximate extremal subdivision* (aes) has the same structure as an extremal subdivision, but the c_k of the local extrema are replaced by c_k^- and c_k^+ approximating c_k, i.e. $c_k^- < c_k < c_k^+$, and c_k^-, c_k^+ close to c_k.

The extremal subdivisions are useful with motion planning problems when it is necessary to decide whether two points lie on the same connected component. A particular case is that of curves as solution set, i.e. $m = d - 1$. They occur in relation with retraction methods of motion planning.

3. Calculating approximate extremal subdivisions

Approximate extremal subdivisions may be calculated by iterated subdivision. The given interval of interest ($[0,1]^d$ initially) is partitioned into sub-intervals. The resulting intervals are in turn partitioned, as long as they are *candidates*. A candidate is an interval on which a given *criterion of solution* may still be satisfiable.

In our discussion we use a strategy of orthogonal subdivision resulting in multidimensional orthogonal parallelepipeds. The parallelepipeds are obtained by halving edges into alternating directions of coordinates, resulting in patterns like those in figures 1 and 2. If the candidate cell is sufficiently small, its boundaries in x_d-direction are taken as c_k^- and c_k^+.

4. Criteria of solution

In many practical cases relatively simple criteria based on classical theorems about the solution of implicit equations and the characterization of local extrema are sufficient for obtaining a considerable restriction of the number of subdivisions to be performed:

Inversion.
Let be $f : \mathbb{R}^d \to \mathbb{R}^d$ defined in an environment $U(x_0)$ of a point $x_0 \in \mathbb{R}^d$, so that
(1) $f \in C^1(U(x_0))$,
(2) $\left| \dfrac{\partial f}{\partial x}(x_0) \right| \neq 0$.
Then there exists an environment $\hat{U}(x_0) \subset U(x_0)$ which is mapped by f one-to-one onto an environment $V(y_0)$ of the point $y_0 := f(x_0)$.□

A consequence is that if $f(x_0) = 0$, x_0 is an isolated zero.

Extrema under constraints.
Let be $g : \mathbb{R}^d \to \mathbb{R}$, $f : \mathbb{R}^d \to \mathbb{R}^m$, $m < d$, $g, f \in C^1(U(x_0))$ for an environment $U(x_0)$ of an $x_0 \in \mathbb{R}^d$. Let the rank of the matrix

$$\left(\frac{\partial f}{\partial x}(x_0) \right)$$

be m. If, under these conditions, g has a relative extremum at x_0 among those $x \in U(x_0)$ with $f(x) = 0$, then there exists a $\lambda \in \mathbb{R}^m$ with

$$\frac{\partial g}{\partial x}(x_0) + \left(\frac{\partial f}{\partial x}(x_0) \right) \cdot \lambda = 0.□$$

Now let be $f(x) = 0$, $f : \mathbb{R}^d \to \mathbb{R}^m$, the algebraic system whose solution set is to be described.

<u>$m = d$.</u>
  Criterion 1.

$$f(x) = 0 \text{ and } \left| \frac{\partial f}{\partial x}(x) \right| \neq 0.$$

According to the inversion theorem these points are isolated.
  Criterion 2.

$$f(x) = 0 \text{ and } \left| \frac{\partial f}{\partial x}(x) \right| = 0.$$

The points characterized by criterion 2 need not to be isolated. If they are not, either they are approximately enumerated exhaustively, or refined criteria sketched below have to be applied.

<u>$m \leq d.$</u>
  Criterion 1.

$$f(x) = 0, \text{ the rank of } \left(\frac{\partial f}{\partial x}(x)\right) = m, \text{ and}$$

$$\left(\frac{\partial f}{\partial x}(x)\right) \cdot \lambda = \begin{pmatrix} 0 \\ \cdots \\ 0 \\ 1 \end{pmatrix} \text{ has a solution.}$$

Criterion 2.

$$f(x) = 0 \text{ and rank of } \left(\frac{\partial f}{\partial x}(x)\right) < m.$$

Points satisfying the criterion 2 are not necessarily isolated. The following refined criterion eventually allows to avoid exhaustive enumeration.

Although the rank of the matrix

$$\left(\frac{\partial f}{\partial x}(x)\right)$$

of $f : \mathbb{R}^d \rightarrow \mathbb{R}^m$ is less than m it may happen that the rank of this matrix augmented by a column of 1's,

$$\left(-\frac{\partial f}{\partial x}(x), 1\right),$$

is m. The rows of the augmented matrix are the normal vectors of the m explicit surfaces in $\mathbb{R}^{d+1}$ defined by the component functions f_i of $f = (f_1, \ldots, f_m)$. Since the rank of the augmented matrix is m, the tangent space of the surface of intersection of these m surfaces has dimension $d + 1 - m$. Further, it can be seen that the tangent space lies in parallel to the hyperplane $x_{d+1} = 0$. Hence the set

$$\{f(x) \mid \text{rank } \left(\frac{\partial f}{\partial x}(x)\right) < m, \text{ rank} \left(-\frac{\partial f}{\partial x}(x), 1\right) = m, \; f_i(x) = f_j(x), \; i, j = 1, \ldots, m\}$$

consists of isolated values only. By treating $f(x) = \varepsilon \cdot 1$ for some small $\varepsilon \in \mathbb{R}$, this situation can be avoided.

<u>$m = 1.$</u> This is a special case of the previous one:
  Criterion 1.

$$f(x) = 0 \text{ and } \frac{\partial f}{\partial x_1}(x) = 0, \ldots, \frac{\partial f}{\partial x_{d-1}}(x) = 0, \frac{\partial f}{\partial x_d}(x) \neq 0.$$

Criterion 2.

$$f(x) = 0 \text{ and } \frac{\partial f}{\partial x}(x) = 0.$$

This situation can be avoided by solving $f(x) = \varepsilon$ for some small ε instead of $f(x) = 0$. This comes from the fact that

$$\{f(x) \mid \frac{\partial f}{\partial x_1}(x) = 0, \ldots, \frac{\partial f}{\partial x_d}(x) = 0\}$$

consists of isolated values only. Figure 4 shows the curve of the previous example with a positive and a negative ε. The originial behavior can be deduced from these two situations.

The suggestion to resolve some of the critical cases by replacing 0 by a small ε is based on the assumption that in practical problems the inputs are not exact, but approximative data.

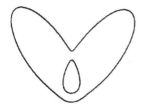 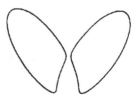

Figure 4: Curves from the same function as in the previous figures, but with positive and negative ε's instead of 0.

Hence, in particular in relation with an approximative numerical evaluation of the solution set, this approach should be feasible. The modification can be done on-line during the run of the subdivision algorithm only in areas where it is recognized necessary. A more general perturbation technique applied in computing arrangements of hyperplanes is presented in [Edelsbrunner, 1987, sect. 9.4].

If an exact algebraic solutions is the goal, and in situations where the argument with the tangent vector does not work, algebraic methods have to be involved in order to get refined criteria. Such criteria can for example be deduced from the following theorem which is taken from [Collins, 1975] and which the cad is based.

Projection.
Let be $\mathcal{A} := \{f_i(x_1,\ldots,x_d) \mid i = 1,\ldots m\}$, $f_i : \mathbb{R}^d \to \mathbb{R}$, a set of polynomials, $d \geq 2$. Let be

$$\mathcal{B} := \{\mathrm{red}^k(f) \mid f \in \mathcal{A}, \ k \geq 0, \ \deg(\mathrm{red}^k(f)) \geq 1\},$$

$$\mathcal{C} := \{\mathrm{ldcf}(g) \mid g \in \mathcal{B}\},$$

$$\mathcal{D}_1 := \{\mathrm{psc}_k(g,g') \mid g \in \mathcal{B}, \ 0 \leq k \leq \deg(g')\},$$

$$\mathcal{D}_2 := \{\mathrm{psc}_k(g_1,g_2) \mid g_1,g_2 \in \mathcal{B}, \ 0 \leq k \leq \min(\deg(g_1),\deg(g_2))\},$$

$$\mathcal{P} := \mathcal{C} \cup \mathcal{D}_1 \cup \mathcal{D}_2.$$

If every element of $\mathcal{P}$ is invariant over $S \subset \mathbb{R}^{d-1}$, then the roots of $\mathcal{A}$ are delineable on S.

"deg" gives the degree of a polynomial, "ldcf" is its leading coefficient, "red_k" is the polynomial obtained by removing the first k terms of highest degree w.r.t to the variable x_d, "f'" means the partial derivative w.r.t the variable x_d, psc_k means the k-th principal subresultant of two polynomials which is a special sub-determinant of the Silvester resultant of them. A polynomial g is "invariant" over a set S, if either $g(\mathbf{x}) > 0$ for all $\mathbf{x} \in S$, or $g(\mathbf{x}) = 0$ for all $\mathbf{x} \in S$, or $g(\mathbf{x}) < 0$ for all $\mathbf{x} \in S$. Intuitively formulated, a set of polynomials is called "delineable" over a set S if their graphs have non-empty intersections over S. For a more precise explanation of this theorem cf. [Collins, 1975].

5. Bernstein-Bézier representation of polynomials
The Bernstein-Bézier representation of system of polynomials has the form

$$\mathbf{f}(x_1,\ldots,x_d) := \sum_{i_1=0}^{n_1} \sum_{i_2=0}^{n_2} \cdots \sum_{i_d=0}^{n_d} \mathbf{b}_{i_1,i_2,\ldots,i_d} B_{i_1}^{n_1}(x_1) B_{i_2}^{n_2}(x_2) \cdots B_{i_d}^{n_d}(x_d),$$

$$B_i^n(x) := \binom{n}{i} x^i (1-x)^{n-i}, \ i = 0,\ldots,n, \text{ the Bernstein polynomials,}$$

$$\mathbf{b}_{i_1,i_2,\ldots,i_d} \in \mathbb{R}^m, \ x_1,x_2,\ldots,x_d \in [0,1], \ n_1,n_2,\ldots,n_d \in \mathbb{N}.$$

Each polynomial function can be converted into this form since Bernstein polynomials B_i^n, $i = 0, \ldots, n$, form a basis of the polynomials of degree n.

Since

$$B_i^n(x) \cdot B_j^m(x) = \frac{\binom{n}{i}\binom{m}{j}}{\binom{n+m}{i+j}} \cdot B_{i+j}^{m+n}(x),$$

the usual polynomial arithmetic almost immediately carries over to the Bernstein-Bézier representation.

The coefficients $b_{i_1,i_2,\ldots,i_d}$ define a set of points

$$C := \{(i_1/n_1, \ldots, i_d/n_d, b_{i_1,i_2,\ldots,i_d}) \mid i_1 = 0, \ldots, n_1; \ldots; i_d = 0, \ldots, n_d\}$$

with the property that the graph of $\mathbf{f}$,

$$\{(x_1, \ldots, x_d, \mathbf{f}(x_1, \ldots, x_d)) \mid x_1, \ldots, x_d \in [0,1]\}$$

lies totally within the convex hull of C. The consequence of the so-called *convex hull property* for the evaluation of the criteria of the previous section is that the non-existence of a zero of a system of polynomials can be easily checked. For example, if the rectangular hyper-parallelepiped induced by the extremal coordinates of the points in C does not intersect the d-dimensional plane spread by the first d axes of coordinates in $\mathbb{R}^{m+d}$ we are sure that $\mathbf{f} \neq \mathbf{0}$ in the interval $[0,1]^d$. The convex hull property also holds for each projection of the graph of $\mathbf{f}$ and the set C.

The Bernstein-Bézier representation is only useful if it can be easily maintained during subdivision. This is indeed possible by the so-called de Casteljau algorithm. The algorithm of de Casteljau applied for subdividing in one coordinate direction works as follows [cf. e.g. Lasser, 1985].

Input. The coefficients $b_{i_1,\ldots,i_d}$, $i_1 = 0, \ldots, n_1$; $\ldots$; $i_d = 0, \ldots, n_d$ of a polynomial in Bernstein-Bézier representation, an integer $k \in \{1, \ldots, d\}$, $t \in [0,1]$.

Output. The Bézier-coefficients $b_{i_1,\ldots,i_d}^-$, $i_1 = 0, \ldots, n_1$, $\ldots$, $i_d = 0, \ldots, n_d$ respectively $b_{i_1,\ldots,i_d}^+$, $i_1 = 0, \ldots, n_1$, $\ldots$, $i_d = 0, \ldots, n_d$, of the restriction of the original polynomial onto the two d-dimensional intervals obtained by dividing $[0,1]^d$ in coordinate direction k with the ratio $t : (1-t)$. The parametrization of both resulting polynomials is again over $[0,1]^d$.

Description.
```
begin
  for (i₁, ..., iₖ₋₁, iₖ₊₁, ..., i_d)
      ∈ {0, ..., n₁} × ... × {0, ..., nₖ₋₁} × {0, ..., nₖ₊₁} × ... × {0, ..., n_d}
  do
  begin
    for j = 0 to nₖ do
      for iₖ = j to nₖ do
        if j = 0 then bʲ_{i₁,...,i_d} := b_{i₁,...,i_d}
        else bʲ_{i₁,...,i_d} := (1 − t) · bʲ⁻¹_{i₁,...,iₖ₋₁,iₖ₋₁,iₖ₊₁,...,i_d} + t · bʲ⁻¹_{i₁,...,iₖ₋₁,iₖ,iₖ₊₁,...,i_d};
      for iₖ = 0 to nₖ do
      begin
        b⁻_{i₁,...,iₖ,...,i_d} := bⁱᵏ_{i₁,...,iₖ,...,i_d};
        b⁺_{i₁,...,iₖ,...,i_d} := bⁿᵏ⁻ⁱᵏ_{i₁,...,nₖ,...,i_d};
      end
  end
end.
```

The time complexity of the algorithm of de Casteljau is $O(n_1 \cdots n_{k-1} n_k^2 n_{k+1} \cdots n_d)$. The coefficients obtained by iteratively applying the subdivision by de Casteljau converge quadratically towards the given polynomial [Dahmen, 1986].

For $x_k = 1/2$ the new coefficients can be calculated by integer arithmetic only if fixed point numbers of the form $(a, 2^i)$ with integers a and i are used for the coefficients. The length of a increases at most by 1 in each step of subdivision so that the coefficients stay reasonably small.

With respect to floating point arithmetic, the algorithm of de Casteljau is also extremely reliable since only convex combinations are to be evaluated. Contributions to the discussion of the numerical stability of algorithms in Bernstein-Bézier representation can be found in [Farouki, Rajan, 1988].

In the criteria of the previous section partial derivatives have to be evaluated. The partial derivative $\partial f/\partial x_k(\mathbf{x})$, $k = 1, \ldots, d$, of a polynomial function $\mathbf{f} : \mathbb{R}^d \to \mathbb{R}^m$ in Bernstein-Bézier representation is

$$\frac{\partial f}{\partial x_k}(\mathbf{x}) = n_k \cdot \sum_{i_1=0}^{n_1} \cdots \sum_{i_k=0}^{n_k-1} \cdots \sum_{i_d=0}^{n_d}$$

$$(\mathbf{b}_{i_1,\ldots,i_{k-1},i_k+1,i_{k+1},\ldots,i_d} - \mathbf{b}_{i_1,\ldots,i_{k-1},i_k,i_{k+1},\ldots,i_d}) \cdot B_{i_1}^{n_1}(\mathbf{x}) \cdots B_{i_k}^{n_k-1}(\mathbf{x}) \cdots B_{i_d}^{n_d}(\mathbf{x}).$$

Hence the Bézier coefficients of the derivative are obtained as differences of consecutive coefficients in direction k, multiplied by n_k. This calculation can be performed easily algebraically exact as well as numerically reliable.

Testing the criteria also involves checking certain determinants

$$|g_{i,j}(x_1,\ldots,x_d)|_{i,j=1,\ldots,m}$$

with polynomial entries whether they can become 0 over a rectangular interval like $[0,1]^d$. If $b_{i,j}^-, b_{i,j}^+ \in \mathbb{R}$ are the smallest respectively greatest of the coefficients of $g_{i,j}$ in Bernstein-Bézier representation, it must be tested whether the interval determinant

$$|[b_{i,j}^-, b_{i,j}^+]|_{i,j=1,\ldots,m}$$

contains 0. The test can be done by evaluating this determinant using the rules of interval arithmetic [Alefeld, Herzberger, 1983].

6. Summary

We have sketched how the Bernstein-Bézier representation can be used to calculate a description of the solution set of an algebraic system. Although the aim in practice is a numerically approximate description of the solution set, an algebraically exact representation can also be obtained under certain conditions after a sufficiently high number of iterations. Algebraically exact means that the extremal values are represented by disjoint intervals. The number of iterations necessary can be estimated by theorems of algebra. Both the numerical as well as the algebraically exact evaluation of polynomials can be efficiently and reliably performed in Bernstein-Bézier representation.

We have used the so-called *tensor product representation* of Bézier polynomials. Alternatively, the triangular representation [cf. e.g. Farin, 1990]

$$f(x_1,\ldots,x_d) := \sum_{i_0+i_1+\ldots+i_d=n,\, i_0,i_1,\ldots,i_d \geq 0} b_{i_0,i_2,\ldots,i_d} B_{i_0,i_1,\ldots,i_d}^n (1 - x_0 - \ldots - x_d, x_1,\ldots,x_d),$$

$$x_1 + \ldots + x_d \leq 1,\ x_1,\ldots,x_d \geq 0,$$

with the generalized Bernstein polynomials

$$B_{i_0,i_1,\ldots,i_d}^n(x_0,\ldots,x_d) := \frac{n!}{i_0! i_1! \ldots i_d!} x_0^{i_0} x_1^{i_1} \ldots x_d^{i_d},$$

$$i_0 + i_1 + \ldots + i_d = n, \; i_0, i_1, \ldots, i_d \geq 0.$$

can be applied. Here the interval of analysis is of triangular instead of rectangular shape.

References

G. Alefeld, J. Herzberger (1983) Introduction to interval computation, Academic Press, New York

E.L. Allgower, G. Georg (1982) Predictor-corrector and simplicial methods for approximating fixed points and zero points of nonlinear mappings, in: Mathematical Programming: The State of the Art, Springer-Verlag, Berlin, 15–56

D.S. Arnon (1983) Topologically reliable display of algebraic curves, Computer Graphics 17(3), 219–227

J. Bloomenthal (1988) Polygonization of implicit surfaces, Computer Aided Geometric Design 5, 341–355

G.E. Collins (1975) Quantifier elimination for real closed fields by cylindrical algebraic decomposition, in: Lecture Notes in Computer Science 33, Springer-Verlag, New York, 134–183,

W. Dahmen (1986) Subdivision algorithms converge quadratically, Journal of Computational and Applied Mathematics 16,145–158

H. Edelsbrunner (1987) Algorithms in combinatorial geometry, Springer-Verlag, 1987

G.E. Farin (1990) Curves and surfaces for computer aided geometric design, Academic Press, San Diego, 2nd ed.

R.T. Farouki, V. Rajan (1988) Algorithms for polynomials in Bernstein form, Computer Aided Geometric Design 5, 1–26

D. Lasser (1985) Bernstein-Bézier representation of volumes, Computer Aided Geometric Design 2 (1985) 145–150

A. Morgan, V. Shapiro (1987) Box-bisection for solving second-degree systems and the problem of clustering, ACM Transactions on Mathematical Software 13,152–167

J.T. Schwartz, M. Sharir, J. Hopcroft, ed. (1987) Planning, geometry and complexity of robot motion, Ablex Publishing Corporation, Norwood, New Jersey

J.T. Schwartz, C.-K. Yap, ed. (1987) Algorithmic and geometric aspects of robotics, Lawrence Erlbaum Ass., Hillsdale, New Jersey

T.W. Sederberg (1984) Planar piecewise algebraic curves, Computer Aided Geometric Design 1, 241–255

T.W. Sederberg (1985) Piecewise algebraic surface patches, Computer Aided Geometric Design, 2, 53–59

T.W. Sederberg and D.C. Anderson and R.N. Goldman (1984) Implicit representation of parametric curves and surfaces, Computer Vision, Graphics and Image Processing 28, 72–84,

T.W. Sederberg and R.J. Meyers (1988) Loop detection in surface patch intersections, Computer Aided Geometric Design 5, 161–171

XYZ:

A project in experimental geometric computation

Jurg Nievergelt, Peter Schorn, Michele De Lorenzi, Christoph Ammann, Adrian Brüngger
Informatik, ETH, CH-8092 Zurich

Abstract

The project XYZ (eXperimental geometrY Zurich) aims to develop practically useful software for geometric computation, and to test it in a variety of applications. In pursuing these goals we emphasize the following points, each of which is described in one section of this paper:

1. Exploit recent progress in computational geometry through a systematic study to determine classes of algorithms that lend themselves to robust and practically efficient programs. Our program library contains many standard algorithms for 2-d problems, several for restricted 3-d problems, and a few for d-dimensional geometry.
2. Verify and evaluate algorithms experimentally: We study the problem of consistency in the presence of numerical errors, and emphasize robust programs that handle all degenerate cases; we often implement and compare different algorithms for the same problem.
3. Use state-of-the-art software engineering techniques in a workbench that supports the development of a library of production-quality programs: The XYZ GeoBench (written in Object Pascal for the Macintosh) is a loosely coupled collection of modules held together by a class hierarchy of geometric objects and common abstract data types.
4. The GeoBench is being used in education as a programming environment for rapid prototyping and visualization of geometric algorithms. Two other application projects are underway: Software for terrain modeling, and interfacing the GeoBench as a "geometry engine" to a spatial data base system.

Contents

0. Software for geometric computation, and the project XYZ

Geometry merged with automatic computation in the late fifties (during the early days of computer graphics, e.g. Sutherland's Sketchpad) to create the discipline of geometric computation. The appearance of computer-aided design (CAD) systems in the sixties greatly widened its range of applications, and gave increased importance to the nagging problem of correctly treating degenerate configurations; whereas a picture can tolerate an occasional error, an engineering design cannot. Computer scientists with a practical orientation working on graphics and CAD pushed the field forward. They developed many interesting algorithms, such as for visibility, and entire classes of related algorithms, such as scan-line algorithms. They collected experimental evidence for comparing the efficiency of different algorithms, and discovered the tantalizing and tough problems of how to compute reliably with degenerate configurations in the presence of round-off errors. Practitioners laid the foundation for a discipline of geometric computation.

In the mid seventies, led by Shamos' pioneering Ph.D. thesis, theoretically oriented researchers took over. They brought the finely honed tools of algorithm design and analysis to bear on geometric algorithms and created a new theoretical discipline of computational geometry. The well-defined, conceptually simple algorithmic problems of geometry, and the highly developed techniques of algorithm analysis, proved to be a perfect match. Computational geometry has now enjoyed a decade and a half of spectacular progress. It turned a field characterized by trial-and-error as recently as the seventies into a discipline where no programmer can work competently in ignorance of theory.

Today's research community in computational geometry still focuses most of its attention on theoretical problems. Research often stops short of investigating practical issues of implementation, so readers are left wondering whether a proposed optimal algorithm is useful in practice or not – a question that rarely has an easy answer. But this question must be answered by the computational geometry research community, and not be left to the applications programmer. It has become abundantly clear that the development of robust and efficient software for geometric computation calls for specialists with a broad range of experience that ranges from algorithm design and analysis to numerics and program optimization.

Even a prototype implementation of just one sophisticated geometric algorithm is an arduous endeavor if attempted without the right tools, such as: A library of abstract data types (e.g. dictionary, priority queue) and corresponding data structures, reliable geometric primitives (e.g. intersection of 2 line segments), and visualization aids. What the applications programmer needs, but cannot find today, are reliable and efficient reusable software building blocks that perform the most common geometric operations. Geometric modelers, the core of CAD systems, do not address his problems – they are typically monoliths from which an applications programmer cannot extract any useful part for his own program.

We are aware of few projects whose main aim is to alleviate the problems faced by implementors of geometric algorithms. The most visible ones are the program library LEDA [MN 89] and the Workbench for Computational Geometry WOCG [ES 90]; both exhibit some similarities and some differences with our project XYZ. The goal common to all three projects is to develop practically useful software for geometric computation that is accessible to a wide range of applications programmers. Differences include: The systems software and programming language chosen as a basis of implementation, the scope of services and functions provided by the system, and range of algorithms and data structures included.

The project XYZ (eXperimental geometrY Zurich) presented here and in the companion paper [Sch 91a] aims at a broad range of goals all of which are essential for turning geometric computation from a specialty into a widely-practiced discipline:

1. Technology transfer: Exploit recent progress in computational geometry through a systematic study to determine classes of algorithms that lend themselves to robust and practically efficient programs.
2. Verify and evaluate algorithms experimentally: We study the problem of consistency in the presence of numerical errors, and emphasize robust programs that handle all degenerate cases; we implement and compare different algorithms for the same problem, and execute them using different number systems.
3. Use state-of-the-art software engineering techniques in a workbench that supports the development of a library of production-quality programs: The XYZ GeoBench (written in Object Pascal for the Macintosh) is a loosely coupled collection of modules held together by a class hierarchy of geometric objects and common abstract data types.
4. Test the software developed by exposing it to the rigors of a number of applications. The GeoBench is being used in education as a programming environment for rapid prototyping and visualization of geometric algorithms. Two other application projects are underway: Software for terrain modeling, and interfacing the GeoBench as a "geometry engine" to a spatial data base system.

This paper is an overview of the entire project, with examples of activities and results in each of these four categories, and particular emphasis on experimentation. [Sch 91a] describes the GeoBench in more detail.

1. Criteria for selection, types of algorithms, and the program library

There is no shortage of algorithms for inclusion in a program library for geometric computation. The problem is one of selection, whereby we emphasize the following criteria.

Robustness. A library routine must yield meaningful results for any geometric configuration, including highly degenerate ones. Unlike random data where degenerate configurations are rare, many practical applications generate a lot of highly degenerate configurations – degeneracy comes from the regularity that is inherent in man-made artifacts. The effort to guarantee correct results under all circumstances accounts for the lion's share of programmer time.

Practical efficiency. We strive for programs that are efficient in practice, that is, outperform competing programs on realistic input data. Example: An optimal algorithm can often be modified to run faster on a battery of realistic test data, even though worst-case optimality is no longer guaranteed. This may occur, for example, by replacing a balanced tree implementation of a dictionary by an array implementation. The XYZ library leaves such choices of data structure to the user.

Standard problems of geometric computation. A program library is never comprehensive enough to solve most users' problems directly. We limit ourselves to basic problems that serve as building blocks for advanced geometric programs.

Well understood and elegant algorithms. We select algorithms that stand out by virtue of their elegant simplicity and can be implemented in a straightforward manner. Even if they are not asymptotically optimal, these tend to do better than their complicated counterparts with respect to robustness and practical efficiency. Some "optimal" algorithms are just too complicated for a reliable, robust implementation.

Start with 2-d geometry. The difficulties posed by 2-d problems must be solved completely before venturing into higher dimensions, so we have concentrated on accumulating a sufficient number of representative 2-d algorithms. We approach 3-d geometry by first studying restricted 3-d problems using layered objects (see also section 4). Just to show that the structure of the GeoBench is not restricted to low-dimensional space we have implemented an algorithm that computes the minimal area disk enclosing a set of points in d-space.

Types of algorithms

As a guide to selection, and in order to benefit from any similarities that might be found, we attempt to classify the multitude of published algorithms. The majority we have investigated fall into one of the following categories:
- Incremental algorithms
  - sweeps (mostly in the plane, occasionally through space)
  - in random order
- Boundary traversal
- Recursive data partitioning, i.e. divide and conquer.

These classes of algorithms differ significantly with respect to their data access patterns. Whether this access pattern is irregular, or simple and predictable, has an effect on efficiency, in particular if data is processed off disk (see also section 4). Some observations:
- Randomized incremental algorithms play an important role in theory because the assumption of randomly ordered data is favorable for average case analysis. In practice they do not appear to be superior to sweeps.
- Sweeps, on the other hand, require (lexicographically) sorted data; this orderly access pattern leads naturally to efficient implementations with simple data structures.
- Algorithms that follow boundaries (e.g. of a convex polygon) exhibit a spatial locality principle. They can be implemented efficiently in central memory using list structures, but are less efficient for data stored on disk.
- Algorithms that partition their data in recursively generate the most irregular access patterns. In section 2 we demonstrate experimentally that divide-and-conquer algorithms are typically less efficient than their plane sweep counterparts.

In conclusion, we favor plane-sweep algorithms as a simple and efficient general purpose skeleton for most 2-d geometric problems.

The 2-d algorithms currently in the library include:

- Convex hull (Graham's scan, divide and conquer)
- Diameter and intersection of convex polygons
- Tangents common to two convex polygons
- Boolean operations (union, intersection, difference) on polygons
- Contour of a set of rectangles
- Winding number
- Intersection of line segments (sweep line for the first intersection and for reporting all intersections, sweep line for the special case of horizontal and vertical line segments)
- Closest pair (sweep line, simplified sweep line, probabilistic)
- All nearest neighbors (sweep line, simplified sweep line, extraction from Voronoi diagram)
- All nearest neighbors in a sector
- Voronoi diagram (sweep line, divide and conquer)
- Euclidean minimum spanning tree (EMST)
- Traveling salesman heuristics (nearest neighbor, EMST, convex hull, tour optimizer)

The presence of distinct algorithms for solving the same problem reflects our concern for experimental assessment and comparison, as discussed in the next section.

2. Verify and evaluate algorithms experimentally

Efficiency is at the heart of algorithm design, but it is neither easy to define nor measure. Algorithm analysis typically interprets the term 'efficiency' as asymptotic complexity, in the worst case or as an average over some data space. Although useful for mathematical analysis, this approach neglects many practical considerations such as: Simplicity and robustness of the implementation, constant factors, and whether data encountered in practice is well modeled by the randomness assumptions that go into the theory (it is usually not). For practical applications, the most promising algorithms must be implemented in a variety of ways and evaluated experimentally. We describe some of the experiments and results obtained. Time is measured in seconds as executed on a Macintosh IIfx.

Efficiency measurements. The GeoBench facilitates experimental verification and evaluation, e.g. by timing and displaying on demand each operation executed. Executing all programs on a common platform has the advantage that geometric primitives are implemented the same way in both programs, making time measurements more meaningful. For example, most algorithms for computing the convex hull are likely to use some means for detecting a 'left turn'. There are many ways to implement this primitive, and they result in different running times, but the peculiarities of this implementation is an issue of program optimization more than of algorithm design. The GeoBench provides about 20 geometric primitives such as 'left turn', and the library programs all use them to the greatest extent possible.

Example: Sweeps and other incremental algorithms vs. divide-and-conquer. Fortune's sweep [Fo 87] for computing the Voronoi diagram is much easier to implement than the standard divide-and-conquer algorithm (e.g. [PS 85]). It uses memory more efficiently, enabling us to compute larger Voronoi diagrams, and is about three times faster, as the following measurements on random point sets show. This and other experiments have convinced us that the divide-and-conquer Voronoi algorithm is not competitive and does not belong in a program library.

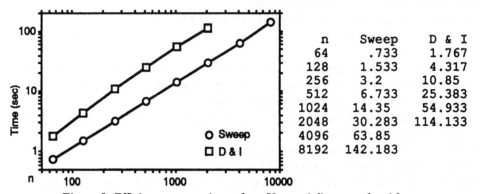

n	Sweep	D & I
64	.733	1.767
128	1.533	4.317
256	3.2	10.85
512	6.733	25.383
1024	14.35	54.933
2048	30.283	114.133
4096	63.85	
8192	142.183	

Figure 1: Efficiency comparison of two Voronoi diagram algorithms

As a second example we consider the computation of the convex hull of a set of points in the plane. The Graham scan [Gr 7], an incremental algorithm, is about twice as fast as Preparata and Hong's [PH 77] divide-and-conquer. The divide-and-conquer code is 60% longer than the Graham scan implementation, which reflects the fact that degenerate cases are more difficult to handle.

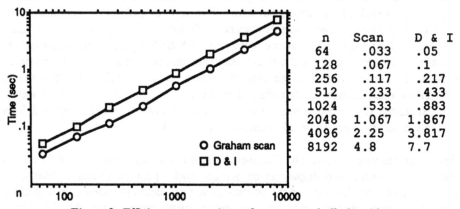

n	Scan	D & I
64	.033	.05
128	.067	.1
256	.117	.217
512	.233	.433
1024	.533	.883
2048	1.067	1.867
4096	2.25	3.817
8192	4.8	7.7

Figure 2: Efficiency comparison of two convex hull algorithms

We conclude that incremental algorithms, in particular sweeps, are usually superior to divide and conquer algorithms. One reason is that most divide and conquer algorithms compute much information that is not part of the final solution, whereas incremental algorithms tend to compute only information that is part of the final solution. A second problem: Divide-and-conquer algorithms tend to use more memory, especially during the last merge where two large objects are combined into the final solution, an operation that can rarely be done in place.

Parametrized arithmetic. Executing a program repeatedly on the same data using different arithmetic is an effective experiment for assessing the robustness of an implementation. The GeoBench contains a floating point arithmetic package where the user specifies, at run time, radix and precision. Low precision, say 2 decimal digits in the mantissa, brings any numerical problems quickly to the observer's attention. The program source code is independent of the choice of arithmetic.

Example: Robustness of a sweep algorithm

The figure below shows the results of executing the all-nearest-neighbors-to-the-left-sweep of [HNS 90] in a low-precision floating point arithmetic (radix 10, two-digit mantissa, equivalent to about 6 bits) and in a 32-bit floating point system. The rightmost window shows the two results superposed using XOR graphics, thus canceling the common part and showing where they differ. The comparison shows that when low precision arithmetic identifies a wrong neighbor, the distances are nevertheless close.

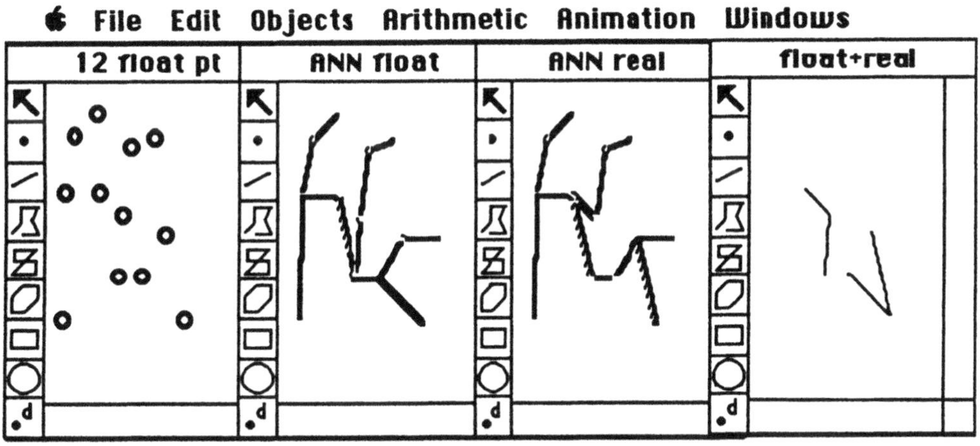

❤ File Edit Objects Arithmetic Animation Windows

12 float pt	ANN float	ANN real	float+real

Figure 3: The effect of using different arithmetic systems

Different implementations of abstract data types, and instrumentation

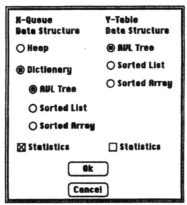

Figure 4: Dialog box for choosing the implementation of abstract data structures

As we may postpone the choice of arithmetic until run time, we may also explore the influence of different implementations of the same abstract data type on the performance of a given algorithm. The dialog box in figure 4 shows how a sweep can be tuned to use various data structures for a dictionary and priority queue. When the user selects 'Statistics', without changing a single line of source code, GeoBench records the maximal number of elements in the data structures and the number of insertions and deletions performed.

Example: Data structures in the all-nearest-neighbors sweep

Figures 5 and 6 show the influence of different implementations of the X-priority-queue and the Y-dictionary in the all-nearest-neighbors sweep [HNS 90]. The experiment involved sets of 8192 and 64 random points uniformly distributed in a square. For 8192 points, the choice of X-queue and Y-table implementations are independent: A heap is best for the X-queue, a balanced tree for the Y-table. It is perhaps surprising that the best and worst time differ by only a factor of 2. For 64 points we have a more complex picture, but the combination heap + balanced tree is still best.

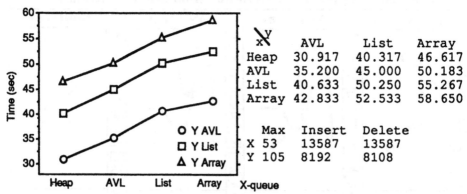

x \ y	AVL	List	Array
Heap	30.917	40.317	46.617
AVL	35.200	45.000	50.183
List	40.633	50.250	55.267
Array	42.833	52.533	58.650

	Max	Insert	Delete
X	53	13587	13587
Y	105	8192	8108

Figure 5: Efficiency comparison of different data structures (n = 8192)

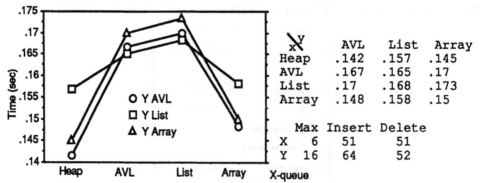

x \ y	AVL	List	Array
Heap	.142	.157	.145
AVL	.167	.165	.17
List	.17	.168	.173
Array	.148	.158	.15

	Max	Insert	Delete
X	6	51	51
Y	16	64	52

Figure 6: Efficiency comparison of different data structures (n = 64)

Test data generation. GeoBench generates two main types of test data, random and degenerate.

Random configurations: Random configurations consist of basic geometric data types such as [d-]dimensional point, line segment, polyline, polygon, convex polygon, circle and rectangle. The user specifies the number of objects (for points, line segments, circles and rectangles) or the number of vertices (for poly lines, polygons and convex polygons). d-dimensional points can be uniformly distributed in a d-dimensional square or in a d-dimensional circle.

Degenerate configurations, including: More than two collinear points, more than three points on a circle, more than two line segments with a common intersection point, horizontal and vertical segments, point sets that lie on a rectangular grid, point sets with equal x- or y-coordinate, coinciding objects. With integer arithmetic most of these degeneracies are exact, that is, the corresponding test polynomials evaluate to zero. In the case of floating point arithmetic the test polynomials either evaluate to zero or to a value close to zero, corresponding to nearly degenerate configurations. Both cases are useful for debugging and testing.

Example: Approximate generalized Voronoi diagrams
GeoBench provides the facility to cover objects consisting of straight lines (e.g. line segments, polygons) or circles with evenly spaced points. This feature can be used in two ways: To create degenerate configurations as explained above, or to approximate linear objects by a sequence of points, as the following example illustrates.
Figure 7 shows, from left to right, line segments, their point covers, and a Voronoi diagram of these points. By omitting the many parallel lines in this diagram we obtain an approximate generalized Voronoi diagram of the original set of line segments.

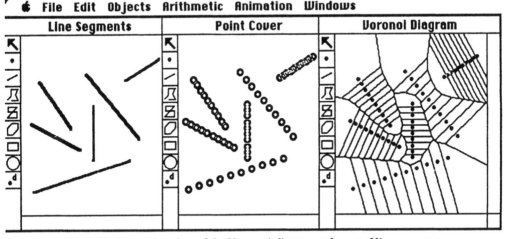

Figure 7: Approximation of the Voronoi diagram of a set of line segments

3. The XYZ software packages

The main goal of the project XYZ is to make available to the practitioner a loosely coupled collection of carefully crafted software packages, in particular:

- The XYZ GeoBench, a programmer's workbench
- The XYZ Program Library,
  an open-ended collection of geometric algorithms
- The XYZ Grid File, a package for managing spatial data on disk.

The relationship among these three packages and the class hierarchy that defines all common object types is shown in the figure below, where an arrow indicates the relationship "is_based_on".

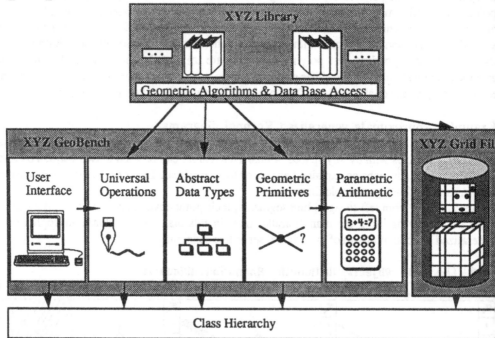

Figure 8: The relationship among the XYZ software packages

The GeoBench ([Sch 91a] and [Sch 91b]) is the programmer's workbench and run-time environment that holds all the library programs and the disk storage grid file package together, keeping them data-compatible. Its components serve the following functions. The user interface manages windows for interactive data generation and algorithm animation, as illustrated in several pictures above. A collection of the most important geometric primitives and abstract data types, and various implementations thereof, saves the programmer a lot of time-consuming detail work. A parametrized arithmetic package supports experimental program validation by providing floating point arithmetic of varying precision and base.

The library is an open-ended collection of geometric algorithms that work in central memory, and of disk access procedures that pack geometric objects into grid files and perform queries on them. The grid file disk management package provides multidimensional data access for an arbitrary number of dimensions, each of which is individually measured by one of the types integer, long-integer, or real.

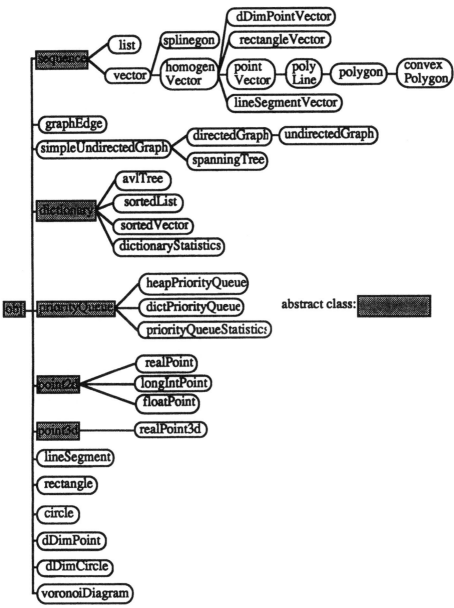

Figure 9: Class hierarchy of the XYZ GeoBench

Figure 9 shows the backbone of this software: The class hierarchy that defines the common data types and serves as interface between all software components. This tree describes the "is_a" relationships among the classes currently in the GeoBench. Algorithms are methods associated with the class on which they operate. For example, the Voronoi algorithm is a method in the class 'pointVector' that yields an object of type 'voronoiDiagram'. The principle of inheritance insures that all methods for a given class are also available for their descendants. Thus the 'method' Voronoi diagram is also applicable to 'polyLine', 'polygon', and 'convexPolygon', for each of which it may have its own implementation.

4. Uses and applications

During its entire period of development, the XYZ GeoBench and program library have served as a useful demonstration package and tool for algorithm animation in courses on algorithms and data structures. Now that the GeoBench is essentially complete, we are also using it as a programming environment for term projects and in a course on computational geometry. Work on realistic applications started only recently; we briefly describe two ongoing projects.

Layered objects and triangulated surfaces for terrain modeling.

Layered objects are an attempt to reconcile two conflicting facts of geometric computation: On the one hand, it is well known that 2-dimensional geometric algorithms are usually a lot simpler and more efficient (often $O(n \log n)$) than their 3-d counterparts (usually $O(n^2)$ or higher); on the other hand, most applications call for 3-d geometry. Fortunately, the objects to be processed in any one application are often subject to restrictions that make it profitable to define various restricted classes of 3-d objects with special properties that yield to simpler algorithms than unrestricted 3-d objects.

Layered objects are a striking example of the benefits and limitations of this approach: A 3-d object is represented or approximated as a vector of layers (parallel slices orthogonal to the z-axis), where each layer is defined by its thickness and a 2-d contour. Important classes of real-world objects are naturally modeled as layered objects, such as terrain (using contour lines), certain semiconductor devices (perhaps using projections and cross-sections), and, in general, objects whose shape is defined by one or more functions of the type $z = f(x, y)$. Layered models are particularly appropriate in CAD systems for stereolithography, a new manufacturing technology that "grows" 3-d objects one layer at a time. Each layer is defined by tracing its outline with a laser and marking the part that is to remain; the latter hardens on top of the preceding layer when exposed to light.

Layered objects are particularly effective, as is the case with maps, when the number L of layers needed to achieve a desired accuracy is small compared to the complexity n of the 2-d figures in each layer. Operations on layered objects typically trigger a sequence of calls to 2-d algorithms, one for each of L layers, and thus work in time $O(L \, n \log n)$. This often compares favorably to the complexity $O(N^2)$ of a quadratic algorithm on a comparable unrestricted 3-d object of complexity N, where a fair comparison suggests $n < N < L \, n$.

Although layered objects greatly simplify the problems of processing 3-d objects, they do not eliminate the necessity to consider the 3rd dimension explicitly. When computing the visible surface, for example, we wish to look at a layered object from an arbitrarily chosen point of view (not necessarily at infinity). This results in clipping a stack of layers against a pyramid in arbitrary position, a true 3-d problem. A second example is the problem of correct treatment of all degenerate configurations: Layered objects exhibit new types of degeneracies beyond those that occur in 2-d.

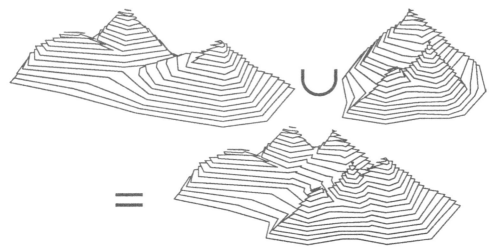

Figure 10: Union of two layered objects

In addition to visibility, we have implemented boolean or set-theoretic operations, as shown in Figure 10.

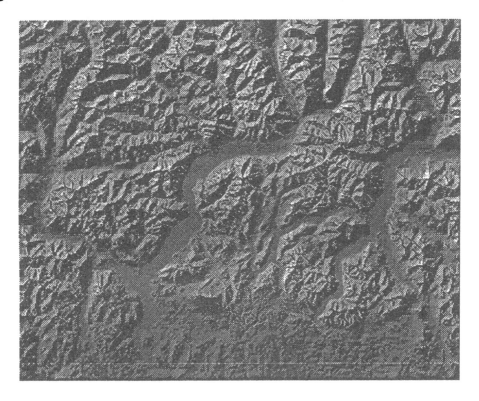

Figure 11: Grey level map of the southern slope of the Alps
(Ticino and Lombardia) automatically generated from geometric data

But the stair-case shape of layered objects make them unsuitable models for the graphic representation of smooth objects, so we are also introducing triangulated surfaces as an alternative 3-d model. As terrain modeling requires seemingly realistic images, we experiment with the automatic generation of synthetic images, with grey-levels or color-shading (a graphics problem rather than a geometric problem). As an example, the terrain image in figure 11 is generated automatically from geographic (x, y, z)-data that represents Switzerland on a 250m × 250m grid.

The interplay between geometric computation and spatial data bases

Most algorithms of computational geometry, and their complexity results, are based on the "random-access-machine model of computation". This model provides realistic performance predictions as long as all the data fits in central memory, where access time to any data element is approximately constant. When large data configurations must be processed off disk, on the other hand, disk access usually becomes the bottleneck. The efficiency of computation is then determined primarily by two issues: 1) How data is stored on disk, and 2) in what order algorithms access data, and, of course, the interplay between 1) and 2).

1) How data is stored on disk. The development of computational geometry has spawned growing interest in spatial data bases, for which efficient data access is perhaps the major issue. [Ni 89] and [Wi 91] are general surveys of spatial data structures. The XYZ GeoBench interfaces to a Grid File [NHS 84] as a general-purpose multidimensional data structure for storing geometric objects on disk. [NH 87] describes how a broad range of proximity queries are answered efficiently on large collections of objects stored as points in parameter space.

2) In what order algorithms access data. In section 2 we classified geometric algorithms according to their data access pattern as follows:
- Sweeps: data is accessed in an order (e.g, of increasing x) known a priori.
- Boundary traversal: data access obeys a spatial locality principle, but is only known at run time.
- Recursive partitioning: data access is usually random.

The advantages of a predictable data access pattern, and the disadvantages of random access, are even more pronounced when the data is processed off disk. Thus the preponderance of sweep algorithms in the XYZ library, and the choice of layered objects for 3-d modeling, are a consequence of our aim at applications such as terrain modeling that require efficient processing of very large data volumes.

A geometry engine as front end to a spatial data base

The tight coupling between the XYZ GeoBench and its Grid File allows us to focus on algorithms that interact in a particularly efficient way with the grid file data structure. The typical situation in applications that process large volumes of spatial data, however, is different. Usually, the user's data is organized and stored in some commercial spatial data management system that provides a few types of spatial queries only – clearly a spatial data base system cannot anticipate the access patterns of all algorithms its users might run on its data. Thus it is an open question in spatial data base research as to how efficiently geometric algorithms interface with typical built-in queries. In order to explore this issue, we started a joint project with the database research group at ETH (H.-J. Schek) where the GeoBench is used as a front end to a spatial data base system built on DASDBS [DSW 90]. The first experiments aim to use the GeoBench as a powerful user interface for retrieving data from the data base, perform geometric operations on it, and finally store (modified) objects back in the database. In a second phase we aim at a tighter coupling based on the extension capabilities of DASDBS, i.e. the ability to manage arbitrary geometric objects provided a certain set of (geometric) operations is supplied [DSW 90].

In conclusion, we have presented an overview of a research project that attacks the practical problems of software development for geometric computation on a broad front. The XYZ GeoBench in particular has proven its usefulness in numerous implementations of geometric algorithms. Its animation capability is regularly used for demonstrating algorithms in courses at ETH. Experiments have led to some surprising insights about efficiency and robustness of well-known algorithms. Other projects have just started, in particular the interaction between a "geometry engine" and spatial data bases.

Acknowledgments. Among the students who have contributed to the development of the XYZ software, we acknowledge in particular Beat Fawer, Markus Furter, Peter Lippuner, and Peter Skrotzky. The XYZ Grid File is based on a grid file package written by Hans Hinterberger and adapted to the GeoBench by Björn Beeli.

References

[DSW 90] G. Dröge, H.-J. Schek, A. Wolf: Erweiterbarkeit in DASDBS, Informatik Forschung und Entwicklung 5, 4 (special issue on Nicht-Standard-Datensysteme), pp. 162-176, 1990.

[ES 90] P. Epstein, A. Knight, J. May, T. Nguyen, J. Sack: A workbench for Computational Geometry (WOCG), Tech. report, Carleton University, 1990.

[Fo 87] S. Fortune: A Sweepline Algorithm for Voronoi Diagrams, Algorithmica 2, pp. 153-174, 1987.

[Gr 72] R. Graham: An efficient algorithm for determining the convex hull of a finite planar set, Information Processing Letters 1, pp. 132-133, 1972.

[HNS 88] K. Hinrichs, J. Nievergelt, P. Schorn: Plane-Sweep Solves the Closest Pair Problem Elegantly, Information Processing Letters 26, pp. 255-261, 11 Jan. 1988.

[HNS 90] K. Hinrichs, J. Nievergelt, P. Schorn: An all-round sweep algorithm for 2-dimensional nearest-neighbor problems, submitted.

[MN 89]   K. Mehlhorn, S. Näher: LEDA, A Library of Efficient Data Types and Algorithms, preliminary version, Universität des Saarlandes, 1989.

[NH 87]   J. Nievergelt, K.H. Hinrichs: Storage and access structures for geometric data bases. Proc. Kyoto 85 Intern. Conf. on Foundations of Data Structures (eds. Ghosh et al.), Plenum Press, pp. 441-455, NY 1987.

[NHS 84]   J. Nievergelt, H. Hinterberger, K. Sevcik: The Grid File: An adaptable, symmetric multikey file structure. ACM Trans. on Database Systems, Vol. 9, No. 1, pp. 35-45, 1984.

[Ni 89]   J. Nievergelt, 7 ± 2 criteria for assessing and comparing spatial data structures, in A. Buchman et al. eds.: Design and Implementation of Large Spatial Databases, invited paper at 1st Symp. SSD'89, UC Santa Barbara, Lecture Notes CS 409, Springer, pp. 3-27, 1990.

[PH 77]   F. Preparata, S. Hong: Convex hulls of finite sets of points in two and three dimensions, Comm. ACM 2 (20), pp. 87-93, Feb. 1977.

[PS 85]   F. Preparata, M. I. Shamos, Computational Geometry: an Introduction, Springer, 1985.

[Sch 91a] P. Schorn: Implementing the XYZ GeoBench: A programming environment for geometric algorithms, in these proceedings.

[Sch 91b] P. Schorn: Robust Algorithms in a Program Library for Geometric Computation, ETH PhD Dissertation 9519, to appear 1991.

[We 90]   E. Welzl: A fast randomized algorithm for computing the minimal area disk enclosing a set of points in d-space, presentation at the Workshop on Computational Geometry, Dagstuhl, Oct 1990.

[Wi 91]   P. Widmayer: Datenstrukturen für Geodatenbanken, Tech. Report, Univ. of Freiburg, 1991.

Implementing the XYZ GeoBench:
A programming environment for geometric algorithms

Peter Schorn, schorn@inf.ethz.ch
Informatik, ETH, CH-8092 Zurich

Abstract: The XYZ GeoBench (eXperimental geometrY Zurich) provides a comprehensive infrastructure for rapid prototyping of geometric algorithms and the implementation of production-quality library programs. This paper introduces the components of this programming environment and gives some implementation details. The system is implemented in an object oriented extension of Pascal on the Apple Macintosh computer. We report our experience with object oriented programming in the context of geometric algorithms and give some advice on building a programming environment for geometric computation.

Contents

188

1 The role of a programmer's workbench in a project for experimental geometric computation

The XYZ project [NSD 91] has the goal to produce production-quality software for geometric computation. This paper describes the design and implementation of the XYZ GeoBench, the major development tool.

The following figure shows the GeoBench and its components, as well as its relation to the other software packages of our system.

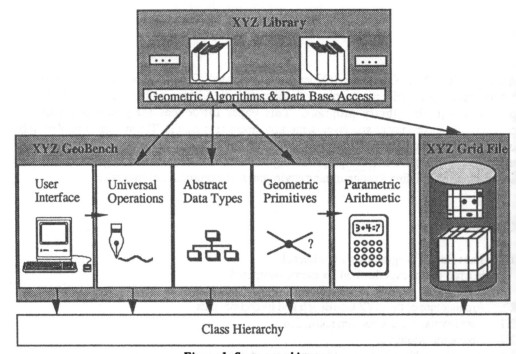

Figure 1: System architecture

The XYZ GeoBench provides the following functions:
1) A programming environment and tool kit to facilitate rapid prototyping of geometric programs
2) A run-time environment for software testing and the experimental validation of geometric algorithms
3) A tool for education, featuring algorithm animation for demonstration purposes

The XYZ GeoBench is written in Object Pascal for the Macintosh. Currently the whole system consists of more than 70 modules with a total source code size of approximately 1000 KB. The author is responsible for the design and architecture of the system and has written most of the code, although other people have contributed substantially.

The paper is organized as follows. Section 2 describes the components in more detail whereas section 3 gives a random sample of some interesting implementation details. Section 4 relates our experience with object oriented programming in the area of geometric algorithms. We assume that the reader is somewhat familiar with object oriented programming techniques in general (e.g. see [M 87] for an introduction).

2 Architecture of the XYZ GeoBench and its components

In the following sections we examine the components in turn and give important details that are useful to implementors of similar systems. We claim that the issues addressed by these components must be solved by any system that enables the user to perform research in experimental geometric computation.

2.1 User interface and algorithm animation

GeoBench uses the Macintosh conventions. The user finds an *info window* containing useful information, such as available memory, the coordinates of the cursor, time taken by the last operation and the type of the currently selected object. Selecting an object for input can either be done by using the palette attached to each geometry window or by using the *objects menu* that also allows the creation of a random instance of any of the currently directly accessible objects point, line segment, circle, rectangle, poly line, polygon, convex polygon and d-dimensional point.

Computation takes place in *geometry windows*: The user creates a new one, enters geometric objects, selects them and chooses the desired operation from the operations menu. The *operations menu* shows only operations which are legal for the selected objects. Performing an operation creates a new geometry window which contains the result of the operation already selected for a subsequent operation. For example one could enter some points using the mouse, select them, compute the Voronoi diagram, compute a Euclidean minimum spanning tree using the Voronoi diagram and finally compute a traveling salesman tour from the spanning tree.

The geometric transformation operations translate, rotate, scale and reflect can be found in the *edit menu* which also provides commands for changing the viewing transformation: Zoom in and zoom out.

In the *animation menu* the user selects which algorithms to animate while the *arithmetic menu* governs which kind of arithmetic to use for newly created objects. Since arithmetic is bound to objects and not to operations, various kinds of conversion operations are available in the operations menu.

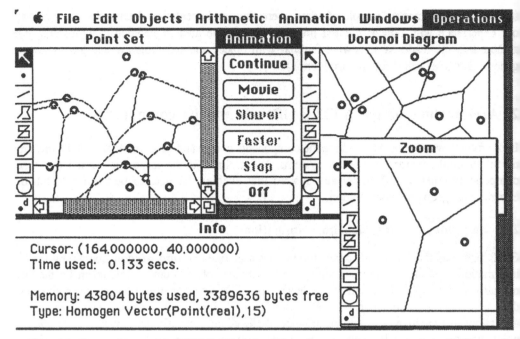

Figure 2: Screendump of the XYZ GeoBench while animating the computation of a Voronoi diagram using Fortune's sweep [F 87]

Algorithm animation

Algorithm animation is used for demonstrating and debugging. We have chosen a simple yet powerful approach to animation. There is only one version of an implementation into which code pertaining to the animation is included via conditional compilation. This code checks whether animation for this particular algorithm is turned on. If yes, it updates the currently visible state of the algorithm and waits for the user to let it proceed. In situations where speed is crucial, we avoid the slight overhead of repeatedly checking whether animation is turned on by setting the appropriate compile time variable to 'false'. Animation code has the following general structure.

```
...
{ Geometric algorithm changing internal state. }
{$IFC myAlgAnim }
    if animationFlag[myAlgAnim] then
       { Update graphical state information, usually draw some objects. }
       waitForClick(animationFlag[myAlgAnim]);
       { Update graphical state information, usually erase some objects. }
    end;
{$ENDC }
...
```

The procedure 'waitForClick' provides an interface between the user and the algorithm currently animated. It supports single step mode and a movie mode with user selectable speed (see the 'Animation' dialog box in the previous screen dump). Updating the visualization of the internal state is facilitated by the convention that all drawing on the screen is done using XOR graphics which has the benefit that erasing is the same as drawing. Animating an algorithm consists of choosing a representation of the internal state (e.g. position of the sweep line, objects in the y-table, deactivated objects, etc.) and determining appropriate locations in the program where this information needs to be updated. Algorithm animation is implemented for all non trivial geometric algorithms.

2.2 Geometric primitives

The type 'point2d' (section 2.3) is the basic building block of our geometry. Therefore all geometric primitives are methods in this class and are usually implemented in three different ways taking advantage of the respective arithmetic. The overwhelming part of our library is based on the following primitives.

```
function whichSide (p, q, r: point2d): (-1, 0, +1);
```
(* Determines on which side of the directed line segment from p to r the point q lies. *)

```
function crossProduct (p, q, r, s: point2d): real;
```
(* Computes the cross product $(p - q) \times (s - r)$. *)

```
function distance (p, q: point2d): real;
```
(* Computes the Euclidean distance between p and q. *)

```
function squaredDistance (p, q: point2d): real;
```
(* $(p_x - q_x)^2 + (p_y - q_y)^2$ *)

```
function squaredDx (p, q: point2d): real;
```
(* $(p_x - q_x)^2$ *)

```
function squaredDy (p: point2d): real;
```
(* $(p_y - q_y)^2$ *)

```
function circleCenter (p, q, r: point2d): point2d;
```
(* Computes the center of the circle through p, q and r. *)

```
function intersectLineSegment
    (p, q, r, s: point2d): [point2d, point2d, boolean];
```
(* Tests whether the segments pq and rs intersect and computes the intersection. *)

```
function xPlusY (p: point2d): real;
```
(* $p_x + p_y$ *)

```
function xMinusY (p: point2d): real;
```
(* $p_x - p_y$ *)

In total we need about 15 primitives for all geometric algorithms implemented so far and 'whichSide' turned out to be the most common one.

2.3   Interchangeable arithmetic and parameterized floating point arithmetic

The choice of arithmetic may have a significant impact on how an implementation of an algorithm behaves in the case of degenerate or nearly degenerate configurations. Since we want to experiment with different arithmetics, easily interchangeable arithmetic is needed. This means that in the best case no line of code of an implementation must be modified in order to try out a different model of arithmetic. Since points are a basic building block of geometry, we achieve this goal by defining an abstract 'point2d' class which has no instance variables for the coordinates but specifies an interface with access procedures to the coordinates and various geometric primitives (see also the previous section 2.2).

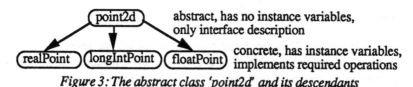

Figure 3: The abstract class 'point2d' and its descendants

From this abstract 'point2d' we derive concrete point objects having instance variables and implementing the geometric primitives in their respective arithmetic. Algorithms using only the functions and procedures specified by the abstract type 'point2d' can be run in any of the three kinds of arithmetic currently supported.

In order to study not only the built-in floating point arithmetic (as used in 'realPoint' where the x- and y-coordinates are of type *real*), we have implemented a software floating point package with arbitrary base (including odd bases) and precision which is used for the coordinates of the object 'floatPoint'. The idea here is of course not to simulate high precision floating point arithmetic but on the contrary low precision arithmetic in order to make rounding errors and other problems of floating point arithmetic more pronounced.

2.4 Abstract data types

Efficient geometric algorithms organize data in lists, dictionaries, queues, etc. We describe design decisions and implementation details for some abstract data types.

Sequence: The abstract class 'sequence', realizing collections of arbitrary objects, is implemented as a linked list structure and as a (dynamic) array. The list implementation is useful when no a priori bound on the number of elements is known and sequential processing of the elements is feasible. The array implementation provides greater flexibility and functionality.

Dictionary: In a dictionary based on the reference concept (see also section 3.1) 'find' is the only one key-based operation. 'find' takes an object o to be found and a key comparison function and returns, if possible, a reference to an object with the same key value. If none is found, 'find' returns a direction d and a reference to an object p such that o can be inserted as p's direct neighbor in direction d. Performing the insertion yields a reference for o. All other operations like 'delete' or 'swap' require references as arguments.

This separation between key-based operations and operations changing the data structure has two advantages: 1) Different objects with identical key values are easier to handle (see the example in 3.1) and 2) operations just changing the data structure can be implemented more efficiently.

If a dictionary is implemented as an AVL-tree, the delete operation takes constant time whenever rebalancing is not needed. In an ordinary implementation we would have to find the object first before we can delete it, making delete an O(log n) operation most of the time. Other implementations realize a dictionary as a sorted list and as a sorted vector.

Priority queue: The abstract data type priority queue is implemented 1) based on a heap and 2) based on a dictionary. We describe in section 3.2 how to implement a heap with an efficient delete operation using the reference concept. In our experience, a heap is sufficient most of the time and a general 'find' operation on a priority queue is rarely needed.

2.6 Universal operations

Object oriented programming allows the specification of a common interface that is understood by all objects in the system by placing this interface at the root of the object hierarchy. All objects in our system are descendants of the root object 'obj' and therefore share a large set of common methods that we call 'universal operations'. A typical example is the ability for an object to display itself on the screen. In the following we discuss the kinds of universal operations provided.

Memory management. We provide methods for the creation, destruction and duplication of objects. When creating an object the difficulty arises that for its proper initialization additional parameters might be necessary, e.g. the length of a dynamic array. In this case we specify that the parameterless initialization method 'init', which must always be called after an object is created, may use additional global variables as implicit parameters.

A second problem is the treatment of objects that contain other *dependent* objects either explicitly as instance variables or implicitly like in a list object. The effect of destruction or duplication of such an object is unclear since the dependent objects might be destroyed or duplicated or not. We solve this problem by explicitly stating at each object declaration for an object o whether the dependent objects of o belong to the *internal state* of o or not. Dependent objects belonging to the internal state are destroyed and copied just like regular instance variables whereas dependent objects that do not belong to the internal state are left intact. The following example shows the difference.

```
lineSegment = object(obj)  (* Derived from the root class 'obj'. *)
    p, q: point
    (* 'point' is a class. 'p' and 'q' belong to the internal state of a 'lineSegment'. *)
end;

vector = object(obj)  (* Derived from the root class 'obj'. *)
    length:   0..maxLength;  (* Length of the vector *)
    elements: array [1..maxLength] of obj
    (* Elements of the vector do not belong to the internal state. *)
end;
```

The decision whether a dependent object belongs to the internal state or not is a purely pragmatic one and must be documented when declaring such an object in order to specify the semantics of destruction and duplication. In the case of 'lineSegment' we prefer completely independent copies whereas in the 'vector' example a (recursive) copy of the vector's elements is usually undesirable. We provide additional methods for cases where the recursive destruction or duplication of all dependent object is required.

Interactive input/output: Interactive input/output is needed for experimentation and demonstration. We support the following tasks.

- Display a geometric object on the screen. We use the XOR mode for drawing which has the advantage that drawing and erasing are the same operations.
- Display an object in a highlighted fashion which is useful for giving a visual feedback in algorithm animation or when an object has been selected by the user.
- Flash an object. This operation is used when animating an algorithm.
- Let the user interactively enter an object. This is usually done by dragging with the mouse. We have implemented a universal dragging method based on the following two primitive operations: 1) Construct the object given the location where the mouse button was first pressed and the current location of the mouse. 2) A method that displays the object in XOR mode on the screen.

File input/output: When processing geometric objects, we are interested in saving them permanently on secondary storage. File input/output becomes therefore another area where operations applicable to all objects must be provided. As file format we normally use a byte stream which contains geometry data and some type information in a prefix format. The introduction of lists and arrays makes this format recursive. A second file format provides a human readable LISP like notation which is useful for debugging or for data exchange with other programs.

Geometric transformations: We provide the following transformations: Translation, rotation, scaling and mirroring of a geometric configuration with respect to a given line.

Type computations: Sometimes we wish to enquire the type of an object, e.g. when testing whether all members of a collection are of the same type. Another example is the creation of a byte, specifying the object's type for storage purposes. These and similar type computations are implemented as universal operations since they must be available for all objects in the system.

Random instances: Generation of random instances is used for the rapid creation of test data. We provide a method that changes the internal state randomly, preserving the object invariant. A global variable, usually a rectangle, specifies the boundaries in which the random change takes place.

Class description and method execution: Each class should be able to describe itself to the user, meaning that given a list of arbitrary objects, a class should deliver all the methods it can execute on these objects in human readable form. Assuming a list of three points, the 'point2d' class should offer a method for creating a circle, the 'pointVector' class should offer a method for computing a closest pair and a Voronoi diagram and so on. After all classes have given such a description, it must be possible to execute a specific method of a specific class on the given list of objects.

3 Implementation aspects: A sample of interesting details

3.1 The reference concept

In the context of key-based data structures the reference concept helps us distinguish between operations that require a key (e.g. 'find') and those that change a data structure without needing a key (e.g. 'swap'). We motivate this distinction with a problem that occurs when implementing the plane sweep algorithm for finding all intersecting line segments.

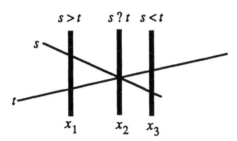

Figure 4: The difficulty of accessing line segments by their y-value

The line segments form a total order along the front and their key value is essentially their y-value when evaluated at the position of the front. At position x_2 we must exchange s and t, which is difficult using their key value: There are two objects with the same key value and we might access the same segment twice.

As a solution, we prohibit key based access in this case. Instead, we provide means for performing the required 'swap' without using key values. We associate with each object o in the data structure a unique *reference* that is supplied whenever o is known to participate in an operation. For easier understanding, imagine a reference to be a pointer into a data structure although the internal representation of a reference is hidden from the user. The idea of a reference resembles the notion of *items* introduced in LEDA [MN 89] as an abstraction of pointers and locations.

3.2 A heap with an efficient delete operation

This section demonstrates how to implement a priority queue as a heap using the reference concept such that an efficient delete operation is supported. A heap is a partially ordered binary tree such that the value associated with each internal node is surpassed or reached by any of its children. The figure below shows a heap together with its standard breadth first array representation.

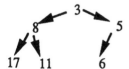
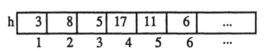

Data structure invarariant: $h[\,i] \le h[2{\cdot}i] \wedge h[i] \le h[2{\cdot}i+1]$

Figure 5: Tree and array representation of a heap

The textbook [AHU 83, p. 163] states that a heap does not allow an efficient delete operation and implementors of priority queues with delete have abandoned the heap and used more complicated data structures such as AVL-trees [B 81]. The reference concept is used to add an efficient delete operation to a heap. The two key observations are: 1) Any element in a heap can efficiently be deleted if we know its position and 2) we can fix this position as soon as an element is inserted. The following figure shows how this works.

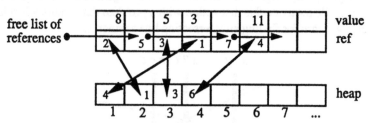

Figure 6: Implementing a heap with the reference concept

Data structure invariant:
value[heap[i]] $\leq$ value[heap[$2 \cdot i$]] $\wedge$ value[heap[i]] $\leq$ value[heap[$2 \cdot i+1$]] $\wedge$ ref[heap[i]] = i.

When an element is inserted into the heap we use the free list of references to assign it a location in the array 'value'. This location is the reference and is never changed. What we change instead is the appropriate double arrow between 'ref' and 'heap' which determines the element's position in the heap. The operation for insertion ('sift' or 'pushDown') is basically the same as for an ordinary heap while the 'delete' operation is a simple generalization of the standard 'deleteMin' operation: The element to be deleted is exchanged with the last element in the heap which is a leaf node. Then we perform the 'pushDown' operation on this sub-heap followed by the 'pushUp' operation reestablishing the data structure invariant. Therefore the cost of delete is O(log n) where n is the number of elements in the heap.

The code is simple.

```
type
    relation  = (less, equal, greater);
    seqIndex  = 1..maxN;
    seqIndex0 = 0..maxN;
    reference = ^integer;

HPQ = object(obj)              (* Heap Priority Queue *)
    length:    seqIndex0;      (* Number of elements *)
    freeList:  seqIndex;       (* Points to first empty element in 'ref' *)
    ref, heap: array [seqIndex] of seqIndex0;
    value:     array [seqIndex] of obj;

    procedure init; (* Establishes the data structure invariant after creation. *)
    function insert   (x: obj;       function compare(a,b:obj):relation
              ): reference;
    procedure delete  (x:reference; function compare(a,b obj):relation);
    procedure swap    (p, q: seqIndex);
    procedure pushUp  (i:seqIndex;  function compare(a,b:obj):relation);
    procedure pushDown(i:seqIndex;  function compare(a,b:obj):relation);
end;
```

```
procedure HPQ.init;
begin   length := 0; freeList := 1; ref[1] := 0  end;

procedure HPQ.swap(p, q: seqIndex); var pm, qm: seqIndex;
begin
   pm := heap[p]; qm := heap[q]; ref[pm] := q; ref[qm] := p;
   heap[p] := qm; heap[q] := pm
end;

procedure HPQ.pushUp(i:seqIndex; function compare(a,b:obj):relation);
begin (*              ↓ conditional and *)
   while (i > 1) & (compare(value[heap[i]],value[heap[i div 2]])=less) do
   begin   swap(i, i div 2); i := i div 2   end
end;

procedure HPQ.pushDown(i:seqIndex; function compare(a,b:obj):relation);
var j, length2: seqIndex0; continue: boolean;
begin
   continue := true; length2  := length div 2;
   while continue and (i <= length2) do
   begin
      j := 2 * i; (* ↓ conditional and *)
      if (j<>length) & (compare(value[heap[j]], value[heap[j + 1]]) > less)
      then j := j + 1;
      if compare(value[heap[i]], value[heap[j]]) = greater
      then begin  swap(i, j); i := j  end
      else continue := false
   end
end;

function HPQ.insert
   (x: obj; function compare(a, b: obj): relation): reference;
var i: seqIndex0;
begin
   length := length + 1; i := freeList;
   if ref[freeList] = 0
   then begin  freeList := freeList + 1; ref[freeList] := 0  end
   else freeList := ref[freeList];
   insert := reference(i); ref[i] := length;
   value[i] := x heap[length] := i; pushUp(length, compare)
end;

procedure HPQ.delete(x:reference; function compare(a,b:obj):relation);
var startPosition: seqIndex0;
begin
   startPosition := ref[seqIndex(x)]; swap(startPosition, length);
   ref[seqIndex(x)] := freeList; freeList := seqIndex(x);
   length := length - 1;
   if startPosition <= length then
   begin
      pushDown(startPosition,compare); pushUp(startPosition,compare)
   end
end;
```

More careful analysis shows that 'pushUp' and 'pushDown' can be implemented more efficiently by remembering the element that is common to consecutive swap operations. In [S 91] and [NSD 91] we present experimental evidence that confirms the superior efficiency of this implementation of a priority queue as compared to an AVL-tree implementation.

A caveat as a final remark: This kind of priority queue cannot be used when the element to be deleted is only known by its key value. In this case we know no efficient way to locate the element in a heap. In our experience implementing plane sweep algorithms however, this kind of priority queue was sufficient in all cases.

3.3 Implementing plane sweep in an object oriented fashion

Another advantage of object oriented design is the uniform treatment of plane sweep algorithms. When implementing plane sweep object oriented, we model events as objects, derived from an abstract event class that provides an execute method. The typical structure is given by the following declarations.

This decomposition has the advantage that all code for handling a certain event is concentrated in one place, making the implementation clearer and easier to understand. Furthermore the main program needs no change in the case new events become necessary (for example when we transform the sweep for testing whether two line segments intersect to the sweep that additionally finds all such intersections).

```
program genericPlaneSweep;

var
   xQueue: priorityQueue;
   yTable: dictionary;

type
   event = object
     procedure process              (* Abstract method *)
   end;

   event1 = object (event)          (* Derive the interface from the abstract class *)
     procedure process; override    (* Real implementation *)
   end;
   event2 = object (event)          (* Derive the interface from the abstract class *)
     procedure process; override    (* Real implementation *)
   end;
   (* Other event types *)

begin
   initialize xQueue; initialize yTable;
   while not xQueue.isEmpty do
     xQueue.extractMinimum.process  (* Let the event process itself. *)
end;
```

3.4 Helpful hints for increasing the reliability of a geometric library

In order to create a reliable system we have tried to catch programming errors early. We recommend the following methods which were used successfully.

Show the dynamic memory allocated. The 'info' window contains the number of bytes currently allocated on the heap for dynamic variables (mostly objects). This number gives a first hint whether memory management is working correctly. A constantly growing number of allocated bytes is usually something to worry about.

Use assertions. Especially geometric algorithms often depend on the truth of certain assertions (e.g. the value of some denominator should be different from zero) and a defensive programmer introduces checks (assertions) at appropriate places which give warnings when the assertion fails. Even errors in algorithms can be detected earlier this way.

Data structures should have a method checking their invariant. This is especially helpful while testing non trivial data structures like AVL trees or even a heap. One tests the data structure by different operations, and after each operation the invariant is checked which increases the confidence in its correctness. Having program code actually check the invariant (as opposed to doing it by hand) also makes automatic testing feasible: One could randomly insert or delete a random element in a dictionary and check the invariant after each operation.

Write and keep test programs for the central data structures. We have written test programs to test the abstract data type 'dictionary' which has the additional advantage that it can be used for all concrete implementations of the type 'dictionary'. The alternative to test programs is to actually use the data structure in an algorithm which has the two disadvantages that 1) in the case of an error one does not know precisely the source of the error and 2)one algorithm rarely uses all methods offered by a data structure designed for universal applicability.

Create geometric test data and keep it for reference in a test suite. Creating good test data is not trivial and the results of this effort should be kept. A new program solving the same problem might be written, or more often the current implementation changes. In both cases one increases the confidence in the reliability by checking with the test suite. It should contain configurations ranging from no degeneracies at all to multiple ones.

Use algorithm animation for geometric algorithms. Algorithm animation should be used to present an algorithm's essential state information in a graphical way. In the debugging phase, this facilitates the detection of inconsistencies before an incorrect result occurs.

Do not remove the code used for debugging. The code used for debugging should be left in the programs since future modifications might benefit from it. Using conditional compilation removes any run time overhead if necessary.

4 Experience with object oriented programming

Techniques from object oriented programming have proven to be useful for implementing a library for geometric computation with extensive support for experimentation. We use the concept of abstract classes together with dynamic binding in order to implement interchangeable arithmetic and data structures. Furthermore the tree structure of the class hierarchy serves as an aid in structuring the whole system in an understandable way. Polymorphism helps in implementing universal data structures. Inheritance is mostly used in the form of interface inheritance; actual code is almost only inherited as methods in abstract classes that can be written in terms of other methods. In the following we discuss the points mentioned in the previous summary in more detail.

4.1 Dynamic binding and abstract classes

An abstract class is a class of objects with the property that no instances of the class itself are created, but only instances of derived classes. Dynamic binding means that the type of an object and the methods to be executed are determined at run time. Consider the example of the class 'point2d' which has descendants 'realPoint', 'floatPoint' and 'longIntPoint' (see also section 2.3). 'point2d' defines a common interface implemented in three different ways using three different kinds of arithmetic. Programs using only variables of the abstract class 'point2d' can be instantiated at run time to work with all three different concrete implementations.

We have used abstract classes and dynamic binding for two purposes: 1) Offering interchangeable arithmetic and interchangeable implementations of abstract data types and 2) factoring out common code. Factoring out common code is best illustrated by the universal dragging routine mentioned in section 2.5. An abstract class provides a method M_0 which is solely implemented in terms of other methods M_1, M_2, ..., M_k of the abstract class. As soon as M_1, M_2, ..., M_k are implemented in a derived class, a working implementation of M_0 is available. The abstract implementation of M_0 factors out the common code.

4.2 The class hierarchy and inheritance

Object oriented design allows us to take a class and derive new classes with enriched functionality from it, modeling the 'is_a' relation. This relationship is rare among geometric objects, causing the class hierarchy to be relatively flat in most places. The most prominent counterexample is the sequence 'convexPolygon' is_a 'polygon' is_a 'polyLine' is_a 'pointVector'.

Inheritance goes hand in hand with derivation: A method M defined in a class is automatically available for each of its subclasses and ideally needs not to be implemented in a different way ('overridden'). This is rarely the case in geometry: The best way for solving the same problem for a class and a derived class can be totally different as the trivial example of computing the convex hull of a polygon and a convex polygon shows.

Having a class hierarchy with a single root has the advantage that universal data types can be created. For example we need a dictionary that can hold any other object. Even objects of different types should be admissible as long as we can define an order relation. The requirement that a dictionary can hold objects of the type of the root class achieves this goal.

As a conclusion, the concept of a class hierarchy is a useful tool for structuring the library and creating universal abstract data types. It is of less use for saving implementation effort.

4.3 Design concepts for classes

Deriving a new class from an existing one should model the 'is_a' relation since otherwise the structure of the class hierarchy would be confusing and the semantics of inherited operations could become unclear. This rule determines where to place a new class in the hierarchy. For example the class 'lineSegment' should not be derived from 'point2d' by adding another point because a line segment is not a point, but is composed of two points. In this case composition is the more appropriate construction principle.

During the evolution of a class hierarchy one often finds two classes which contain some similar methods. An example were the class for linked lists and the class for vectors for which various methods like displaying itself on the screen were similar. In this case we factor out common methods by introducing a common abstract ancestor class which realizes the common behavior.

5 Conclusion

We have described the basic building blocks that are available to the implementor of geometric algorithms working in the XYZ GeoBench programming environment.

A comprehensive set of universal operations applicable to all objects in the system covers a wide range of tasks: Memory management, interactive input/output, binary and text file I/O, geometric transformations, type computations, creation of random instances, class description and method execution.

Geometric algorithms can be implemented in an arithmetic independent way. This allows the user to experiment with different kinds of arithmetic: Ordinary built-in floating point arithmetic, a floating point arithmetic featuring arbitrary basis and precision and standard integer arithmetic.

A set of reliable geometric primitives and common universal data structures like sequence, dictionary and priority queue facilitate the implementation effort. The reference concept on which dictionaries and priority queues are based allows efficient delete operations in AVL trees and heaps.

A variety of already implemented geometric algorithms covers a wide range of 2-dimensional Computational Geometry and serves as the building block for further implementation efforts.

The XYZ GeoBench serves also as an interactive front end to the XYZ Library of geometric algorithms. A smooth Macintosh style user interface is used to access most implemented algorithms, often with animation of their execution, and to perform a wide variety of experiments. Typical examples are efficiency measurements, testing the influence of different kinds of arithmetic, testing the effect of different implementations of abstract data types, running an algorithm on random or degenerate geometric configurations, etc.

Currently the most active application area of the XYZ GeoBench is in education. On the one hand we can demonstrate a wide variety of geometric algorithms in the class room due to the built-in animation capabilities and on the other hand we give our students the opportunity to implement geometric algorithms for themselves. The latter is greatly facilitated by the programming environment our workbench provides.

Acknowledgements

I thank J. Nievergelt for his support of the XYZ project and for commenting on an earlier draft of this paper. I am grateful to C. Ammann and M. Furter for their dedication in improving the user interface.

References

[AHU 83]   A. Aho, J. Hopcroft, J. Ullman: Data Structures and Algorithms, Addison Wesley, 1983.

[B 81]   K. Brown: Comments on "Algorithms for reporting and counting geometric intersections", IEEE Trans. Comput. vol. C-30, pp. 147-148, Feb. 1981.

[F 87]   S. Fortune: A Sweepline Algorithm for Voronoi Diagrams, Algorithmica 2, pp. 153-174, 1987.

[NSD 91]   J. Nievergelt, P. Schorn, C. Ammann, A. Brüngger, M. De Lorenzi: XYZ: A project in experimental geometric computation, in these proceedings, 1991.

[M 87]   B. Meyer: Object-Oriented Software Construction, Prentice Hall, 1987.

[MN 89]   K. Mehlhorn, S. Näher: LEDA, A Library of Efficient Data Types and Algorithms, preliminary version, Universität des Saarlandes, 1989.

[S 91]   P. Schorn: Robust Algorithms in a Program Library for Geometric Computation, ETH PhD Dissertation 9519, to appear 1991.

[W 90]   E. Welzl: A fast randomized algorithm for computing the minimal area disk enclosing a set of points in d-space, presentation at the Workshop on Computational Geometry, Dagstuhl, Oct 1990.

Computing the Rectilinear Link Diameter of a Polygon[*]

Bengt J. Nilsson[†]   Sven Schuierer[†]

Abstract

The problem of finding the diameter of a simple polygon has been studied extensively in recent years. $O(n \log n)$ time upper bounds have been given for computing the geodesic diameter and the link diameter for a polygon.

We consider the rectilinear case of this problem and give a linear time algorithm to compute the rectilinear link diameter of a simple rectilinear polygon. To our knowledge this is the first optimal algorithm for the diameter problem of non-trivial classes of polygons.

1 Introduction

In many versions of motion planning problems the large cost involves making turns. An example is broadcasting a radio signal through a beam. At particular points relay stations must be erected to reflect the beam in a new direction. This leads to a cost measure for paths which is commonly known as the *link metric*.

The problem of finding link optimal paths between points inside a simple polygon has been solved by Suri [Sur87]. He gives a linear time algorithm for this problem. In addition he shows that the *link diameter*, the furthest link distance between any two points in a polygon, can be computed in $O(n \log n)$ time and linear storage (n will in the following always denote the number of polygon edges). A similar result is obtained by Ke [Ke89].

In many cases we may want to restrict the paths to be rectilinear, i.e., all segments of the path are axis parallel. This particular restriction has been studied for simple rectilinear polygons by de Berg [B89]. He devises a data structure which allows to compute, given two query points inside the polygon, the shortest path between the points in $O(\log n + l)$ time, where l is the number of links of the path. The data structure can be constructed in $O(n \log n)$ time and requires $O(n \log n)$ storage. Furthermore, he shows how to compute the rectilinear link diameter in $O(n \log n)$ time with a simple divide and conquer approach.

In the more general case when the obstacles do not form the boundary of a simple polygon but instead are just rectilinear line segments in the plane, a generalization of the shortest path problem has been studied by de Berg et al. [BKNO90].

In this paper we give a linear time algorithm for computing the rectilinear link diameter of a simple rectilinear polygon. To our knowledge this is the first optimal algorithm found for the diameter problem of non-trivial classes of polygons. The algorithm is based on a divide and conquer algorithm presented in [B89]. The idea is to split the polygon into two subpolygons of approximately equal size, compute the diameter of the two subpolygons recursively, and to compute the longest distance between any pair of vertices in different subpolygons. The maximum of these three values is the diameter. This algorithm runs in $O(n \log n)$ time. In order to reduce the running time to linear it is possible to show that one only needs to recur on one of the subpolygons. For the other subpolygon one can show that the diameter is either smaller than the diameter of the full polygon or the diameter of the subpolygon can be computed explicitly without recursion and in linear time.

The paper is organized as follows. In the next section we state our definitions and give some preliminary results for simple rectilinear polygons. We show in Section 3 how to compute the link

---

[*]This work was supported by the Deutsche Forschungs Gemeinschaft under Grant No. Ot 64/5–4.
[†]Institut für Informatik, Universität Freiburg, Rheinstr. 10–12, D-7800 Freiburg, Fed. Rep. of Germany.

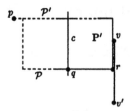

Figure 1: *One of v and v' is at least as far from p as q is.*

diameter in optimal time. In Section 4 we present a simple algorithm to compute an approximate link center, i.e., a point from which all other points are reachable using at most $R(\mathbf{P}) + 1$ links, in linear time ($R(\mathbf{P})$ is the rectilinear link radius of polygon $\mathbf{P}$).

2 Definitions

Let P be a Jordan curve consisting of n axis parallel line segments such that no two consecutive segments are collinear. We define a *simple rectilinear polygon* $\mathbf{P}$ to be the union of P and its interior. A *(rectilinear) path* $\mathcal{P}$ is a curve that consists of (a finite number of) axis parallel line segments inside $\mathbf{P}$. The length of $\mathcal{P}$, denoted by $\lambda(\mathcal{P})$, is the number of line segments it consists of. From now on we will only consider rectilinear polygons and rectilinear paths in a polygon $\mathbf{P}$. Hence, whenever we talk of polygons, we mean simple rectilinear polygons and whenever we talk of paths, we mean rectilinear paths in $\mathbf{P}$.

Let p and q be two points in a polygon $\mathbf{P}$ and e be an axis parallel line segment in $\mathbf{P}$. The *(rectilinear) link distance* between p and q, denoted by $d(p,q)$, is defined as the length of the shortest path connecting p and q. We say a polygonal path $\mathcal{P}$ from p to line segment e is *admissible* if it is rectilinear and the last link of $\mathcal{P}$ is orthogonal to e. We define the link distance of p and e, again denoted by $d(p,e)$, to be the length of the shortest admissible path from p to a point of e.

The *(rectilinear) link diameter* denoted $D(\mathbf{P})$ is defined as the maximum rectilinear link distance between any two points in $\mathbf{P}$. The diameter gives a classification on the amount of winding in the polygon. There is an interesting relationship between the link diameter and the *(rectilinear) link radius* of $\mathbf{P}$, denoted by $R(\mathbf{P})$, which is defined as the minimum integer k for which there is some point in $\mathbf{P}$ from which all other points can be reached by k links. It can be shown that $\lceil D(\mathbf{P})/2 \rceil \leq R(\mathbf{P}) \leq \lceil D(\mathbf{P})/2 \rceil + 1$ and that these bounds are tight. We will make use of this fact in Section 4 to compute a point that can reach all points in $\mathbf{P}$ with at most $R(\mathbf{P}) + 1$ links. Such a point can viewed as an approximation to the link center of $\mathbf{P}$. The *link center of* $\mathbf{P}$ is the set of all points that can reach all other points using at most $R(\mathbf{P})$ links.

In the rest of the paper the letter v with various subscripts and superscripts will always denote vertices of $\mathbf{P}$. In the same way the letters p, q, and r denote points of $\mathbf{P}$.

In the following lemma it is proven that there is always a vertex which is a furthest neighbour of a point of the polygon. This fact is used to discretize the problem, i.e., our link diameter algorithm only needs to concern itself with the vertices of $\mathbf{P}$.

LEMMA 2.1 For any two points $p, q \in \mathbf{P}$, there is a vertex v such that $d(p,q) \leq d(p,v)$.

PROOF: Let $k = d(p,q)$ and let $\mathcal{P}$ be a shortest path from p to q. Let r be the point of intersection between the boundary of $\mathbf{P}$ and the extension of the last link l of $\mathcal{P}$. Let $\mathcal{P}'$ be a shortest path from p to r and suppose furthermore that $\lambda(\mathcal{P}') < k$. Obviously, q can not be a point on $\mathcal{P}'$. If the last link of $\mathcal{P}'$ is collinear with l, q can be reached from p with less than k links, a contradiction. On the other hand, if l is orthogonal to the last link of $\mathcal{P}'$, then follow the two paths $\mathcal{P}$ and $\mathcal{P}'$ from r until they join again. The path traced like this bounds a simple polygon $\mathbf{P}'$ interior to $\mathbf{P}$. The point q is obviously a point on the boundary of $\mathbf{P}'$. Also, since $\mathcal{P}$ is a shortest path, the chord c of $\mathbf{P}'$ through

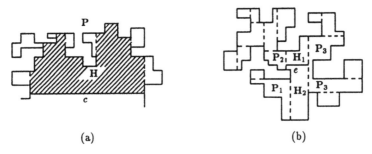

Figure 2: *An example of a maximal histogram and a histogram partition.*

q and orthogonal to l has one end point on the path $\mathcal{P}'$ and, hence, by following the path $\mathcal{P}'$ to the intersection point with c and by continuing along the chord, a shorter path from p to q has been constructed. This proves that $d(p,r) \geq k$; see Figure 1.

The point r lies on a boundary edge of $\mathbf{P}$ with vertices v and v'. Suppose both v and v' have distance less than k from p. Then, so does r since the line segment through r orthogonal to the edge vv' will join one of the paths from p to v or from p to v', a contradiction. $\qquad\square$

Hence, the diameter definition can be equivalently stated in terms of the vertices (v and v') of $\mathbf{P}$.

$$D(\mathbf{P}) = \max_{v,v' \in \mathbf{P}} d(v, v').$$

A *horizontal (vertical) histogram* is a rectilinear polygon such there is one side that can be reached with one vertical (horizontal) line segment from any point inside the polygon. This side is called the *base* of the histogram. We define a *maximal histogram* $\mathbf{H}$ inside a rectilinear polygon $\mathbf{P}$ having an axis parallel chord c in $\mathbf{P}$ as its base to be the maximum area histogram interior to $\mathbf{P}$ with c as its base (see Figure 2a). The definition does not yield a unique maximal histogram since $\mathbf{H}$ can extend to either side of c. However, if we specify a direction orthogonal to c the ambiguity can be resolved. A *window* is a maximal segment of the boundary of a histogram which is not also part of the boundary of $\mathbf{P}$. A *histogram partition* of a polygon $\mathbf{P}$ w.r.t. some axis-parallel line segment e in $\mathbf{P}$ is defined as follows. If $\mathbf{P}$ is a histogram with base e, then the histogram partition of $\mathbf{P}$ w.r.t. e just consists of $\mathbf{P}$ itself. Otherwise, let $\mathbf{H}_1$ and $\mathbf{H}_2$ be the two maximal histograms defined by e. $\mathbf{P} \setminus (\mathbf{H}_1 \cup \mathbf{H}_2)$ consists of a number of simple polygons $\mathbf{P}_i$, $1 \leq i \leq k$, each of which has one edge e_i in common with $\mathbf{H}_1 \cup \mathbf{H}_2$. The histogram partition of $\mathbf{P}$ consists then of $\mathbf{H}_1$ and $\mathbf{H}_2$ together with the histogram partitions of $\mathbf{P}_i$ w.r.t. e_i, $1 \leq i \leq k$ (see Figure 2b).

We will make use of the following construction. For an axis parallel line segment e and a point p in $\mathbf{P}$, let $e(p, d)$ denote the part of e that can be reached from p with an admissible path of length at most d. de Berg shows in [B89] that, for any edge e of $\mathbf{P}$, the set of distances $\{d(v, e) \mid v \in \mathbf{P}\}$ and the sets of subedges $\{e(v, d(v, e)) \mid v \in \mathbf{P}\}$ and $\{e(v, d(v, e) + 1) \mid v \in \mathbf{P}\}$ can be computed in linear time. It is important to note that $e(p, d)$ is a connected set, for any point p and any integer d. Because of this property, we will often refer to the subedges as *intervals*. Note that, in particular, the sets of subedges have the property that their intersection is non-empty if and only if the pairwise intersection of any two subedges is non-empty.

Since we rely heavily on these results we give informal arguments as to the truth of these statements. First, note that given the line segment e it is possible to partition the polygon into histograms such that e is an interval on the base of two such histograms; see [Lev87] for the algorithm. The partitioning can be computed in linear time given the vertex-edge pairs of [Cha90].

The path from a vertex to e will pass through the bases of all the histograms on the way and will make one turn in each histogram (except in the first). Thus, the distance $d(v, e)$ is easy to find using the dual graph of the histogram partition. Refer to Figure 3 for an example.

Moreover, de Berg shows in [B89] that for any vertex v with $d(v, e) > 2$, there is a vertex v_{next} such that the distance from v_{next} to e is one less than $d(v, e)$ and such that v and v_{next} have the same

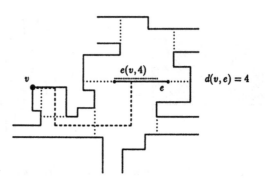

Figure 3: *A path from v to e of four links.*

interval on e. The reason for this is that the second link of of a shortest path between v and e can always be chosen so that it touches the boundary of $\mathbf{P}$. In other words, this means that the set of intervals $\{e(v, d(v, e)) \mid v \in \mathbf{P}\}$ only depends on the vertices that are exactly two links away from e and these can be computed in linear time. By keeping a reference at each vertex v to v_{next} we can obtain the intervals for all the vertices. For the set of intervals $\{e(v, d(v, e) + 1) \mid v \in \mathbf{P}\}$ a similar argument applies.

To see that the subedges are intervals it is enough to consider the histogram having the chord through e as base. A path from a point outside the histogram to e enters the histogram through a window and can then make a turn to reach the base. This turning point must lie within the interval between the window and the "opposite" side of the histogram.

The following lemma is also proven in [B89].

LEMMA 2.2 (DE BERG) *Let c be a chord which cuts $\mathbf{P}$ into two subpolygons such that p and q are points in different subpolygons and d_p be the distance from p to c and d_q the distance from q to c. Then, we have*

$$\begin{aligned} d(p, q) &= d_p + d_q - 1 && \text{when } c(p, d_p) \cap c(q, d_q) \neq \emptyset, \\ &\geq d_p + d_q && \text{otherwise.} \end{aligned}$$

This lemma will prove useful later which is why we restate it here.

3   The Rectilinear Link Diameter

Our starting point is the following divide-and-conquer algorithm by de Berg [B89]. It makes use of the following fact. If we are given a chord c that splits $\mathbf{P}$ into two parts $\mathbf{P}_1$ and $\mathbf{P}_2$, then we have that the diameter is spanned either by two vertices in $\mathbf{P}_1$ or two vertices in $\mathbf{P}_2$ or one vertex in $\mathbf{P}_1$ and the other in $\mathbf{P}_2$. In the latter case the shortest path between the two vertices has to cross c. Hence, if we are given the intervals that the vertices in $\mathbf{P}_1$ and $\mathbf{P}_2$ can reach on c, then we can compute the maximal distance from a vertex in $\mathbf{P}_1$ to a vertex in $\mathbf{P}_2$ by Lemma 2.2.

(i) If $\mathbf{P}$ is a rectangle, then $D(\mathbf{P}) = 2$ and stop.

(ii) Compute a chord c of $\mathbf{P}$ that cuts $\mathbf{P}$ into two subpolygons $\mathbf{P}_1$ and $\mathbf{P}_2$ such that $|\mathbf{P}_1|, |\mathbf{P}_2| \leq 3/4\,n + 2$.

(iii) Compute $M = \max\{d(v, w) \mid v \in \mathbf{P}_1, w \in \mathbf{P}_2\}$ and remember the vertices v, w with $d(v, w) = M$.

(iv) Compute $D(\mathbf{P}_1)$ and $D(\mathbf{P}_2)$ recursively.

(v) Let $D(\mathbf{P}) := \max(M, D(\mathbf{P}_1), D(\mathbf{P}_2))$.

de Berg shows how Steps (ii) and (iii) can be carried out in linear time. This immediately leads to an $O(n \log n)$ algorithm. Our algorithm differs from the above one only in Step (iv). It is based on the following observation. For $i = 1, 2$, let d_i be the maximal distance from a vertex in $\mathbf{P}_i$ to c, i.e., $d_i = \max_{v \in \mathbf{P}_i} d(v, c)$. In the following discussion we assume that $d_1 \geq d_2$. There is one vertex v_i in each subpolygon $\mathbf{P}_i$ with $d(v_i, c) = d_i$, for $i = 1, 2$. For these we have $d(v_1, v_2) \geq d_1 + d_2 - 1$ again by Lemma 2.2. This gives us a lower bound of $d_1 + d_2 - 1$ on M. On the other hand, there is a trivial upper bound on the diameter of $\mathbf{P}_2$ which is $2d_2 + 1$ since all points of $\mathbf{P}_2$ can reach a point on c with d_2 links. Hence, if we can show that $D(\mathbf{P}_2) \leq d_1 + d_2 - 1$, we can avoid to recur on $\mathbf{P}_2$. Unfortunately, we cannot guarantee that $D(\mathbf{P}_2)$ is smaller than M. But we know that the interesting values of $D(\mathbf{P}_2)$ are in the range from $d_1 + d_2 - 1$ to $2d_2 + 1$. In the next section we will show that we can decide in linear time (in the size of $\mathbf{P}_2$) if $D(\mathbf{P}_2) \leq 2d_2 - 1 \leq M$ and, furthermore, if $D(\mathbf{P}_2) > 2d_2 - 1$, the exact value of $D(\mathbf{P}_2)$ can also be computed in linear time. As an immediate consequence we only have to recur on $\mathbf{P}_1$ in Step (iv).

3.1 Computing Bounds on the Diameter of a Polygon

For the rest of this section suppose that we are given a polygon $\mathbf{P}$ and one edge e of it. We denote the maximal distance of a vertex to e by d_e, i.e., $d_e = \max_{v \in \mathbf{P}} d(v, e)$. The problem we solve in this section is to show how to decide whether $D(\mathbf{P}) \leq 2d_e - 1$ and, if not, how to compute the exact value of $D(\mathbf{P})$, both in linear time. To this end we need the two sets of intervals $\mathcal{I}_1 = \{e(v, d_e) \mid v \in \mathbf{P}\}$ and $\mathcal{I}_2 = \{e(v, d_e + 1) \mid v \in \mathbf{P}\}$. Recall both of them can be computed in linear time [B89, p. 10/11].

The idea of the algorithm is to make use of the information that is supplied by the sets $\mathcal{I}_1$ and $\mathcal{I}_2$. In order to do so we make a few observations which are stated in the following three lemmas. The first lemma gives an upper bound on the distance between two points if the intervals on e that the points can reach with a given number of links overlap.

LEMMA 3.1 *If p and q are two points in $\mathbf{P}$ and d_p and d_q are two integers such that $e(p, d_p) \cap e(q, d_q) \neq \emptyset$, then $d(p, q) \leq d_p + d_q - 1$.*

PROOF: Let s be a point in $e(p, d_p) \cap e(q, d_q)$. Consider two admissible paths $\mathcal{P}$ from p to s and $\mathcal{Q}$ from q to s with d_p and d_q links, respectively. Since the last link is orthogonal to e in both cases, we can join these two paths into one by leaving aside the shorter of the two last links. The new path obviously has length $d_p + d_q - 1$. Hence, $d(p, q) \leq d_p + d_q - 1$. $\qquad\square$

An immediate consequence of Lemma 3.1 is that if $\bigcap \mathcal{I}_1 \neq \emptyset$,[1] then $D(\mathbf{P}) \leq 2d_e - 1 \leq M$ since all the intervals $e(v, d_e)$ with $v \in \mathbf{P}$ overlap. The second lemma gives us a lower bound on the distance between two points if their intervals do not overlap.

LEMMA 3.2 *If p and q are two points in $\mathbf{P}$ and d_p and d_q are two integers such that $e(p, d_p)$ and $e(q, d_q)$ are non-empty and $e(p, d_p) \cap e(q, d_q) = \emptyset$, then $d(p, q) \geq d_p + d_q - 1$.*

PROOF: W.l.o.g. assume that e is vertical, that the interior of $\mathbf{P}$ is to the right of e, and that $e(p, d_p)$ is above $e(q, d_q)$. Let s_p be the lower end point of $e(p, d_p)$ and s_q be the upper end point of $e(q, d_q)$. Further, let $\mathcal{P}$ be an admissible path of length d_p from p to s_p with last link l_p and $\mathcal{Q}$ be an admissible path of length d_q from q to s_q with last link l_q (see Figure 4).

Since we cannot move l_p lower resp. l_q higher, there is a horizontal upward (downward) edge of $\mathbf{P}$ that overlaps l_p (l_q). Since we allow more than two vertices to be collinear, l_p (l_q) may overlap a number of horizontal edges. We choose e_p (e_q) to be the leftmost of these. Let r_p be the left end point of e_p and r_q be the left end point of e_q. Furthermore, let h_p be the horizontal line segment from r_p to e and h_q the horizontal line segment from r_q to e. We denote the part of $\mathbf{P}$ above[2] h_p by $\mathbf{P}_1$ and the part of $\mathbf{P}$ below h_q by $\mathbf{P}_2$.

---

[1] We denote the intersection of all sets in a family $\mathcal{F}$ by $\bigcap \mathcal{F}$.

[2] If we speak of polygons, in this case $\mathbf{P}_1$ and $\mathbf{P}_2$, being *above* or *below* a line segment, we mean that the interior of the curve is locally above or below the line segment.

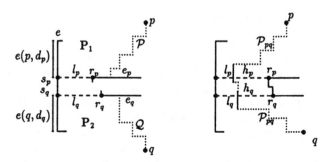

Figure 4: *A path between p and q has at least $d_p + d_q - 1$ links.*

Our first claim is that p is in $\mathbf{P}_1$. To see this assume the contrary. Since the right end point of l_p is contained in $\mathbf{P}_1$ and the boundary of $\mathbf{P}_1$ is a Jordan curve, $\mathcal{P}$ crosses h_p. Let l be the first vertical link that intersects h_p. Since there is no upward edge of $\mathbf{P}$ that overlaps h_p, we can connect a point of l slightly below h_p with one horizontal link to e which contradicts the fact that we chose s_p to be the lowest point on e that can be connected to p with d_p or less links.

Since we cannot reach any point below s_p from p with d_p links, $\mathcal{P}$ cannot cross h_p and, hence, p is in $\mathbf{P}_1$. For similar reasons q is in $\mathbf{P}_2$. Therefore, a shortest path $\mathcal{P}_{pq}$ from p to q has to cross both h_p and h_q. Let l_1 be the first vertical link from p that intersects h_p. The part of $\mathcal{P}_{pq}$ from to p to l_1 can have no less than $d_p - 1$ links since, otherwise, p can reach points below s_p with d_p links as we argued above. Let l_2 be the first vertical link from q that intersects h_q. For the same reason as above, the part of $\mathcal{P}_{pq}$ from q to l_2 has at least $d_q - 1$ links. Hence, the length of $\mathcal{P}_{pq}$ is at least $(d_p - 1) + (d_q - 1) + 1$. □

An immediate consequence of Lemma 3.2 is that if $\bigcap \mathcal{I}_2$ is empty, then $D(\mathbf{P}) = 2d_e + 1$. To see this first note that, for a family $\mathcal{I}$ of intervals, we have $\bigcap \mathcal{I} = \emptyset$ if and only if there are two intervals I_1, $I_2 \in \mathcal{I}$ with $I_1 \cap I_2 = \emptyset$. Hence, there are two vertices v and v' in $\mathbf{P}$ with $e(v, d_e + 1) \cap e(v', d_e + 1) = \emptyset$ and Lemma 3.2 yields that $d(v, v') \geq 2d_e + 1$. On the other hand, we noted in the previous section that $D(\mathbf{P}) \leq 2d_e + 1$ and, thus, we obtain that $D(\mathbf{P}) = 2d_e + 1$.

Unfortunately, it is not always true that either $\bigcap \mathcal{I}_1$ is non-empty or $\bigcap \mathcal{I}_2$ is empty. So from now on suppose that $\bigcap \mathcal{I}_1 = \emptyset$ and $\bigcap \mathcal{I}_2 \neq \emptyset$. What are the remaining cases? Consider two points p and q. Since $\bigcap \mathcal{I}_1 = \emptyset$, we can assume that $e(p, d_e) \cap e(q, d_e) = \emptyset$. Furthermore, we have that $e(p, d_e + 1) \cap e(q, d_e + 1) \neq \emptyset$ because $\bigcap \mathcal{I}_2$ is non-empty. Hence, we have to look at the intersections $e(p, d_e) \cap e(q, d_e + 1)$ and $e(p, d_e + 1) \cap e(q, d_e)$. Either both of them contain at least one point or at least one of the two intersections is empty. The former case (both intersections are non-empty) is dealt with in the following lemma. As we will show later the latter case can also be treated with the tools we have developed so far.

LEMMA 3.3 *If p and q are two points in $\mathbf{P}$ and d_p and d_q are two integers such that*

(i) $e(p, d_p)$ *and* $e(q, d_q)$ *are non-empty,*

(ii) $e(p, d_p) \cap e(q, d_q) = \emptyset$,

(iii) $e(p, d_p + 1) \cap e(q, d_q) \neq \emptyset$, *and*

(iv) $e(p, d_p) \cap e(q, d_q + 1) \neq \emptyset$,

then $d(p, q) = d_p + d_q - 1$.

PROOF: By Property (ii) and Lemma 3.2 we have that $d(p, q) \geq d_p + d_q - 1$. Hence, we only have to show the reverse inequality. Before we start proving the claim we give a precise definition of the

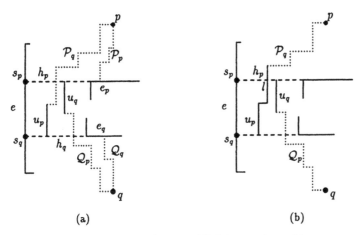

Figure 5: *There is a horizontal link between h_p and h_q.*

situation we consider. As before we assume that e is vertical, that the interior of the polygon is to the right of e, and that $e(p, d_p)$ is above $e(q, d_q)$. We again denote the lowest point of $e(p, d_p)$ by s_p and the highest point of $e(q, d_q)$ by s_q. Further, let $\mathcal{P}_p$ be an admissible path from p to s_p of length d_p. Since we cannot reach any point below s_p with at most d_p links, the last link of $\mathcal{P}_p$ overlaps with some horizontal edges. We denote the leftmost one of these by e_p. Let h_p be the horizontal line segment from e_p to s_p. Furthermore, let $\mathcal{Q}_q$, e_q and h_q be defined analogously for q. Let $\mathcal{P}_q$ be an admissible path from p to s_q (having length $d_p + 1$) and $\mathcal{Q}_p$ be an admissible path from q to s_p (of length $d_q + 1$). Note that these exist since $e(p, d_p + 1) \cap e(q, d_q)$ is non-empty as is $e(p, d_p) \cap e(q, d_q + 1)$. As we have argued in Lemma 3.2, $\mathcal{P}_q$ crosses h_p and $\mathcal{Q}_p$ crosses h_q. It is our aim to show that $\mathcal{P}_q$ and $\mathcal{Q}_p$ can be joined to yield a path of length $d_p + d_q - 1$ that connects p and q. Let u_p be the d_p^{th} link of $\mathcal{P}_q$ and u_q be the d_q^{th} link of $\mathcal{Q}_p$. Clearly, u_p and u_q are vertical. For illustration see Figure 5a.

The proof is now organized as follows. First we show that we can assume that the second to last links u_p and u_q of $\mathcal{P}_q$ and $\mathcal{Q}_p$ both span the part from h_p to h_q and that $\mathcal{P}_p$ and $\mathcal{Q}_q$ are shortest paths. In the second part we show how to construct a path of length $d_p + d_q - 1$ if the rectangle $\mathbf{R}$ spanned by the upper end point s_1 of u_p and the upper end point s_2 of u_q is empty, i.e., contains no points of the boundary of $\mathbf{P}$. Finally, we proof that the rectangle $\mathbf{R}'$ spanned by s_1 and the right end point r of the last link of $\mathcal{P}_p$ is empty. Since it can be easily seen that $\mathbf{R}'$ contains $\mathbf{R}$, this completes the proof.

Part 1

First assume that one of the paths, say $\mathcal{P}_q$, has a horizontal link between h_p and h_q as shown in Figure 5b. Let l be the first vertical link of $\mathcal{P}_q$ that intersects h_p. Since there is a horizontal link of $\mathcal{P}_q$ between h_p and h_q, l is at least the fourth link from the end of $\mathcal{P}_q$. Hence, if we connect u_q with one horizontal link to l, the resulting path has length at most $\lambda(\mathcal{P}_q) - 3 + \lambda(\mathcal{Q}_p) - 1 + 1 = d_p - 2 + d_q + 1 = d_p + d_q - 1$.

We now show that $\mathcal{P}_p$ is a shortest path; for suppose that this is not the case, say we can reach s_p with an admissible path $\mathcal{P}'$ consisting of $d_p - 1$ links. Then, we just connect u_q to $\mathcal{P}'$ by extending the last (horizontal) link l of $\mathcal{P}'$ towards u_q. This is always possible since either l contains the last link of $\mathcal{P}_p$ and, hence, connects to u_q or l is contained in the last link of $\mathcal{P}_p$ and then the horizontal line segment from the end point of u_q to l does not cross the boundary of $\mathbf{P}$. This again yields a path of length $\lambda(\mathcal{P}') + \lambda(\mathcal{Q}_p) - 1 = d_p - 1 + d_q + 1 - 1 = d_p + d_q - 1$.

So we can assume that both u_p and u_q intersect h_p and h_q and that $\mathcal{P}_p$ and $\mathcal{Q}_q$ are shortest paths. Since we cannot reach s_q from p with d_p links, $\mathcal{P}_q$ is also a shortest path and so is $\mathcal{Q}_p$.

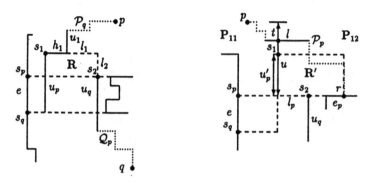

Figure 6: R *has to be empty.*

Part 2

Next we show how to construct a path of length $d_p + d_q - 1$. W.l.o.g. assume that u_p is to the left of u_q. Let s_1 be the upper end point of u_p and let s_2 be the upper end point of u_q. Note that by our assumption on u_p s_1 is above h_p. For now suppose that the horizontal line segment l_1 from s_1 to the x-coordinate of s_2 and the vertical line segment l_2 from s_2 to the y-coordinate of s_1 do not cross the boundary of **P**. We construct a path from p to q of length $d_p + d_q - 1$ in the following way. We take the part of $\mathcal{Q}_p$ from q to u_q and append l_2 to it which yields a path of length d_q. Clearly, we can connect the end point of l_2 to the $(d_p - 2)^{\text{nd}}$ link u_1 of $\mathcal{P}_q$ with one link. Hence, the resulting path has length $d_q + (d_p - 2) + 1 = d_p + d_q - 1$ (see Figure 6).

Part 3

Hence, we only have to prove that $l_1 \cup l_2$ does not cross the boundary of **P**. Let **R** be the rectangle spanned by s_1 and s_2. Clearly, the boundary of **P** does not cross $l_1 \cup l_2$ if **R** contains no boundary points of **P** in its interior.

To see this consider the last link l_p of $\mathcal{P}_p$. Let u'_p be the part of u_p above h_p. Since s_2 is on l_p, **R** is contained in the rectangle **R'** spanned by the right end point r of l_p and s_1. Therefore, it suffices to show that there is no point of the boundary of **P** in **R'**.

As we have argued in the proof of Lemma 3.2, p is in the part $\mathbf{P}_1$ of **P** that is above h_p. We extend u'_p to a vertical segment u that touches the boundary of $\mathbf{P}_1$ at both end points. u splits $\mathbf{P}_1$ into two parts, $\mathbf{P}_{11}$ to the left of u and $\mathbf{P}_{12}$ to the right of u. It is easy to see that p is in $\mathbf{P}_{12}$ since the right end point r of l_p is already in $\mathbf{P}_{12}$ and, hence, $\mathcal{P}_p$ would cross u twice if p were in $\mathbf{P}_{11}$. Obviously, $\mathcal{P}_q$ does not cross u either since, if it does, we can shorten $\mathcal{P}_q$ by at least one link.

Next we turn to proving that no boundary point of **P** intersects the interior of **R'**. The proof is again by contradiction. So suppose that there is a boundary point x of **P** in the interior of **R'** (see Figure 7). We will show that either $\mathcal{P}_p$ or $\mathcal{P}_q$ is not a shortest path in this case. We choose x to be one of the leftmost points of the boundary of **P** in **R'** (which may be on u'_p). We denote the horizontal line segment from x to u'_p by h_x (if x is on u'_p, then h_x degenerates to x). h_x partitions $\mathbf{P}_{12}$ into two parts, $\mathbf{P}_a$ above h_x and $\mathbf{P}_b$ below h_x. We distinguish two cases according to the location of p.

Case 1 p is in $\mathbf{P}_a$.

We show that we can shorten $\mathcal{P}_p$ in this case. $\mathcal{P}_p$ crosses h_x since r is in $\mathbf{P}_b$. Let l be the first vertical link of $\mathcal{P}_p$ that intersects h_x. Since x is one of the leftmost points of the boundary of **P** in **R'**, we can prolong l down so that it reaches l_p which forms the lower boundary of **R'**. On the other hand, we need at least one more link to reach r from l. This contradicts the fact that we assumed $\mathcal{P}_p$ to be a shortest path.

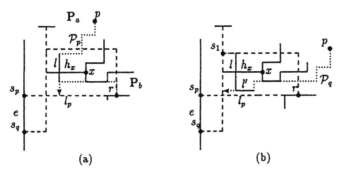

Figure 7: *There is a boundary point in* $\mathbf{R}'$.

Case 2 p *is in* $\mathbf{P}_b$.

We show that then $\mathcal{P}_q$ can be shortened. $\mathcal{P}_q$ crosses h_x, since the upper end point s_1 of u_p belongs to $\mathbf{P}_a$. Let l be again the first vertical link of $\mathcal{P}_q$ that intersects h_x. x is in this case clearly to the right of u'_p since otherwise l would be collinear with u_p which immediately contradicts the fact that $\mathcal{P}_q$ is a shortest path.

Clearly, the horizontal link l' before l on $\mathcal{P}_q$ is above l_p, i.e., it is (partially) contained in $\mathbf{R}'$. Since there are no boundary points of $\mathbf{P}$ to the left of x and, therefore, to the left of l, we can extend l' to the left until it intersects u_p. The resulting path from p to s_q has at least two links less than $\mathcal{P}_q$ which contradicts the fact that $\mathcal{P}_q$ is a shortest path from p to s_q.

Hence, the assumption that the boundary of $\mathbf{P}$ intersects the interior of $\mathbf{R}'$ leads to a contradiction in both cases. This, in particular, implies that the two line segments l_1 and l_2 that we need for the construction of a path of length $d_p + d_q - 1$ from p to q do not cross the boundary of $\mathbf{P}$. Thus, the construction of the path is possible and we can connect p and q with $d_p + d_q - 1$ links as claimed in the lemma. This completes the proof. ☐

With the help of the above three lemmas we are now in a position to attack the problem of computing the claimed bounds on $D(\mathbf{P})$. We already have considered the case that $\bigcap \mathcal{I}_1$ is non-empty and that $\bigcap \mathcal{I}_2$ is empty; we have shown that we obtain the desired bounds on $D(\mathbf{P})$ in these cases. So suppose that we have that $\bigcap \mathcal{I}_1 = \emptyset$ and at the same time that $\bigcap \mathcal{I}_2 \neq \emptyset$. In the following lemma we summarize the results and also deal with the two remaining cases that can occur.

LEMMA 3.4 *If* $\mathbf{P}$ *is a polygon,* e *is an edge of* $\mathbf{P}$*, and* $d_e = \max_{v \in \mathbf{P}} d(v, e)$*, then we have that*

(i) *if* $\bigcap \mathcal{I}_1 \neq \emptyset$*, then* $D(\mathbf{P}) \leq 2d_e - 1$*;*

(ii) *if* $\bigcap \mathcal{I}_2 = \emptyset$*, then* $D(\mathbf{P}) = 2d_e + 1$*;*

(iii) *if* $\bigcap \mathcal{I}_1 = \emptyset$ *and* $\bigcap \mathcal{I}_2 \neq \emptyset$*, then*

    (a) *if, for all* v_1*,* $v_2 \in \mathbf{P}$*, we have that* $e(v_1, d_e) \cap e(v_2, d_e + 1) \neq \emptyset$*, then* $D(\mathbf{P}) = 2d_e - 1$*;*

    (b) *if there are two points* v_1*,* $v_2 \in \mathbf{P}$ *with* $e(v_1, d_e) \cap e(v_2, d_e + 1) = \emptyset$*, then* $D(\mathbf{P}) = 2d_e$*.*

PROOF: Claims (i) and (ii) have been dealt with previously. So we immediately turn to proving Claims (iii.a) and (iii.b). First consider Claim (iii.a). By Lemma 3.2 we have that $D(\mathbf{P}) \geq 2d_e - 1$ since $\bigcap \mathcal{I}_1 = \emptyset$ implies that there is a pair of vertices v' and v'' with $e(v', d_e) \cap e(v'', d_e) = \emptyset$. Now let v_1 and v_2 be two arbitrary vertices in $\mathbf{P}$. We have to show that we can connect them with a path of length at most $2d_e - 1$. If $e(v_1, d_e) \cap e(v_2, d_e) \neq \emptyset$, then Lemma 3.1 yields the desired result. If $e(v_1, d_e) \cap e(v_2, d_e) = \emptyset$, then we can apply Lemma 3.3 since by Condition (iii.a) we have that

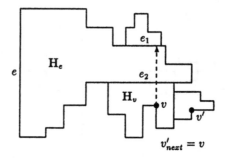

Figure 8: *How to obtain a sorted interval list.*

$e(v_1, d_e) \cap e(v_2, d_e + 1) \neq \emptyset$ and $e(v_1, d_e + 1) \cap e(v_2, d_e) \neq \emptyset$. This yields that there is a path of length $2d_e - 1$ that connects v_1 and v_2.

Now consider Claim (iii.b). Let v' and v'' be two vertices in **P** with $e(v', d_e) \cap e(v'', d_e + 1) = \emptyset$ which exist by assumption, then Lemma 3.2 yields that $d(v', v'') \geq 2d_e$ and, therefore, $D(\mathbf{P}) \geq 2d_e$. So now let v_1 and v_2 be two arbitrary vertices in **P**. We want to show that $d(v_1, v_2) \leq 2d_e$. If $e(v_1, d_e) \cap e(v_2, d_e + 1) \neq \emptyset$, then we have by Lemma 3.1 that $d(v_1, v_2) \leq 2d_e$. Hence, assume that $e(v_1, d_e) \cap e(v_2, d_e + 1) = \emptyset$. Since $\cap \mathcal{I}_2 \neq \emptyset$, we have $e(v_1, d_e + 1) \cap e(v_2, d_e + 1) \neq \emptyset$, and, further, by the definition of d_e we have $e(v_1, d_e) \neq \emptyset$ and $e(v_2, d_e + 2) = e$ so that $e(v_1, d_e) \cap e(v_2, d_e + 2) \neq \emptyset$. Hence, the prerequisites of Lemma 3.3 are met if we consider $p = v_1$, $q = v_2$, $d_p = d_e$ and $d_q = d_e + 1$. Thus, we have $d(v_1, v_2) = d_p + d_q - 1 = d_e + d_e + 1 - 1 = 2d_e$. □

Given the two sets of intervals $\mathcal{I}_1$ and $\mathcal{I}_2$, we can easily compute their intersections and test whether $\cap \mathcal{I}_1 \neq \emptyset$ or $\cap \mathcal{I}_2 = \emptyset$. If none of the two conditions hold, we have to test for the cases of Lemma 3.4. Note that there are two vertices v_1 and v_2 such that $e(v_1, d_e) \cap e(v_2, d_e + 1) = \emptyset$ if and only if there exists a vertex v such that $e(v, d_e) \cap \cap \mathcal{I}_2 = \emptyset$. Hence, this test can also be carried out in time proportional to the number of intervals in $\mathcal{I}_1$ and $\mathcal{I}_2$ which yields an overall linear time algorithm.

3.2 Computing a Sorted Interval List

In order to execute Step (iii) of the link diameter algorithm presented previously, we do not only need to compute all the intervals $e(v, d(v, e))$, for edge e, but we also need them sorted according to the upper and lower end points, for all the vertices at a given distance.

In the following we show how to obtain a sorted list in linear time. W.l.o.g. assume that e is a vertical edge of **P** and that we have computed a histogram partition of **P** w.r.t. e. de Berg [B89] shows that, for each vertex v at distance $d(v, e) > 2$, there exists a vertex v_{next} that has a distance of one less to e and that has the same interval as v. Furthermore, he shows that v_{next} can be computed in linear time, for *all* vertices v of **P**.

In order to obtain an ordered list of all the intervals at e, we compute at each vertex v in **P**, a list $\mathcal{L}_v$ of the vertices v' with $v'_{next} = v$. This can be done by adding each vertex v to $\mathcal{L}_{v_{next}}$. Since $\mathcal{L}_{v_{next}}$ is unsorted, this costs only a constant amount of time per vertex and, hence, linear time in total.

Let $\mathbf{H}_v$ be the histogram of the partition that contains vertex v. For a vertex v that needs only two links to reach e, de Berg computes the edge e_1 of $\mathbf{H}_e \cup \mathbf{H}_v$ that can be reached from v by an axis parallel line segment where $\mathbf{H}_e$ is the maximal histogram induced by e (in **P**) (see Figure 8). W.l.o.g. assume that e_1 is a north edge of $\mathbf{H}_e$. e_1 obviously determines the upper end point of the interval $e(v, 2)$. At e_1 we keep a list $\mathcal{L}_{e_1}$ of all the vertices (at distance two to e) that have the y-coordinate of e_1 as the upper end point of their interval. Let e_2 be the edge that is collinear with the base of $\mathbf{H}_v$. We also keep a list $\mathcal{L}_{e_2}$ of all the vertices that have the y-coordinate of e_2 as the lower end point of their interval. Clearly, we can compute the lists $\mathcal{L}_{e'}$, for all edges e' of $\mathbf{H}_e$, in time linear in the number of vertices of **P**.

In order to obtain a list of the intervals that is sorted according to the lower end points we now proceed as follows. We traverse the histogram $\mathbf{H}_e$ from bottom to top. Since $\mathbf{H}_e$ is monotone w.r.t. the y-axis we encounter the horizontal edges in sorted order. At each horizontal edge e' we follow the list $\mathcal{L}_{e'}$ of the vertices at distance 2 whose intervals have the y-coordinate of e' as the lower end point of their interval. For each vertex v in $\mathcal{L}_{e'}$, we follow the vertices in the list $\mathcal{L}_v$ backward until the vertices at a given distance are found which are then added to the sorted list together with their intervals. To obtain a list of the intervals that is sorted according to upper end points we proceed analogously.

Since the computation of the intervals $e(v, d(v, e)+1)$ (see [B89]) is similar to the above described procedure we can also sort them in linear time if we keep the list $\mathcal{L}_v$ of backward references introduced above.

The algorithm to compute the link diameter of a polygon $\mathbf{P}$ can now be stated as follows.

(i)   If $\mathbf{P}$ is a rectangle, then $D(\mathbf{P}) = 2$ and stop.
(ii) Compute a chord c of $\mathbf{P}$ that cuts $\mathbf{P}$ into two subpolygons $\mathbf{P}_1$ and $\mathbf{P}_2$ such that $|\mathbf{P}_1|, |\mathbf{P}_2| \leq 3/4\, n + 2$.
(iii) Compute $m = \max\{d(v, v') \mid v \in \mathbf{P}_1, v' \in \mathbf{P}_2\}$ and remember the vertices v, v' with $d(v, v') = m$.
(iv) Compute d_1 and d_2. If $d_1 < d_2$, then exchange $\mathbf{P}_1$ and $\mathbf{P}_2$.
(v)   Compute $M = \max(m, D(\mathbf{P}_2))$:
      (v.a) Compute $\mathcal{I}_1 = \{e(v, d_2) \mid v \in \mathbf{P}_2\}$ and $\mathcal{I}_2 = \{e(v, d_2 + 1) \mid v \in \mathbf{P}_2\}$.
      (v.b) if $\bigcap \mathcal{I}_2 = \emptyset$
            then $M := \max(m, 2d_2 + 1)$
            else if there is a vertex v in $\mathbf{P}_2$ with $e(v, d_2) \cap \bigcap \mathcal{I}_2 = \emptyset$
                 then $M := \max(m, 2d_2)$
                 else $M := m$;
            (* Note that we only have to check for the Cases (ii) and (iii.b) of Lemma 3.4
               since in Cases (i) and (iii.a) we have that $D(\mathbf{P}_2) \leq 2d_2 - 1 \leq m$. *)
(vi) Compute $D(\mathbf{P}_1)$ recursively.
(vii) Let $D(\mathbf{P}) := \max(M, D(\mathbf{P}_1))$.

The function $T(n)$ for the number of steps which are needed by the above algorithm for an n-vertex polygon satisfies the following recurrence.

$$
\begin{aligned}
T(n) &\leq T(3/4\, n) + O(n) \\
T(1) &= O(1)
\end{aligned}
$$

which yields that $T(n) = O(n)$. Thus, we have proven the following theorem.

THEOREM 1 *The rectilinear link diameter of a simple rectilinear polygon can be computed in time linear in the number of vertices of the polygon.*

4   Computing an Approximation of the Link Center

In this section we show how to find a chord c in $\mathbf{P}$ that contains points which can reach any other point in $\mathbf{P}$ with at most $\lceil D(\mathbf{P})/2 \rceil + 1 \leq R(\mathbf{P}) + 1$ links. A good candidate for c is the chord c_m through the middle link l_m of a path between two vertices that span the diameter D of $\mathbf{P}$. So let v_1 and v_2 be two vertices with $d(v_1, v_2) = D$ and let $\mathcal{P}$ be a shortest path between them. We will only concern ourselves with the case that D is odd. The case that D is even can be treated in essentially the same way. So assume from now on that $D = 2k + 1$.

Let l_m be the $(k+1)^{\text{st}}$ link of $\mathcal{P}$ as seen from v_1 and c_m the maximal chord through l_m. c_m splits $\mathbf{P}$ into a number of subpolygons $\mathbf{P}_1, \ldots, \mathbf{P}_s$. We have $d(v_1, c_m) = k$ and $d(v_2, c_m) = k$. Clearly, $c_m(v_1, k)$ and $c_m(v_2, k)$ do not intersect for otherwise we have that $d(v_1, v_2) \leq k + k < D$ by Lemma 2.2. We claim that any point in $\mathbf{P}$ can reach c_m with $k + 2$ links.

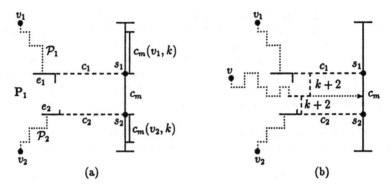

Figure 9: *How to find a point on c_m close to the link center.*

For assume there is a vertex v that needs $k + 3$ or more links. If v is in a different subpolygon than v_1, then we have by Lemma 2.2 or Lemma 3.2 that $d(v_1, v) \geq k + k + 3 - 1 \geq 2k + 2$ which contradicts the fact that the diameter of $\mathbf{P}$ is $2k + 1$. The same argument of course applies if v and v_2 are in two different subpolygons.

So we have to consider the case that v_1, v_2, and v are all in the same subpolygon, say $\mathbf{P}_1$. W.l.o.g. assume that c_m is vertical, $\mathbf{P}_1$ is to the left of c_m, and that $c_m(v_1, k)$ is above $c_m(v_2, k)$. Let s_1 be the lowest end point of $c_m(v_1, k)$ and s_2 be the highest end point of $c_m(v_2, k)$. Furthermore, let $\mathcal{P}_i$ a shortest path from v_i to s_i, $i = 1, 2$. Let l_i be the last link of $\mathcal{P}_i$. As in the proof of Lemma 3.2 we can show that there is a horizontal upward (downward) edge e_1 (e_2) that intersects l_1 (l_2). Let c_i be the horizontal chord from e_i to c_m. The situation is illustrated in Figure 9a. If v is in the part of $\mathbf{P}_1$ below c_2, then v needs at least $k + 2$ links to reach c_2 since $d(v, c_m) \geq k + 3$. On the other hand, v_1 needs at least $k + 1$ links to reach c_2 since otherwise the distance from v_1 to v_2 is less than $2k + 1$. Hence, we have $d(v_1, v) \geq k + 2 + k + 1 - 1 = 2k + 2$ in this case which contradicts the fact that the maximal distance between two vertices is $2k + 1$. For an analogous reason v cannot be above c_1. So suppose that v is below c_1 and above c_2. It can be easily seen that in this case v cannot reach both c_1 and c_2 with $k + 2$ links for otherwise we can prolong the $(k + 1)^{st}$ link of one of the two path to c_m (see Figure 9b). But if, say $d(v, c_1) \geq k + 3$, then $d(v_1, v) \geq k + k + 3 - 1 = 2k + 2$ by Lemma 2.2 which is again a contradiction. Hence, v is at most $k + 2$ links from c_m.

Now let v' and v'' be two arbitrary vertices in $\mathbf{P}$. We want to show that $c_m(v', k + 2)$ and $c_m(v'', k + 2)$ intersect. If v' and v'' are on two different sides of c_m, then Lemma 2.2 implies that $d(v'v'') \geq k + 2 + k + 2 = 2k + 4 > D(\mathbf{P})$ if $c_m(v', k + 2)$ does not intersect $c_m(v'', k + 2)$. If they are on the same side of c_m, then an empty intersection implies that $d(v', v'') \geq k + 2 + k + 2 - 1 = 2k + 3$ by Lemma 3.2 again in contradiction to fact that $D(\mathbf{P}) = 2k + 1$. Hence, the intervals $c_m(v', k + 2)$ and $c_m(v'', k + 2)$ have a non-empty intersection, for any two vertices v' and v'' in $\mathbf{P}$. This implies that $I = \bigcap \{c_m(v, k + 2) \mid v \in \mathbf{P}\}$ is non-empty and any point in I can reach all the vertices in $\mathbf{P}$ with $k + 2 = \lceil D(\mathbf{P})/2 \rceil + 1 \leq R(\mathbf{P}) + 1$ links. Since computation of $\{c_m(v, k + 2) \mid v \in \mathbf{P}\}$ can be done in linear time as well as the taking the intersection, I can computed within the same time bound.

It should be noted that there is an algorithm which also runs in linear time that computes all points of the exact link center [NS91].

5 Conclusions

In this paper we have studied the concept of rectilinear link distance within simple rectilinear polygons. We treat the problem of computing the rectilinear link diameter and solve it in optimal linear time. In fact, the algorithm computes a diametral pair of vertices of the polygon. To our knowledge this is the first optimal algorithm for the diameter problem of a non-trivial class of polygons.

Open questions are, of course, to generalize our approach for polygons and paths having more than two (fixed) orientations. It is also possible to define the rectilinear link diameter in general polygons and we conjecture that for this problem our result still works with minor modifications. A problem in this setting is that the size of the partitioning of the polygon into histograms (actually, histogram-like polygons) is no longer of the order of the size of the polygon. In fact, it can become arbitrarily large. Our algorithm would then run in time proportional to the size of this partition instead of the size of the polygon.

Another possibility is to use a different metric. In the rectilinear case, the L_1 metric would be a natural choice. There exists no result in this context.

References

[B89] MARK DE BERG. *On Rectilinear Link Distance.* Technical Report RUU-CS-89-13, Department of Computer Science, University of Utrecht, P.O.Box 80.089, 3502 TB Utrecht, the Netherlands, May 1989.

[BKNO90] M.T. DE BERG, M.J. VAN KREVELD, B.J. NILSSON, M.H. OVERMARS. Finding Shortest Paths in the Presence of Orthogonal Obstacles Using a Combined L_1 and Link Metric. In *Proc. 2nd Scandinavian Workshop on Algorithm Theory*, pages 213–224, 1990.

[Cha90] BERNARD CHAZELLE. Triangulating a Simple Polygon in Linear Time. In *Proc. 31th Symposium on Foundations of Computer Science*, pages 220–230, 1990.

[Ke89] YAN KE. An Efficient Algorithm for Link-distance Problems. In *Proceedings of the Fifth Annual Symposium on Computational Geometry*, pages 69–78, ACM, ACM Press, Saarbrücken, West Germany, June 1989.

[Lev87] CHRISTOS LEVCOPOULOS. *Heuristics for Minimum Decompositions of Polygons.* PhD thesis, University of Linköping, Linköping, Sweden, 1987.

[NS91] B.J. NILSSON, S. SCHUIERER. An Optimal Algorithm for the Rectilinear Link Center of a Rectilinear Polygon. In *2nd Workshop on Algorithms and Data Structures*, Lecture Notes in Computer Science, 1991.

[Sur87] SUBHASH SURI. *Minimum Link Paths in Polygons and Related Problems.* PhD thesis, Johns Hopkins University, Baltimore, Maryland, August 1987. pages 213–224, 1990.

Layout of Flexible Manufacturing Systems - Selected Problems

Hartmut Noltemeier

University of Würzburg

Extended Abstract

Summary: In this paper a selection of important problems, which occur in the layout of Flexible Manufacturing Systems (FMS), is considered and analyzed from a more computational geometry and knowledge engineering point of view. This includes the representation of proximity properties as well as applications in the layout of assembly lines, in machine layout and in robot vision/ motion planning problems. Some recent results on monotonous bisector trees are given and their applicability in these fields are outlined. New results on motion planning in dynamic scenes are summarized.

Introduction

The layout of Flexible Manufacturing Systems (FMS) has been recognized as a central problem of modern production systems in a rapidly growing number of different branches of economies. A lot of remarkable work (scientific papers as well as extensive field studies) has been done in the last two decades. For short we refer only to a very few of many papers relevant to this field ([KusWi], [SteSu], [BuzYa], [BiSkiZa], [MyrKähLu]).

We got involved into this field massively, when we had to design an assembly line for the production of a large variety of printed circuit boards (PCB) in 1983 (see [DeNo]).

Let us pick up this problem as a (simplified) starting example for our considerations:

Let $E = \{e_1,\ldots,e_m\}$ denote the set of types of (electronic) elements which have to be put onto specific PCB-layers (by pick-and-place devices along the line), and let $G = \{g_1,\ldots,g_n\}$ denote the set of types of boards which have to be produced in the next period with expected frequencies $f_j (j = 1,\ldots,n)$ respectively.

We assume that the production data are given by a weighted 'Gozinto' graph ('bill of material');this wheighted bipartite graph indicates which electronic element (chip, resistance,...) how often has to be put into each board. A further distinction (applicability of specific robot operations within a set R of robot operations) may induce an additional weight-vector r on each element $e \in E$; thus we finally get a bipartite graph with weighted edges (w_{ij}) and weighted nodes (f_j for $g_j \in G$ resp. r_i for $e_i \in E$).

The layout problem roughly is to distribute the elements on 'tables' (manufacturing units (cells), which can be handled by a person or a robot) along the assembly line such that 'highly parallel' manufacturing is possible and the 'output' (in some precise sense) is maximized (obeying limited amounts of investment, inventory cost etc.).

Thus we have to find a suitable family of subsets of E with each subset representing those elements of E which are stored and handled on a table - and in practice additionally easily allows the introduction of a few of new types of boards within the next period.

While other authors were mainly concerned with optimal sequencing of insertion operations [BaMa], with component and tool allocation in a dual delivery placement machine [AhGroJoh] or with predicting performance of board-manufacturing by means of simulation and regression [ShaTa] - in each case for only one specific type or a very small number of types of boards, we were faced with more than thousand different types of boards (n ≈ 2000), which on the other hand had to be produced with low frequencies - and with about m = 15000 different types of electronic elements altogether.

The many facets of our complex problem showed some common features and caused us to address (sub)problems in a more geometric oriented setting, e.g. to find suitable proximity properties

(with respect to robot operation characteristics/

to locations in a cell/

to locations along the line etc.)

and to use them to solve at least important subproblems.

These kinds of proximities are usually quite more general than those used in simulated annealing approaches, for example to solve the Standard Cell Placement-Problem (see [MaGro]) in Board Design.

This paper is organized as follows: in section 2 we will introduce monotonous bisector trees and will state some theorems about the effort to construct these tools representing proximity properties in a very general setting.

Section 3 gives some hints to the applications in the layout of assembly lines and in robot vision/motion planning.

In section 4 only a few comments are listed about our vision control devices and methods, while in section 5 most recent algorithms and results for motion planning problems in dynamic scenes are sketched. A list of selected references completes this extended abstract.

2. Representation of proximity properties by means of monotonous bisector trees

Proximity properties in a set E of objects frequently are induced solely by binary relations. Typical examples are demonstrated in Facilities Layout Problems [FouGibGif], in Machine Layout Problems [HerKus] and in Clustering Problems (see f.e. [HanJaufra], [HanJauMu]). But in most of those problems we are concerned with, geometric attributes are involved and yield types of spaces which manifest some kind of coherence. For simplicity we will restrict ourselves in the following to quasi-metric spaces and we will assume a

distance function ('dissimilarity function', 'quasi metric',...) $d:E^2 \rightarrow \mathbb{R}_+$, which is reflexive, symmetric, does obey the triangle inequality, but there may be different elements $e, e' \in E$ with distance $d(e,e') = 0$.

First let us introduce Monotonous Bisector Trees (MBT) in a very familiar space.

Definition 1: Let $S \subset \mathbb{R}^m$ be a nonempty, finite set of n points in $\mathbb{R}^m$ ($m \in \mathbb{N}_+$: dimension), $P_0 \in S$ and
$d: \mathbb{R}^m \times \mathbb{R}^m \rightarrow \mathbb{R}_+$ any quasi-metric.
A Monotonous Bisector Tree MBT (S, P_0) is a binary tree with
- if $1 \leq |S| \leq 2$ then MBT (S, P_0) consists of exactly one node, which contains S.
- if $|S| > 2$ and r is the root of MBT (S, P_0), then r contains exactly two points $P_0, P_1 \in S$ and the subtrees of r are monotonous bisector trees
MBT (S_0, P_0) resp. MBT (S_1, P_1) with
$S_0 \cup S_1 = S$, $S_0 \cap S_1 = \emptyset$ and for $i \neq j$ ($i,j \in \{0,1\}$)
$\{ P \in S \mid d(P, P_i) < d(P, P_j) \} \subseteq S_i \subseteq \{ P \in S \mid d(P, P_i) \leq d(P, P_j) \}$.

Recently the following theorem was proven ([Zi])

Theorem 1: For any finite set $S \subset \mathbb{R}^m$, arbitrarily chosen $P_0 \in S$ and any L_p-metric ($1 \leq p \leq \infty$) a monotonous bisector tree MBT (S, P_0) can be constructed in $O(n \log n)$ time which has logarithmic height.

This is a supplementary theorem to earlier results on Voronoi Trees ([Heu], [No]), where we have at most a triple of bisectors in each node, which yield a monotonous hierarchical partition of the underlying space as well. (If we again have a finite set of points in $\mathbb{R}^m$, the Voronoi Tree can be interpreted as a 'flat extract/mixture' of the order-k-Voronoi diagrams, where k ranges from 1 to m-1 [No]).

As we are usually dealing with 'extended' objects (extended in geometric space or/and additionally in attribute spaces like color, texture etc.) a leaf oriented variant of Monotonous Bisector Trees resp. Voronoi Trees is useful.

<u>Definition 2</u>: Let S be a finite, non empty set of objects,

D a set of splitting values, $p_0 \in D$ and

$d: D \times S \to \mathbb{R}_+$ a distance function.

A <u>Monotonous Bisector* Tree</u> MB*T (S,p_0) is a binary tree with

- if $1 \le |S| \le 2$, then MB*T (S,p_0) consists of exactly one node, which represents S.

- if $|S| > 2$, then MB*T (S,p_0) has root r,

which contains p_0 and another $p_1 \in D-\{p_0\}$,

and r has two sons w_0 and w_1, which are roots of MB*T (S_0,p_0) resp. MB*T (S_1,p_1)

with

$S_0 \cup S_1 = S$, $S_0 \cap S_1 = \emptyset$ and for $i \ne j$ $(i,j \in \{0,1\})$

$\{o \in S | d(p_i,o) < d(p_j,o)\} \subseteq S_i \subseteq \{o \in S \mid d(p_i,o) \le d(p_j,o)\}$.

<u>Theorem 2</u> ([NoRoZi]):

Let S be a finite set of n convex objects in m-dimensional euclidean space which are pairwise disjoint and let $D = \mathbb{R}^m$ be the set of splitting values. Then an (optimal) balanced) Monotonous Bisector* Tree with logarithmic height can be constructed in (optimal) $O(n \log n)$ time and $O(n)$ storage assuming that the distance $d(p,o)$ between an object $o \in S$ and a point $p \in D$ can be computed in constant time.

Similar results hold for Voronoi* Trees too; generalizations can be shown even for nondisjoint objects and for more general metrics (e.g. convex distance functions in $\mathbb{R}^2$).

3. <u>Applications of Bisector Trees resp. Bisector* Trees</u>

Due to limitation of space we have to skip the details of this chapter and will only state two facts:

- *Applications in layout of assembly lines for printed circuit boards*
  As mentioned in the introduction the weighted bipartite graph, which summarizes essential data of production, served as a base for extracting several suitable distance functions on the set E of electronic elements. By simply constructing related Bisector* Trees resp. Voronoi* Trees - representing the set E of electronic elements - one easily can extract suitable subtrees yielding a family of subsets of E. The resulting solutions were very satisfactory (for more detail see [DeNo]).
  This procedure can be transfered to more general cell-layout and machine-layout problems too.

- *Applications in ROBOT VISION and MOTION PLANNING*
  Monotonous Bisector* Trees are an efficient and elegant tool to store and represent large scenes of 'geometric objects', which optimizes search costs by

means of the logarithmic height and additionally supports ('coherence'-)queries along 'walks', if the splitting values are chosen carefully.

Alternative approaches for large extended (frequently overlapping) objects are given in [GueNo].

ROBOT VISION

To support flexibility in assembling (a large variety of types of printed circuit boards on one line e.g.) or in more general manufacturing, one usually is faced with the need for a VISION CONTROL, which identifies the work pieces transported by any kind of Material Handling System.

We have developed a robot vision system, which allows to identify types of boards online by a low cost system based on a video camera and a knowledge based segmentation, feature extraction and structural matching process [GoeHoeNo]. More general information on ROBOT CONTROL and INFERENCE SYSTEMS are given in a lot of papers and books (see e.g. [WoPu]).

MOTION PLANNING IN DYNAMIC SCENES

Earlier Material Handling Systems (gantry robot, circular operating handling robot, linear operating automated guided vehicles (AGV), vibratory feeders, magazines etc.) are characterized by a small degree of freedom with respect to motion. But recently more advanced applications (autonomous robots,...) need to look for 'geometric scenes', e.g. set of geometric objects, some of them altering their shape and/or position over time.

A number of different subproblems arise immediately:

- the *path existence problem* is to search for the possibility for any motion between two given configurations without collision with the objects in a surrounding working space,
- the *optimal path detection problem* is to calculate an optimal feasible path (collisionfree and with shortest time, shortest length e.g.),
- the *collision detection problem* assumes a given path and has to decide whether this path is collisionfree in a (dynamically changing) scene,
- the *proximity detection problem* is to decide whether an object moving along a given path does obey obstacle specific tolerance distances to all the objects along this path.

  In certain cases this problem can be reduced to the intersection of circles, which can be decided efficiently by means of power diagrams (see [Au]).
- the *scene representation problem* is how to store large scences of geometric objects.

Among the *nonrigid objects*, the arm of a manipulator is the most important one.
Planning motions of robot arms is usually done by simplifying them to moving
chains of arm segments. But even in 3-space resp. 2-space motion planning of chains
of line segments with a fixed number of joints was shown to be NP-hard and PSPACE-
complete (Reif (1979); Hopcroft, Joseph, Whitesides (1985)). But for more specialized and frequently occuring scenes,algorithms which avoid the double expo-
nential growth (in the degree of freedom; [SchwSha]) meanwhile have been developed
([AbMü 1,2]).Yet in very large scenes these algorithms are not really practicable. To
overcome this difficulty we looked for a special class of pathes in large dynamic
scenes, which are 'maximal' with respect to collision avoidance, namely 'locally
safest' pathes. In a static scene 'locally safest' pathes are essentially given
by the edges (surfaces) of the Voronoi diagram. Recently we have presented a new
algorithm for planning the motion of a disc in a dynamic scene of n (with constant
velocities) moving sites in $\mathbb{R}^2$, which is based on the corresponding sequence of
Voronoi diagrams [RoNo]. This sequence can be maintained efficiently, as the number
of topological events, which cause an update of the Voronoi diagram, has a
nearly cubic upper bound $O(n^2\lambda_s(n))$, where $\lambda_s(n)$ is the maximum length of a (n,s)-
Davenport-Schinzel-sequence and s is a constant depending on the motion of sites
([GuiMiRo]). Each topological event uses only $O(\log n)$ time which is worst-case
optimal.
Using adequate data structures (monotonous bisector* trees, see section 2) and efficient algorithms for computing and maintaining Voronoi diagrams of a large number of
moving sites, we got really practicable algorithms to generate feasible pathes in
the class of 'safest pathes'.
The problem to get an optimal (shortest time resp. shortest length) path within
this class is a very interesting kind of 'minmax'-problem and still unsettled.

References

[AbrMül] S. Abramowski, H. Müller:
Collision avoidance for nonrigid objects
in: Special issue "Computer Science and Operations Research"
(ed. H. Noltemeier), ZOR, 32, 3/4, 165-186, 1988

[AbrMü2] S. Abramowski, H. Müller:
Collision avoidance for nonrigid objects
in: Computational Geometry and its Applications
(ed. H. Noltemeier), LNCS 333, 168-179, Springer 1988

[AhGroJoh] J. Ahmadi, S. Grotzinger, D. Johnson:
Component allocation and partitioning for a dual delivery placement
machine
Operations Research, 36, 2, 176-191, 1988

[Au] F. Aurenhammer:
Improved algorithms for discs and balls using power diagrams
J. Algorithms, 9, 151-161, 1988

[BaMa] M.O. Ball, M.J. Magazine:
Sequenzing of insertions in printed circuit board assembly
Operations Research, 36, 2, 192-201, 1988

[BiSkiZa] G.R. Bitran, W. Skinner, W.I. Zangwill (ed.):
"Operations Research in Manufacturing"
Operations Research, vol. 36, 2, March-April 1988

[BuzYa] J.A. Buzacott, D.D. Yao:
Flexible Manufacturing Systems: A Review of Analytic Models
Man. Sci., vol. 32, 7, 890-905, July 1986

[DeNo] F. Dehne, H. Noltemeier:
Clustering methods for geometric objects and applications to design problems
Visual Computer, vol. 2, 1, 31-38, Springer 1986

[FouGibGif] L.R. Foulds, P.B. Gibbons, J.W. Giffin:
Facilities Layout Adjacency Determination: An experimental comparision
of three graph theoretic heuristics
Operations Research, 33, 5, 1091-1106, 1985

[GoeHoeNo] M. Goerke, K.U. Höffgen, H. Noltemeier:
PICASYS: Ein Bildanalysesystem zur Identifikation von Leiterplatinen
in: Mustererkennung 1989 (ed. Burkhardt, Höhne, Neumann), IFB vol. 219,
391-395, Springer 1989

[GueNo] O. Günther, H. Noltemeier:
Spatial database indices for large extended objects
Proc. 7th IEEE-Conf. on Data Engineering, Kobe, Japan, April 1991

[GuiMiRo] L. Guibas, J.S.B. Mitchell, T. Roos:
Voronoi diagrams of moving points in the plane
Res. Report, Univ. Würzburg, Oct. 1990

[HanJaufra] P. Hansen, B. Jaumard, O. Frank:
An $\Theta(N^2)$ algorithm for maximum sum-of-splits clusterings
Res. Report # 10-89, RUTCOR, 1989

[HanJauMu] P. Hansen, B. Jaumard, K. Musitu:
Weight constrained maximum split clustering
Res. Report # 28-89, RUTCOR, 1989

[HerKus] S.S. Heragu, A. Kusiak:
Machine layout problem in flexible manufacturing systems
Operations Research, 36, 2, 258-268, 1988

[Heu] H. Heusinger:
Cluster algorithms for sets of geometric objects
PhD-thesis (in German), Univ. Würzburg, 1989

[KusWi] A. Kusiak, W.E. Wilhelm (ed.):
Analysis, Modeling and Design of Modern Production Systems
Annals of Oper. Res., 17, Baltzer 1989

[MaGro] S. Mallela, L.K. Grover:
Clustering based Simulated Annealing for Standard Cell Placement
Proc. 25th ACM/IEEE Design Automation Conference, 312-317, 1988

[MyrKähLu] M. Myrup Andreasen, S. Kähler, T. Lund:
Design for Assembly, Springer 1988 (2.ed.)

[No] H. Noltemeier:
Voronoi Trees and Applications
in: Proc. Discrete Algorithms and Complexity, 69-74, Fukuoka/Japan, Nov. 1989

[NoRoZi] H. Noltemeier, T. Roos, C. Zirkelbach:
Partitioning of complex scenes of geometric objects
Res. Report, Univ. Würzburg, Nov. 1990

[RoNo] T. Roos, H. Noltemeier:
Dynamic Voronoi diagrams in motion planning
Res. Report, Univ. Würzburg, Dec. 1990

[SchwSha] J.T. Schwartz, M. Sharir:
On the piano movers problem. II. General techniques for computing
topological properties of real algebraic manifolds
Advances in Applied Math., 4, 298-351, 1983

[ShaTa] J.S. Shang, P.R. Tadikamalla:
Yield Analysis of an automated printed circuit board assembly line
in: [SteSu], 293-298, 1989

[SteSu] K.E. Stecke, R. Suri (ed.):

   Flexible Manufacturing Systems, Elsevier 1989

[WoPu] A.K.C. Wong, A. Pugh (ed.):

   Machine intelligence and knowledge engineering for robotic applications

   NATO ASI Series, vol. 33, Springer 1987

[Zi] C. Zirkelbach:

   Monotonous Bisector Trees and Clustering Problems

   Res. Report, Univ. Würzburg, June 1990

Authors's address: Prof. Dr. Hartmut Noltemeier

                   Computer Science Department

                   University of Würzburg

                   Am Hubland

                   D-8700 Würzburg

                   email: noltemei@informatik.uni-wuerzburg.dbp.de

                   fax: +49-931-888-4600

                   phone: +49-931-888-5054 and -5055

Dynamic Voronoi Diagrams in Motion Planning

(Extended Abstract)

Thomas Roos and Hartmut Noltemeier*
University of Würzburg, Germany

Abstract

Given a set of n points in the Euclidean plane each of which is continuously moving along a given trajectory. At each instant of time, these points define a Voronoi diagram which also changes continuously, except for certain critical instances – so-called *topological events*.

In [Ro 90], an efficient method is presented of *maintaining* the Voronoi diagram over time. Recently Guibas, Mitchell and Roos [GuMiRo 91] improved the trivial quartic upper bound on the number of topological events by almost a linear factor to the nearly cubic upper bound of $O(n^2 \lambda_s(n))$ topological events, where $\lambda_s(n)$ is the maximum length of an (n, s)-Davenport-Schinzel sequence and s is a constant depending on the motion of the sites. Each topological event uses only $O(\log n)$ time (which is worst-case optimal).

Now in this work, we present a new algorithm for planning the motion of a disc in a dynamic scene of moving sites which is based on the corresponding sequence of Voronoi diagrams. Thereby we make use of the well-known fact, that locally the Voronoi edges are the *safest paths* in the dynamic scene. We present a quite simple approach combining *local* and *global strategies* for planning a feasible path through the dynamic scene.

One basic advantage of our algorithm is that only the topological structure of the dynamic Voronoi diagram is required for the computation. Additionally, our *goal oriented approach* provides that we can maintain an existing feasible path over time. This guarantees that we reach the goal if there is a feasible path in the dynamic scene at all. Finally our approach can easily be extended to general convex objects.

1 Introduction

The *Voronoi diagram* is one of the most fundamental data structures in computational geometry. In its most general form, the Voronoi diagram $VD(S)$ of a set S of n objects in a space E is a subdivision of this space into maximal regions, so that all points within a given region have the same nearest neighbor in S with regard to a general distance measure d.

Shamos and Hoey [ShHo 75] introduced the Voronoi diagram for a finite set of points in the Euclidean plane $\mathbb{E}^2$ into the field of computational geometry, providing the first efficient algorithm for its computation. Since then, Voronoi diagrams in many variations have appeared throughout the algorithmic literature; see, for example, [ChEd 87], [Ya 87], [Ro 89] and [Au 90].

One problem of recent interest has been of allowing the set of objects S to vary continuously over time along given trajectories. This "dynamic" version has been studied by [ImSuIm 89], [AuImTo 90] and [Ro 90].

---

*This work was supported by the Deutsche Forschungsgemeinschaft (DFG) under contract No 88/10 - 1) and (No 88/10 - 2).

Recently Guibas, Mitchell and Roos [GuMiRo 91] improved the naive quartic bound on the number of topological events, achieving a nearly cubic upper bound of $O(n^2 \lambda_s(n))$, where $\lambda_s(n)$ is the maximum length of an (n, s)-Davenport-Schinzel sequence and s is a constant depending on the motion of the sites.

Our main result is to present a new algorithm for planning the motion of a disc in a dynamic scene of moving sites (and also general convex objects), which is based on the corresponding sequence of Voronoi diagrams. Thereby we make use of the well-known fact, that locally the Voronoi edges are the *safest paths* in the dynamic scene.

The basis of our algorithm is a generalization of an early result due to Ó'Dúnlaing and Yap [O'DuYa 85] proving that the general path-existence problem in a dynamic scene is equivalent to the path-existence problem restricted to the dynamic Voronoi diagram of the underlying objects. For developing good approximations of shortest feasible paths it is necessary to combine both, *local* and *global strategies*, to avoid unnecessary (and arbitrarily long) detours. We present a quite simple approach which guarantees that the moving center of the disc remains on the dynamic Voronoi diagram thereby approaching the final position (local optimization).

To satisfy the global condition we adopt a static technique by Rohnert [Ro 91] who uses *maximum bottleneck spanning trees* to decide efficiently if there is a feasible path between two arbitrary points in the plane. Indeed, using the *dynamic maximum spanning trees* this result can be generalized to dynamic scenes, too. Thereby we can apply the methodology of dynamic Voronoi diagrams for characterizing the topological stability and topological events of maximum spanning trees analogously.

The basic advantages of our algorithm are its simplicity, its numerical stability (cf. [Sulr 89]) and the fact that only the topological structures of the dynamic Voronoi diagram and the maximum spanning tree are required for the computation. Additionally, our *goal oriented approach* provides that we can maintain an existing feasible path over time. This guarantees that we reach the goal if there is a feasible path in the dynamic scene at all. Finally our approach can easily be extended to convex objects and results in a general *dynamic free space approach* (cf. [AbMü 88] for the static case) by adding some local heuristics. The algorithms have also been implemented with success.

2 The Topological Structure of Voronoi Diagrams

This section summarizes the elementary definitions concerning classical Euclidean Voronoi diagrams, where closeness is defined by the Euclidean distance function d. Given a finite set

$$S := \{P_1, \ldots, P_n\}$$

of $n \geq 3$ points in the Euclidean plane $\mathbb{E}^2$.

First of all let $B(P_i, P_j)$ denote the perpendicular *bisector of P_i and P_j* and $v(P_i) := \{x \in \mathbb{E}^2 \mid \forall_{j \neq i} \; d(x, P_i) \leq d(x, P_j)\}$ the *Voronoi polygon of P_i*. The vertices of the Voronoi polygons are called *Voronoi points* and the bisector parts on the boundary are called *Voronoi edges*. Finally let

$$VD(S) := \{v(P_i) \mid P_i \in S\}$$

denote the *Voronoi diagram of S*.

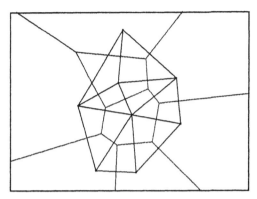

Figure 1: The Voronoi diagram with its straight line dual,
the Delaunay triangulation.

The embedding of the Voronoi diagram provides a planar straight line graph that we call the *geometrical structure* of the underlying Voronoi diagram (see figure 1).

Now we turn our attention to the dual graph of the Voronoi diagram, the so-called *Delaunay triangulation* $DT(S)$. If S is in *general position* – i.e. no four points of S are cocircular and no three points of S are collinear – every bisector part in $VD(S)$ corresponds to an edge and every Voronoi point in $VD(S)$ to a triangle in $DT(S)$. The use of the dual graph not only has numerically advantages, but also allows a clearer separation between geometrical and topological aspects.

We now introduce a *one - point - compactification* to simplify the following descriptions. Therefore we consider the modified basic set $S' := S \cup \{\infty\}$ and obtain the extended Delaunay triangulation

$$DT(S') = DT(S) \cup \{(P_i, \infty) \mid P_i \in S \cap \partial CH(S)\}$$

i.e. in addition to the Delaunay triangulation $DT(S)$, every point on the boundary of the convex hull $\partial CH(S)$ is connected to ∞. We call the underlying graph of the extended Delaunay triangulation $DT(S')$ the *topological structure* of the Voronoi diagram. We obtain the following relation characterizing triangles in $DT(S')$:

$$\{P_i, P_j, P_k\} \in DT(S') \iff v(P_i, P_j, P_k) \text{ is a Voronoi point in } VD(S).$$
$$\{P_i, P_j, \infty\} \in DT(S') \iff P_i \text{ and } P_j \text{ are neighboring points of } S \text{ on}$$
$$\text{the boundary of the convex hull } \partial CH(S).$$

As $DT(S')$ is a complete triangulation of the extended plane $\overline{\mathbb{E}^2}$ – i.e. every triangle is bounded by exactly three edges and every edge belongs to exactly two triangles – *Euler's polyhedron formula* implies that the number of of edges and triangles of the topological structure $DT(S')$ of the Voronoi diagram $VD(S)$ is linear. Furthermore it is easy to see, that the hardest part of constructing a Voronoi diagram is to determine its topological structure, because the geometrical structure of a Voronoi diagram can be derived from it in linear time by a simple flow of the current Delaunay triangles in $DT(S')$. In addition, the geometrical structure is determined only locally by its topological structure, namely in the neighborhood of the corresponding Voronoi point. This implies the possibility of a local update of the Voronoi diagram after a local change of one or more points in S.

3 Voronoi Diagrams of Moving Points in the Plane

In this section we consider the case of *continuously moving points* in the Euclidean plane. For that we are given a finite set

$$S := \{P_1, \ldots, P_n\}$$

of $n \geq 3$ continuous curves in the Euclidean plane $\mathbb{E}^2$, with $P_i : \mathbb{R} \to \mathbb{E}^2, t \mapsto P_i(t)$, under the following *assumptions*:

A   *The points move without collisions, i.e.* $\forall_{i \neq j} \forall_{t \in \mathbb{R}} \ P_i(t) \neq P_j(t)$

B   *There exists a moment $t_0 \in \mathbb{R}$ where $S(t_0)$ is in general position.*

First of all, we summarize the elementary properties of dynamic Voronoi diagrams. The omitted proofs can be found in [Ro 90] and [GuMiRo 91]. Our first theorem describes the *local stability* and the *elementary changes* of the topological structure $DT(S'(t_0))$ – the so-called *topological events*.

Theorem 1   For a finite set S of points in general position, the topological structure of the Voronoi diagram is *locally stable* under sufficiently small continuous motions of the sites.
Elementary changes in the topological structure of the Voronoi diagram $VD(S)$ are characterized by SWAPs of adjacent triangles in $DT(S')$, except for degenerated cases (compare figure 2).

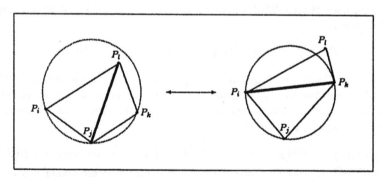

Figure 2: SWAP: an elementary change of the topological structure.

In this connection the *original advantage* of the one - point - compactification becomes apparent. Even changes on the boundary of the convex hull $\partial CH(S)$ can be treated by simple SWAPs of diagonal edges of adjacent extended triangles.

The degenerated cases, where more than four points in $S'(t)$ become cocircular, can be handled without loss of time by the triangulation algorithm presented in [Ag 87] (compare [Ro 90]).

As we have seen, topological events are characterized by moments of cocircularity or collinearity of neighboring points. Therefore it is necessary, that the zeros of the functions $INCIRCLE(\ldots)$ and $CCW(\ldots)$ introduced by [GuSt 85] are calculable.[1] For that we demand the following additional assumption, that is achieved, for example, in the case of piecewise polynomial curves of bounded degree.

C   *The functions $INCIRCLE(P_i, P_j, P_k, P_l)$ and $CCW(P_i, P_j, P_k)$ have at most $s \in O(1)$ zeros.*

Assumption (C) implies that each *quadrilateral* – i.e. each pair of adjacent triangles – generates at most a constant number of topological events.

Now we proceed with a coarse sketch of the *algorithm* which maintains the topological structure over time :

Preprocessing :

1. Compute the topological structure $DT(S'(t_0))$ of the starting position.

2. For every existing quadrilateral in $DT(S'(t_0))$ calculate the potential topological events and build up a balanced SWAP – tree.

Iteration :

1. Determine the next topological event and decide whether it's a SWAP or a RETRIANGULATION.

2. Process the topological event and do an update of the SWAP – tree.

Now our second theorem summarizes on the one hand the time and storage requirements of the algorithm above and on the other hand the recently derived results [GuMiRo 91], concerning the maximum number of topological events. In the meantime, we have implemented dynamic Voronoi diagrams, achieving a very good performance and only $\Theta(n\sqrt{n})$ topological events in the average case.

Theorem 2   Given a finite set $S(t)$ of n continuous curves under the assumptions (A), (B) and (C). The motion of the points requires $O(n \log n)$ preprocessing time and $O(n)$ storage. Every topological event that appears uses $O(\log n)$ time (which is worst-case optimal).

Furthermore there are at most $O(n^2 \lambda_s(n))$ topological events during the entire flow of all points. Thereby $\lambda_s(n)$ denotes the maximum length of a (n, s)-Davenport-Schinzel sequence and s is a constant depending on the complexity of the underlying curves.

In addition, if we move only k points (while leaving the other $n - k$ points fixed), there are at most $O(k n \lambda_s(n) + (n - k)^2 \lambda_s(k))$ topological events, which proves a nearly quadratic upper bound if k is constant.

---

[1]The functions are defined as follows

$$INCIRCLE(P_i, P_j, P_k, P_l) := \begin{vmatrix} x_{P_i} & y_{P_i} & x_{P_i}^2 + y_{P_i}^2 & 1 \\ x_{P_j} & y_{P_j} & x_{P_j}^2 + y_{P_j}^2 & 1 \\ x_{P_k} & y_{P_k} & x_{P_k}^2 + y_{P_k}^2 & 1 \\ x_{P_l} & y_{P_l} & x_{P_l}^2 + y_{P_l}^2 & 1 \end{vmatrix} \quad \text{and} \quad CCW(P_i, P_j, P_k) := \begin{vmatrix} x_{P_i} & y_{P_i} & 1 \\ x_{P_j} & y_{P_j} & 1 \\ x_{P_k} & y_{P_k} & 1 \end{vmatrix}$$

4   Motion Planning in Dynamic Scenes

We consider the following problem: Given a dynamic scene of points in the Euclidean plane and a disk D with center x, we search a collision free path of this disk through the dynamic scene from a (safe) starting position $\tilde{x}$ to a final position $\bar{x}$. It is well-known that locally the Voronoi edges of dynamic Voronoi diagrams are the *safest paths* in the dynamic scene (compare [AbMü 88] for the static case).

Ó'Dúnlaing and Yap [O'DuYa 85] were the first to observe that the general *path-existence problem* in a static scene[2] is equivalent to the path-existence problem restricted to the static Voronoi diagram of the underlying objects. Using the continuity of the dynamic Voronoi diagram, we can present the following extension.

Theorem 3   There is a safe path from the starting point $\tilde{x}$ to the final position $\bar{x}$ in the dynamic scene iff there exists such a path "in the dynamic Voronoi diagram".

Therefore the above path-equivalence holds for dynamic scenes, too. Now we generalize this approach to continuously (piecewise differentiable) moving sites in the plane with the help of the presented dynamic Voronoi diagram. Thereby we can restrict ourselves to a starting position $x(t_0)$ and a final position $x(t_1)$, lying on Voronoi edges, because there are various strategies to get the disk on a bisector in the starting position and vice versa (if the disk has reached a Voronoi edge which belongs to the Voronoi polygon that contains the final position). As we'll see, it is necessary to combine both, *local and global strategies*, to obtain good approximations of feasible paths and to avoid unnecessary (and arbitrarily long) detours. For that, we demand the following *conditions*:

(1)   **moving along locally safest paths**
Guarantee that the center $x(t)$ never leaves the moving Voronoi edges in $VD(S(t))$.

(2)   **avoiding collisions**
Make sure that D avoids any collision with the moving points $S(t)$.

(3)   **goal oriented approach**

(a)   **local strategy**
Guarantee that the center $x(t)$ move towards the final position $\bar{x}$ (if possible).

(b)   **global strategy**
Make sure that the center $x(t)$ reaches the final position $\bar{x}$ if there is a feasible path.

Now, we are going to present an approach by which the demands (1), (2) and (3a) can be controlled easily.

---

[2]They studied only bounded Voronoi diagrams, but their results easily extend to unbounded Voronoi diagrams, if we admit the disk to move between the unbounded Voronoi edges far outside the scene (see also figure 4).

The basic idea of our approach is contained in the following formulation (see also figure 3):

$$x(t) \quad := \quad m(t) + \lambda(t)\, n(t) \quad \text{where}$$

$$m(t) \quad := \quad \frac{P_i(t) + P_j(t)}{2} \quad \text{and}$$

$$n(t) \quad := \quad \begin{pmatrix} P_{i2}(t) - P_{j2}(t) \\ P_{j1}(t) - P_{i1}(t) \end{pmatrix} \perp [P_j(t) - P_i(t)]$$

Thereby $\lambda(t)$ is a *scalar function* that we have to select carefully with respect to condition (1), (2) and (3a). Notice, that this formulation already guarantees that $x(t)$ moves along the current bisector $B(P_i(t), P_j(t))$.

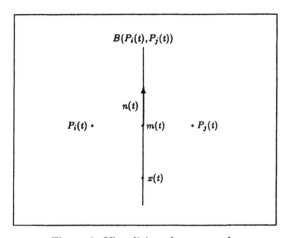

Figure 3: Visualizing the approach.

Now, we are going to check our conditions, one by one. First of all, $\lambda(t_0)$ is determined by the current position $x(t_0)$. To satisfy condition (1) we only have to make the restriction that $x(t)$ moves between the two adjacent (possibly extended) Voronoi points. This leads to the following limitation on $\lambda(t)$:

$$\left. \begin{array}{ll} -\infty & \text{if } P_l = \infty \\[4pt] low(t) & \text{otherwise} \end{array} \right\} \leq \lambda(t) \leq \left\{ \begin{array}{ll} \infty & \text{if } P_k = \infty \\[4pt] high(t) & \text{otherwise} \end{array} \right.$$

where

$$low(t) \quad := \quad \frac{[P_l(t) - P_j(t)]^T [P_l(t) - P_i(t)]}{2\, n(t)^T [P_l(t) - P_i(t)]}$$

$$high(t) \quad := \quad \frac{[P_k(t) - P_j(t)]^T [P_k(t) - P_i(t)]}{2\, n(t)^T [P_k(t) - P_i(t)]}$$

Thereby $\{P_i, P_j, P_k\}$ and $\{P_i, P_j, P_l\}$ are the two triangles which are adjacent to the edge (P_i, P_j) in $DT(S'(t))$ (in the right orientation).

[AuImTo 90] H. Aunuma, H. Imai, K. Imai and T. Tokuyama, *Maximin Locations of Convex Objects and Related Dynamic Voronoi Diagrams*, Proc. of the 6th ACM Symposium on Computational Geometry, Berkeley, 1990, pp 225 - 234

[Au 90] F. Aurenhammer, *Voronoi Diagrams – A Survey of a Fundamental Geometric Data Structure*, Technical Report B 90-09, Fachbereich Mathematik, Serie B Informatik, FU Berlin, Nov. 1990

[ChEd 87] B. Chazelle and H. Edelsbrunner, *An Improved Algorithm for Constructing k-th - Order Voronoi Diagrams*, IEEE Transactions on Computers, Nov. 1987, Vol. C-36, No. 11, pp 1349 - 1354

[GuMiRo 91] L. Guibas, J.S.B. Mitchell and T. Roos, *Voronoi Diagrams of Moving Points in the Plane*, Proc. 17th International Workshop on Graphtheoretic Concepts in Computer Science, Fischbachau, Germany, June 1990, to appear in LNCS

[GuSt 85] L. Guibas and J. Stolfi, *Primitives for the Manipulation of General Subdivisions and the Computation of Voronoi Diagrams*, ACM Transactions on Graphics, Vol. 4, No. 2, April 1984, pp 74 - 123

[ImSuIm 89] K. Imai, S. Sumino and H. Imai, *Minimax Geometric Fitting of Two Corresponding Sets of Points*, Proc. of the 5th ACM Symposium on Computational Geometry, Saarbrücken, 1989, pp 266 - 275

[No 88] H. Noltemeier, *Computational Geometry and its Applications*, Proceedings Workshop CG '88, Universität Würzburg, März 1988, LNCS 333, Springer, 1988

[PrSh 85] F.P. Preparata and M.I. Shamos, *Computational Geometry – An Introduction*, Springer, 1985

[O'DuYa 85] C. Ó'Dúnlaing and C. Yap, *A Retraction Method for Planning the Motion of a Disc*, Journal of Algorithms, Vol. 6, 1985, pp 104 –111

[Ro 91] H. Rohnert, *Moving a Disc Between Polygons*, Algorithmica, Vol. 6, 1991, pp 182 –191

[Ro 89] T. Roos, *k - Nearest - Neighbor Voronoi Diagrams for Sets of Convex Polygons, Line Segments and Points*, Proceedings 15th Intern. Workshop on Graph-Theoretic Concepts in Computer Science WG89, LNCS 411, pp 330 - 340, Springer, 1990

[Ro 90] T. Roos, *Voronoi Diagrams over Dynamic Scenes (Extended Abstract)*, Proceedings 2nd Canadian Conference on Computational Geometry, Ottawa, 1990, pp 209 - 213

[ShHo 75] M.I. Shamos and D. Hoey, *Closest - Point Problems*, Proc. 16th Annual Symp. on FOCS, 1975, pp 151 – 162

[SuIr 89] K. Sugihara and M. Iri, *Construction of the Voronoi Diagram for One Million Generators in Single-Precision Arithmetic*, private communications, 1989

[Ya 87] C.K. Yap, *An $O(n \log n)$ Algorithm for the Voronoi Diagram of a Set of Simple Curve Segments*, Discrete & Computational Geometry, 1987, Vol. 2, pp 365 – 393

Theorem 4   For a set of n points in the Euclidean plane, let v_1 and v_2 be two Voronoi points and $MBST(VD(S))$ be a spanning tree of maximum total width. Then, the bottleneck (i.e. the minimum edge width) on the unique path in $MBST(VD(S))$ from v_1 to v_2 is maximal over all paths from v_1 to v_2 in the Voronoi diagram $VD(S)$.

Therefore, using the maximum bottleneck spanning tree $MBST(VD(S))$ the bottleneck between two Voronoi points can be calculated in $O(n)$ time.[4] Now the question arises whether this approach can be generalized to dynamic scenes, too. Indeed, we can apply the methodology of dynamic Voronoi diagrams for characterizing the topological stability and topological events of maximum (minimum) spanning trees analogously. At first it is easy to see, that maximum spanning trees are locally stable as long as the weights of the edges are different. On the other hand, topological changes in the maximum spanning tree can be described as exchanges of tree edges with non-tree edges at the moment of equal weight providing that both edges lie on a common cycle in the maximum spanning tree (except for degenerated cases). Finally we obtain the following theorem.

Theorem 5   For a set of n continuously moving sites $S(t)$ the maximum bottleneck spanning tree $MBST(VD(S(t))$ of a dynamic Voronoi diagram $VD(S(t))$ can be easily maintained over time. There appear at most $O(n^4)$ topological events which can be handled in $O(n)$ time each.

With that we can now apply a general *depth-first-search algorithm technique* to our problem. Thereby we stay on the current Voronoi edge until we reach an adjacent Voronoi point (in which case we select one of the adjacent Voronoi edges) or the current Voronoi edge disappears due to a topological event. If there is a feasible path at the starting situation, we can keep this feasibility (condition (3b)) during the process by selecting the next Voronoi edge carefully: At any reached Voronoi point there is one adjacent Voronoi edge which is part of the unique safest path generated by the *dynamic maximum bottleneck spanning tree*. Therefore at least one adjacent Voronoi edge preserves condition (3b) thereby possibly breaking condition (3a).

The basic advantages of our algorithm are its simplicity, its numerical stability (cf. [SuIr 89]) and the fact that only the topological structure of the dynamic Voronoi diagram is required for the computation. Our approach can easily be extended to convex objects and results in a general *dynamic free space approach* (cf. [AbMü 88] for the static case) by adding some local heuristics.

References

[Ag 87]   A. Aggarwal, L. Guibas, J. Saxe and P. Shor, *A Linear Time Algorithm for Computing the Voronoi Diagram of a Convex Polygon*, Proc. of the 19th Annual ACM Symposium on Theory of Computing, New York City, 1987, pp 39 - 45

[AbMü 88]   S. Abramowski and H. Müller, *Collision Avoidance for Nonrigid Objects*, in H. Noltemeier (ed.): ZOR - Zeitschrift für Operations Research, Vol. 32, pp 165 - 186

---

[4]Indeed, this can be further improved. Rohnert [Ro 91] uses an additional data structure, the so-called edge tree to speed up the computation of the bottleneck to $O(\log n)$ time.

At next, satisfying condition (2), we demand $d(x(t), P_i(t)) > radius(D)$ to avoid collisions. This is obviously equivalent to :

$$\lambda(t)^2 > \frac{radius(D)^2 - \frac{1}{4} d(P_i(t), P_j(t))^2}{n(t)^T n(t)}$$

On the other hand condition (3a) can be expressed as $x'(t)^T[\bar{x} - x(t)] > 0$ (directional derivative). Therefore we can satisfy this condition at the moment t_0 by a suitable choice of $\lambda'(t_0)$, except for that case where $n(t_0)^T[\bar{x} - x(t_0)] = 0$. But in that special case we select :

$$\lambda(t) := \frac{[\bar{x} - m(t)]^T n(t)}{n(t)^T n(t)} \qquad \text{which minimizes} \qquad \min_{x(t) \in B(P_i(t), P_j(t))} \|x(t) - \bar{x}\|_2^2$$

Altogether there is a choice of $\lambda(t)$ as a linear function (except for the special case above) in such a way that the conditions (1), (2) and (3a) hold in a whole neighborhood of t_0. Additionally, under linear motions of the points[3] these conditions can be checked easily, because they can be expressed as polynomials of degree ≤ 4.

At next, to satisfy the global condition (3b) we adopt a static technique by Rohnert [Ro 91] who uses *maximum bottleneck spanning trees* to decide efficiently if there is a feasible path between two arbitrary points in the plane. Thereby this data structure is defined as follows. Let $Width(e)$ denotes the minimum distance of a Voronoi edge $e \in VD(S)$ to any point of S. Then the maximum bottleneck spanning tree $MBST(VD(S))$ is defined to be a spanning tree of the (extended) Voronoi diagram $VD(S)$ whose edges are weighted by the $Width$ function and which has maximum total width (see figure 4).

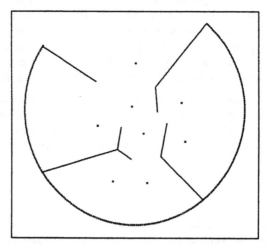

Figure 4: A maximum bottleneck spanning tree $MBST(VD(S))$.

Now, the usefulness of maximum bottleneck spanning trees relies on the following theorem.

---

[3]Realize, that the linear motion of the sites also provides the linearity of the functions $m(t)$ and $n(t)$ and that $n(t) \neq \bar{0}$ using assumption (A).

Generating triangulations of 2-manifolds

Haijo Schipper[*]

July 1991

Abstract

In this paper we will show how for a given triangulation T of a 2-manifold $\mathcal{M}$ a minimal triangulation T_{min} of $\mathcal{M}$ and a sequence S of vertex-splittings can be calculated such that performing S on T_{min} gives T. Our method will work in either $O(n \log n)$ time and linear space or in linear time if $O(n^2)$ space is available, where n is the number of vertices of T.

1 Introduction

A theorem of Steinitz and Rademacher [StRa 34] states that triangulations of the 2-sphere can be generated from a single minimal triangulation, the 2-complex consisting of four vertices, six edges and four triangles, by a sequence of vertex-splittings.
Later analogous results have been found by Barnette [Ba 82] for the projective plane (starting with two minimal triangulations) and by Duke and Grünbaum [DuGr], Rusnak [Ru] and Lavrenchenko [La 87] for the torus (22 minimal triangulations).
Barnette and Edelson [BaEd 88, BaEd 89] proved that all compact 2-manifolds have finitely many minimal triangulations.

In this paper we will show that we can calculate efficiently a minimal triangulation to start with and the sequence of vertex-splittings to generate a given triangulation subject to the constraint that the 2-manifold we are working on has a finite set of minimal triangulations. (In particular all compact 2-manifolds will do.)
The rest of the paper is organized as follows : in section 2 we will give some definitions, in section 3 we will give some lemmata which are needed to prove the complexity bounds, in section 4 we will introduce some data structures, and in section 5 we will give the algorithm.

2 Definitions

Definitions 1 A **2-manifold** or **surface** is a topological space in which every point has a neighbourhood which is topologically equivalent to an open disc.
A 2-manifold is called **triangulable** if it can be obtained from a set of triangles by the identification of edges such that each pair of triangles is identified either at a single vertex, or along a single edge or both triangles are completely disjoint. Also each edge is identified with exactly one other edge and the triangles identified at each vertex can be arranged in a cycle $T_1, T_2, \ldots, T_k, T_1$ such that adjacent triangles are identified along an edge. [He 79]
The 2-complex of the triangles, edges and vertices mentioned above is called the **triangulation** of the 2-manifold.
A **k-cycle** or **k-circuit** in a triangulation consists of k edges $e_1, \ldots, e_k$ such that $e_i \cap e_j$ is a single vertex if and only if $|i - j| = 1$ or $|i - j| = k - 1$ and $e_i \cap e_j = \emptyset$ otherwise.

---

[*]Department of Computer Science, University of Groningen, P. O. Box 800, 9700 AV GRONINGEN, The Netherlands.

A k-cycle is called **planar** *if it bounds a cell (i. e. a subset of the 2-complex which is a topological disc), otherwise it is called* **non-planar**. *A k-cycle is* **facial** *if it bounds a triangle, otherwise it is* **non-facial**.
The **star** *of a vertex v is the union of the simplices (including their boundary components) meeting at v.*
The **link** *of a vertex v consists of the opposite edges of the triangles meeting at v.*
The **dual complex** T^d *of a triangulation T is the 2-complex in which every triangle of T is a vertex in T^d and two vertices of T^d are joined by an edge if and only if the corresponding triangles in T share an edge. Facets of T^d correspond to vertices of T.*

In the latter by a 2-manifold we mean a 2-manifold with finitely many minimal triangulations.

Observations 1

   i) A triangulation does not contain any 1-cycles or 2-cycles.

   ii) If a cycle is facial, then it is planar.

   iii) Every vertex of a triangulation has degree at least three.

   iv) The link of a vertex v can be arranged as a planar cycle and is the boundary of the star of v.

   v) Every vertex of a dual complex has degree 3.

Because there are no 1-cycles nor 2-cycles in a triangulation we can determine an edge uniquely by its endpoints and cycles by the sequence of the endpoints of the consecutive edges of the cycle.

Definitions 2 *Let the link of a vertex v be a cycle C. Let C be the union of two paths A and B whose intersection consists of two vertices x and y. A* **vertex splitting** *at v consists of replacing v by two new vertices v_1 and v_2 joined by an edge, the triangles determined by the star of v by triangles determined by v_1 and edges of A, triangles determined by v_2 and edges of B, and (two) triangles determined by the edge v_1v_2 and the vertices x and y [Ba 82]. See figure I.*
The inverse operation is called **edge shrinking**, *while the dual operations (thus in T^d) are called* **facet splitting** *and the inverse* **facet merging**.

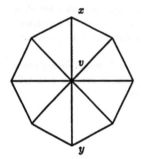 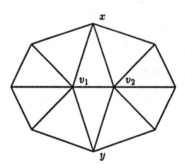

Figure I: Vertex splitting & edge shrinking.

Observation 2 When a vertex splitting is performed the number of vertices increases by one, the number of edges by three, and the number of triangles by two. Hence the Euler characteristic ($V - E + F$) does not change.

While every vertex can be split, not every edge is shrinkable. Suppose an edge e_1 lies on non-facial 3-cycle $e_1e_2e_3$. If we shrink e_1 we get the non-facial 2-cycle e_2e_3. This cycle does not bound a facet so we cannot replace it by a single edge without changing the topological type of the 2-manifold.

Definitions 3 *An edge which produces a triangulation after shrinking is a* **shrinkable edge**, *otherwise the edge is* **non-shrinkable**. *A vertex which is merely incident upon non-shrinkable edges is a* **non-shrinkable vertex**, *otherwise it is a* **shrinkable vertex**.

Note that if we shrink a non-shrinkable edge we still have a 2-complex consisting of triangles, edges and vertices. However the 2-complex will contain an edge which is identified with three other edges. (Thus four triangles meet at one edge.) According to definitions 1 the 2-complex is not a triangulation.

Definition 4 *A triangulation T of a 2-manifold is a* minimal triangulation *if and only if every edge of T is non-shrinkable.*

Of course, a triangulation is minimal if and only if every vertex is non-shrinkable.

3 Some useful lemmata

Important is that the topological type of the surface does not change while shrinking edges or splitting vertices.

Lemma 1 *If T is a triangulation of a 2-manifold M and T' is created from T by splitting a vertex v of T or shrinking an edge e of T then T' is also a triangulation of M.*

Proof It is obvious that T' is a triangulation. What remains is to prove that T' is a triangulation of M.
First consider the case of splitting a vertex v. Let $st\,(\,v)$ be the star of v, and $L\,(\,v)$ the link of v. Then $st\,(\,v)$ is topologically equivalent to a disc. A splitting of v does not make changes outside of $st\,(\,v)$ and also $L\,(\,v)$ remains the same. The changed star remains a topological disc with boundary $L\,(\,v)$, therefore T' is a triangulation of M.
Since edge shrinking is the inverse of vertex splitting both T and T' are triangulations of M.
q.e.d.

Lemma 2 *An edge e of a triangulation T is non-shrinkable if and only if e lies on at least three 3-cycles or T is the triangulation of the sphere with four vertices, six edges and four triangles (T_4).*

Proof Obviously both cases cannot occur simultaneously. So suppose T is T_4. If we reduce an edge e of T then the number of vertices decreases by one, so we will have only three vertices left. But we cannot make a triangulation of the sphere with that few vertices. (See for instance Henle [He 79, §27].) Therefore such a shrinking is not the inverse of a vertex splitting because the manifold does not change upon a vertex splitting nor upon an edge shrinking (lemma 1). So if T is T_4 then all edges are non-shrinkable.
If an edge e is non-shrinkable and T is not T_4 then it lies on at least one non-facial 3-cycle. But every edge lies on exactly two facial 3-cycles, therefore every non-shrinkable edge of a triangulation not equal to the T_4 lies on at least three 3-cycles. If an edge lies on at least three 3-cycles then at least one 3-cycle it lies on is non-facial, since an edge lies on exactly two facial 3-cycles. But then the edge is non-shrinkable, which proves the claim.
q.e.d.

Lemma 3 *If a vertex v in a triangulation T other than T_4 is non-shrinkable, then every non-facial 3-cycle vw_1w_2 (with w_1 and w_2 vertices of T) is non-planar.*

Proof Suppose there is a non-shrinkable vertex v in a triangulation $T \neq T_4$ such that there is at least one planar non-facial 3-cycle which contains v. Let the neighbours of v in counterclockwise order be $w_1, \ldots, w_k$. Define the set of all planar non-facial 3-cycles containing v, $C = \{c_{i,j} = vw_iw_j\}$. Note that if $c_{i,j} = vw_iw_j \in C$ then $c_{j,i} = vw_jw_i \in C$. Because of our assumption C is not empty. Define the size of a cycle $c_{i,j}$ as follows : if $j > i$ then the size is $j - i$ else the size is $k + j - i$. Note that the size is at least 2 and less than $k - 1$. Let c be the (or one of the) 3-cycle(s) of C with smallest size. W. l. o. g. assume $c = vw_1w_r$. Consider all cycles of the form $c_s = vw_2w_s$. Since c is planar $s \leq r$. But then for every c_s we have either that c_s is facial or the size of c_s is smaller than the size of c. Therefore either vw_2 is shrinkable or c is not (one of) the smallest 3-cycle(s) of C. This contradiction proves the lemma.
q.e.d.

A similar lemma as the lemma above was proven by Barnette and Edelson [BaEd 88].

Corollary 4 *If T is a triangulation of the sphere other than T_4 then every vertex of T is shrinkable.*

Proof The Jordan curve theorem implies that every cycle on the sphere is planar. Therefore a vertex of T cannot lie on a non-planar 3-cycle, which is necessary according to lemma 3 for being a non-shrinkable vertex in a triangulation other than T_4.   q.e.d.

Lemma 5 *If v is a non-shrinkable vertex of a triangulation T, other than T_4, then v has degree at least four.*

Proof Suppose T is not T_4 and let the degree of v be three. (Since T is a triangulation v has degree at least three.) Let the neighbours of v be w_1, w_2 and w_3. If v would be non-shrinkable, the edge vw_1 is non-shrinkable, so it has to lie on at least one non-facial 3-cycle. But it lies only on the 3-cycles vw_1w_2 and vw_1w_3, both boundaries of triangles, thus facial. This is a contradiction. Hence either the degree of v is at least four or T is T_4.   q.e.d.

Lemma 6 *If a vertex v is non-shrinkable and if v has degree four, then all the edges of the link of v are non-shrinkable.*

Proof Let the neighbours of v (in counterclockwise order) be w_1, w_2, w_3 and w_4. Suppose the lemma above is not true so there is an edge of the link of v which is shrinkable. W. l. o. g. assume that this is edge w_1w_2. But because v is non-shrinkable there is an edge w_1w_3 such that the cycle vw_1w_3 is non-planar. But then the 3-cycle $w_1w_2w_3$ is also non-planar. Contradiction since w_1w_2 was assumed to be shrinkable.   q.e.d.

Lemma 7 *If a vertex v of a triangulation T is non-shrinkable and an edge shrinking is performed on T then afterwards v still is non-shrinkable.*

Proof Let e be an edge of T. There are three possibilities.

  i) e is incident with v.

  ii) e lies on the link of v.

  iii) e lies outside of the star of v.

In case i) e lies on a non-planar 3-cycle vxy, for some x and y in T, so e cannot be shrunken.
In case ii) e is non-shrinkable (then we are ready) or v has degree at least five. (Lemmata 5 and 6). Let the neighbours of v be (in counterclockwise order) $w_1, \ldots, w_k$, ($k \geq 5$). W. l. o. g. assume $e = w_1w_2$ is shrinkable and shrunken. Note that neither the edge w_1w_3 nor the edge w_2w_k is present (otherwise w_1w_2 would be non-shrinkable or v would be shrinkable). Also note that because v is non-shrinkable for every i, $1 \leq i \leq k$, there is a j, $1 \leq j \leq k$, such that $1 < |i-j| < k-1$ and the 3-cycle vw_iw_j is non-planar. Let $w_{1,2}$ be the new vertex which is induced by shrinking w_1w_2. If $3 \leq i,j \leq k$ then the 3-cycles vw_iw_j have not changed and are still non-planar, thus the edges vw_i, $3 \leq w_i \leq k$ are still non-shrinkable. Otherwise assume $i = 1$. Then $4 \leq j \leq k-1$. But then the 3-cycle $vw_{1,2}w_j$ is present and non-planar, thus $vw_{1,2}$ is non-shrinkable. Therefore v stays non-shrinkable.
In case iii) either e lies on a 3-cycle vxy or it does not. In the first case vxy is non-facial so e is non-shrinkable, in the latter case shrinking e cannot change a 3-cycle of the form vxy so v stays non-shrinkable.   q.e.d.

The following lemma is well-known.

Lemma 8 *If T is a triangulation of the sphere then T has a vertex v with degree at most five.*

Proof Eulers relation tells $V - E + F = 2$. Since T is a triangulation we have $3F = 2E$. Thus

$$V - E + F = 2 \Rightarrow 3V - 3E + 3F = 6 \Rightarrow$$

$$6 = 3V - 3E + 2E = 3V - E \Rightarrow E = 3V - 6 \Rightarrow \frac{E}{V} < 3$$

Because every edge is incident with two vertices the average degree of the vertices is less than six. Therefore at least one vertex has degree at most five.   q.e.d.

A more general result is the following :

Lemma 9 *For every compact 2-manifold $\mathcal{M}$ there is a constant $c(\mathcal{M})$, depending solely on $\mathcal{M}$, such that every triangulation T of $\mathcal{M}$ with $n > c(\mathcal{M})$ vertices has a shrinkable vertex with degree at most 6.*

Proof Let T be a triangulation of $\mathcal{M}$. Let the number of vertices, edges and facets of T be V, E and F. Then $V - E + F = \chi(\mathcal{M}) \leq 2$. $\chi(\mathcal{M})$ depends only on $\mathcal{M}$. Because T is a triangulation we have $3F = 2E$.

Barnette and Edelson [BaEd 89] proved that every compact 2-manifold $\mathcal{M}$ has finitely many minimal triangulations. Therefore every minimal triangulation $T_{\min}$ of $\mathcal{M}$ has at most $k(\mathcal{M})$ vertices. According to lemma 7 every triangulation T of $\mathcal{M}$ has at most $k(\mathcal{M})$ non-shrinkable vertices.

Let $c'(\mathcal{M}) = k(\mathcal{M}) - \chi(\mathcal{M})$. Note that $c'(\mathcal{M})$ is a positive constant depending only on $\mathcal{M}$. Let $V = V_s + V_{ns}$, where V_s is the number of shrinkable vertices of T and V_{ns} is the number of non-shrinkable vertices of T. Thus $V_{ns} \leq k(\mathcal{M})$ and $V - k(\mathcal{M}) \leq V_s \leq V$. Then :

$$V - E + F = \chi(\mathcal{M}), \quad 3F = 2E \Rightarrow 3V - 3E + 3F = 3V - E = 3\chi(\mathcal{M}) \Rightarrow$$

$$E = 3V - 3\chi(\mathcal{M}) = 3V_s + 3V_{ns} - 3\chi(\mathcal{M}) = 3V_s + 3(V_{ns} - \chi(\mathcal{M})) \leq 3V_s + 3c'(\mathcal{M}) \Rightarrow$$

$$\frac{E}{V_s} \leq 3 + 3\frac{c'(\mathcal{M})}{V_s} \leq 3 + 3\frac{c'(\mathcal{M})}{V - k(\mathcal{M})}$$

Let $c(\mathcal{M}) = 6c'(\mathcal{M}) + k(\mathcal{M}) = 7k(\mathcal{M}) - 6\chi(\mathcal{M})$ and let $V > c(\mathcal{M})$, then

$$\frac{E}{V_s} \leq 3\frac{1}{2}$$

(Note that indeed $c(\mathcal{M})$ depends solely on $\mathcal{M}$.) Thus twice the number of edges in the triangulation is less than seven times the number of non-shrinkable edges in the triangulation. Therefore there is a shrinkable vertex with degree at most 6.   q.e.d.

4 Preliminaries

4.1 The incidence graph

Edelsbrunner [Ed 87, Section 8.5.2] introduces the incidence graph of a polytope. In this graph all vertices, edges and facets of a polytope and their incidence relations are represented.

Let $\mathcal{P}$ be the polytope which is represented. Then every edge e, between the vertices v and w, of $\mathcal{P}$ is represented in the incidence graph $\mathcal{I}(\mathcal{P})$ by two directed edges, one with source v and destination w, the other with source w and destination v.

Let $\vec{e}$ be the directed edge (in $\mathcal{I}(\mathcal{P})$) between source v and destination w. Then we have in $\mathcal{I}(\mathcal{P})$ a pointer (i. e. direct access) to :

- $v = orig(\vec{e})$,
- $f = left(\vec{e})$, the facet left of $\vec{e}$ viewed from the outside of $\mathcal{P}$,
- $\vec{e}_s = sym(\vec{e})$, the symmetric copy of $\vec{e}$, from source w and destination v,
- $\vec{e}_o = onext(\vec{e})$, such that $\vec{e}$ and $\vec{e}_o$ have the same origin and $left(sym(\vec{e}_o)) = left(\vec{e})$, and
- $\vec{e}_l = lnext(\vec{e})$ such that $\vec{e}$ and $\vec{e}_l$ have the same facet on their left, and the destination of $\vec{e}$ is the source of $\vec{e}_l$.

Finally we need for a given vertex v direct access to $edge(v)$, an arbitrary edge with source v, and for a given facet f direct access to $edge(f)$, an arbitrary edge with f on its left.

Note that not all the incidence relations are represented by a pointer, but all incidence relations can be found in $\mathcal{I}(\mathcal{P})$.

The following operations can be done in constant time :

- add a facet f (where all directed edges $\vec{e}$ with $left(\vec{e}) = f$ are already in the incidence graph and we have a pointer to one of them),

- add a directed edge $\vec{e}$ (if the edges which become $onext(\vec{e})$ and $lnext(\vec{e})$ are known),

- remove a facet, and

- remove a directed edge.

Observations 3

- We can visit all outgoing edges of a vertex in (counter)clockwise order in constant time per edge visited.

- We can visit all edges and vertices of a certain facet in (counter)clockwise order in constant time per edge and vertex. If $\mathcal{P}$ is a triangulation we can travel around a triangle in constant time.

- The structure uses linear storage.

4.2 The bitvector

The bitvector is defined by Mehlhorn [Me 84, Section III.8.1] as follows :

Let $U = \{1,\ldots,N\}$. Then a set $S \subseteq U$ can be represented by a boolean array $B[1,\ldots,N]$ (bitvector). We have $B[i] = \text{true}$ if and only if $i \in S$. Otherwise $B[i] = \text{false}$. Then the operations insert, delete and member query take constant time. The drawback is that initialization takes time $\Omega(N)$.

However, we can change the structure such that we can initialize the data structure in constant time without a drawback on the time needed for the operations insert, delete and member query.

In stead of the array B we use an array $A[1,\ldots,N]$ and a stack, realized by an array $C[1,\ldots,N]$ and a variable TOP. Initially $TOP = 0$. We will maintain the following invariant :

$$i \in S \text{ if and only if } 1 \leq A[i] \leq TOP \wedge C[A[i]] = i$$

Initially $TOP = 0$ and hence $S = \emptyset$. The test $i \in S$ can be done in constant time using the invariant. We can insert an element in S using the algorithm insert and we can delete an element from S using the algorithm delete.

Algorithm Insert
Suppose we want to insert i in S (i not already present in S).
$TOP := TOP + 1$
$A[i] := TOP$
$C[TOP] := i$
End algorithm

Algorithm Delete
Suppose we want to delete i from S (i present in S).
$A[C[TOP]] := A[i]$
$C[i] := C[TOP]$
$TOP := TOP - 1$
End algorithm

It is easy to see that these algorithms maintain the invariant. Thus initialization, insertion, deletion and member query can all be performed in constant time. The structure uses obviously $O(N)$ storage.

Lemma 10 *If a bitvector is used to represent a set $S \subseteq \{1,\ldots,N\}$ then the minimum k can be found in $O(k)$ time.*

Proof First test if $1 \in S$. If so then we are ready, else test if $2 \in S$, et cetera. Each test takes constant time.   **q.e.d.**

5 The algorithm

5.1 Overview

The general idea of the algorithm is as follows. Since the inverse of vertex splitting is edge shrinking we start with the given triangulation T and shrink edges until we reach a minimal triangulation i.e. we cannot find a shrinkable edge.

Because a vertex stays non-shrinkable once it becomes non-shrinkable we keep track of those vertices of which we do not know yet whether they are shrinkable or not. At each iteration we take a vertex and decide whether it is shrinkable. If so we shrink a shrinkable edge incident upon it such that the current vertex disappears. Otherwise it is non-shrinkable and will be reported as a vertex of the resulting minimal triangulation.

Finally we have to report the resulting triangulation and the inverses of the edge shrinkings in reverse order.

We assume that T is given as an incidence graph $I(T)$ and a list of all the vertices.

5.2 Preprocessing

To perform the algorithm efficiently we need a data structure to perform queries like "is v a neighbour of w" rather fast. Therefore for each vertex v we make a data structure N_v (neighbourhood structure) in which we store its neighbours. Then we only have to perform member queries. Because the neighbours of a certain vertex can change during the algorithm, the neighbourhood structures have to be dynamic. Several data structures will do.

Secondly we need a global structure D (dictionary) which stores the vertices we have not treated yet ordered by their degree. The dictionary is used to get the not treated vertex with lowest degree efficiently. Since during the algorithms vertices disappear or their degree is changed, D also has to allow for insertions and deletions. Again a number of data structures is available.

For the dictionary we will use the following notation for query time (that is to find the minimum of D), insertion time, deletion time and the time needed to initialize the structure : $Q_D(n), I_D(n), D_D(n)$ and $P_D(n)$.
For the neighbourhood structures the corresponding quantities are detoned by : $Q_N(n), I_N(n), D_N(n)$ and $P_N(n)$.
We will use data structures such that all times mentioned above are $O(n)$.

We will use the following preprocessing :

Algorithm Preprocess
initialize D
for all vertices $v \in T$ do
    $degree := 0$
    initialize N_v
    for all edges $vw \in T$ do
        $degree := degree + 1$
        insert w in N_v
    od
    insert v in D using $degree$
od
End algorithm

Lemma 11 *The preprocessing algorithm needs time $O(n(I_N(n)+I_D(n)+P_N(n))+P_D(n))$, where n is the number of vertices in the triangulation.*

Proof We need time $P_D(n)$ to initialize the dictionary. We have to initialize n neighbourhood structures, each taking time $P_N(n)$. Of course we only need a linear number of insertions in the dictionary, each taking $O(I_D(n))$ time. The number of insertions in neighbourhood structures is twice the number of edges in the triangulation which is, for a fixed manifold, linear in the number of vertices. Each insertion takes $O(I_N(n))$, which proves the lemma.   q.e.d.

5.3 Testing an edge

We already know that an edge $e = vw$ of a triangulation T is non-shrinkable if and only if $T = T_4$ or if e lies on at least three 3-cycles (lemma 2). Note that for every 3-cycle e lies on, v and w must have a common neighbour. So if we count the number of vertices which are both in $\mathcal{N}_v$ and in $\mathcal{N}_w$ we can decide whether e is shrinkable or not.

Lemma 12 *The test to decide whether an edge vw is shrinkable or not takes $O(dQ_N(n))$ time, where d is the degree of v.*

Proof The test $T = T_4$ can be done in constant time. So assume $T \neq T_4$. Then for each neighbour of v we need to check if it is a neighbour of w. The number of neighbours of v is d. Each check needs $O(Q_N(n))$ time, so totally we need $O(dQ_N(n))$ time.   q.e.d.

5.4 Testing a vertex

If we test a vertex v we want to decide whether v is non-shrinkable or find a shrinkable edge e incident with v.

Lemma 13 *The test if a vertex v is non-shrinkable or finding a shrinkable edge e incident with v can be done in $O(d^2Q_N(n))$ time, where d is the degree of v.*

Proof The algorithm is straightforward. For every edge incident with v we test if it is shrinkable or not. If we find such an edge then we report it, else we report that v is non-shrinkable. Since the degree of v is d we have to test at most d edges. Each test can be done in $O(dQ_N(n))$ time (lemma 12) which proves our claim.   q.e.d.

5.5 Shrinking an edge

If an edge is shrinkable and it is shrunken then some updates in T have to be made. Since T is given as a incidence graph $\mathcal{I}(T)$ we have to update $\mathcal{I}(T)$. Also we have to update some neighbourhood structures and the dictionary. In fact all what we have to do is merging two vertices, deleting three edges and two triangles, and replacing some incidence relations.
We will use the following algorithm to shrink an edge. See also figure II.

Algorithm Shrinking an edge
Let $e = vw_1$ be the edge between the vertices v and w_1 which is shrunken. Let the neighbours of v be in counterclockwise order $w_1, \ldots, w_d$. Thus the degree of v is d.

delete w_1, w_2, w_d and v from $\mathcal{D}$ if they are present
subtract 1 from the degrees of w_1, w_2 and w_d
delete v from $\mathcal{N}_{w_1}, \mathcal{N}_{w_2}$ and $\mathcal{N}_{w_d}$
delete the edges $vw_1, w_1v, vw_2, w_2v, vw_d$ and w_dv from $\mathcal{I}(T)$
delete the triangles vw_1w_2 and vw_dw_1 from $\mathcal{I}(T)$
for $i := 3$ to $d - 1$ do
    delete v from $\mathcal{N}_{w_i}$
    insert w_i in $\mathcal{N}_{w_1}$
    insert w_1 in $\mathcal{N}_{w_i}$
    replace the edges vw_i and w_iv by the edges w_1w_i and w_iw_1 in $\mathcal{I}(T)$
    add 1 to the degree of w_1
    replace the triangle $w_iw_{i-1}v$ by the triangle $w_iw_{i-1}w_1$ in $\mathcal{I}(T)$

od
replace the triangle $w_d w_{d-1} v$ by the triangle $w_d w_{d-1} w_1$ in $\mathcal{I}(\mathcal{T})$
insert w_1, w_2 and w_d in $\mathcal{D}$ using their new degrees, if they were present in $\mathcal{D}$
End algorithm

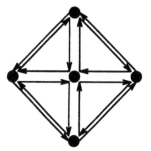 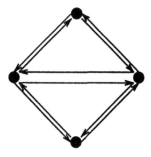

Figure II: Shrinking an edge.

Lemma 14 *Shrinking an edge $e = vw_1$ of a triangulation $\mathcal{T}$ with n vertices can be performed in time $O(D_{\mathcal{D}}(n) + I_{\mathcal{D}}(n) + d(D_{\mathcal{N}}(n) + I_{\mathcal{N}}(n)))$, where d is the degree of v.*

Proof We have to delete at most four vertices from $\mathcal{D}$, and we have to insert at most three vertices in $\mathcal{D}$.
Each deletion of v from a neighbourhood structure can be charged to an edge incident with v, thus d times.
An insertion of w_i in $\mathcal{N}_{w_1}$ and an insertion of w_1 in $\mathcal{N}_{w_i}$ can be charged to the edge vw_i. Then every edge vw_i is charged twice, for $3 \leq i \leq d-1$. Thus $d-3$ times. Therefore there are at most $2d - 6$ insertions in a neighbourhood structure.
Note that the pointers in the $\mathcal{I}(\mathcal{T})$ structure give us enough freedom to perform the updates in this structure in $O(d)$ time, because we walk one time around the vertex v. (See section 4.1.)
q.e.d.

5.6 Reduction algorithm

Now we can give the reduction algorithm. Recall that in the dictionary $\mathcal{D}$ the not treated vertices are stored. Since each vertex has to be treated once – because either it is non-shrinkable or it is shrunken – we can stop if $\mathcal{D}$ is empty.

Algorithm
preprocess $\mathcal{T}$
while $\mathcal{D}$ is not empty **do**
    let v be the (a) vertex of $\mathcal{D}$ with minimal degree.
    if there is a shrinkable edge e incident with v
    then shrink e
    else delete v from $\mathcal{D}$
od
report the resulting minimal triangulation.
End algorithm

Lemma 15 *Reducing a triangulation $\mathcal{T}$ of a 2-manifold $\mathcal{M}$ with n vertices to a minimal triangulation of $\mathcal{M}$ by performing consecutive edge shrinkings can be done in time $O(n(Q_{\mathcal{D}}(n) + Q_{\mathcal{N}}(n) + I_{\mathcal{D}}(n) + I_{\mathcal{N}}(n) + D_{\mathcal{D}}(n) + D_{\mathcal{N}}(n) + P_{\mathcal{N}}(n)) + P_{\mathcal{D}}(n) + C(\mathcal{M}))$, where $C(\mathcal{M})$ depends only on $\mathcal{M}$.*

Proof The preprocessing can be done in time $O(n(I_{\mathcal{D}}(n) + I_{\mathcal{N}}(n) + P_{\mathcal{N}}(n)) + P_{\mathcal{D}}(n))$ (lemma 11).

Let $c(M)$ be as in lemma 9. If $n > c(M)$ then v has degree at most 6. According to lemmata 13 and 14 one iteration takes time $O(Q_D(n) + Q_N(n) + I_D(n) + I_N(n) + D_D(n) + D_N(n))$. This happens $n - c(M) = O(n)$ times. If $n \leq c(M)$ then the degree of v is at most $c(M)$ and we can test v in time $O((c(M))^2 Q_N(c(M))) = O((c(M))^3)$ (lemma 13). Shrinking an edge incident with v takes time $O(D_D(c(M)) + I_D(c(M)) + c(M)(D_N(c(M)) + I_N(c(M)))) = O((c(M))^2)$. Deleting v from $\mathcal{D}$ can be done in $O(c(M))$ time. Since we have to make $O(c(M))$ iterations this will take totally $O((c(M))^4)$ time. $(c(M))^4$ is a constant depending solely on M. If we choose $C(M) = (c(M))^4$ we have proven our claim.   q.e.d.

Corollary 16 *Reducing a triangulation T of a 2-manifold M with n vertices to a minimal triangulation of M by performing consecutive edge shrinkings can be done in time $O(n(Q_D(n) + Q_N(n) + I_D(n) + I_N(n) + D_D(n) + D_N(n) + P_N(n)) + P_D(n))$.*

Proof Because M is fixed, $C(M)$ is a constant.   q.e.d.

5.7 Complexity bounds

Now are we able to prove the following theorems :

Theorem 1 *Let T be a triangulation of a 2-manifold M with n vertices. Then we can compute in $O(n \log n)$ time, and with the use of $O(n)$ space, a minimal triangulation $T_{\min}$ of M and a sequence of vertex splittings S such that performing S on $T_{\min}$ gives T.*

Proof We will use the algorithms as presented above. If we use for both the dictionary and neighbourhood structures balanced trees, e. g. AVL-trees, then we can initialize them in constant time, and inserting in, deleting from and querying in the structures all take $O(\log n)$ time. Using corollary 16 we get the desired time bounds.

For the space bound note that every vertex is stored at most once in the dictionary, so this needs linear storage. For every edge in the triangulation two vertices are stored in some neighbourhood structure. Therefore all neighbourhood structures together take linear storage. This completes our proof.   q.e.d.

Theorem 2 *Let T be a triangulation of a 2-manifold M with n vertices. Then we can compute in $O(n)$ time a minimal triangulation $T_{\min}$ of M and a sequence of vertex splittings S such that performing S on $T_{\min}$ gives T if we have $O(n^2)$ space available.*

Proof We will use the algorithms as presented above. We label all vertices with numbers from 1 to n, so we have a universe N of size n. Also the degrees of the vertices all lie in N. Thus we can use for both the dictionary and neighbourhood structures bitvectors (see section 4.2). We have to use chaining for the dictionary, because there can be vertices with same degree (details omitted). Then initialization of, insertion in, deletion from and querying in the structures can be done in constant time. Using corollary 16 we get the desired time bounds.

For the space bounds note that we have $O(n)$ structures each "claiming" $O(n)$ space, so totally the structure needs $O(n^2)$ space. This completes the proof.   q.e.d.

6 Conclusions and further remarks

In this paper we have shown how one can calculate a minimal triangulation and a sequence of vertex splittings for generating a given triangulation. The time needed for the algorithm is up to a constant term independent of the 2-manifold of the triangulation.

However this term depends on the Euler characteristic of the manifold and on the size of the largest minimal triangulation of the 2-manifold.

Tight lower bounds on minimal triangulations are known. Ringel [Ri 55] and Jungerman and Ringel [JuRi 80] proved the following results for compact surfaces :

Theorem 3 [Ri 55, JuRi 80] *Let M be a compact 2-manifold with Euler characteristic χ. Then there is a triangulation T of M with δ_χ triangles, with*

$$\delta_\chi = \begin{cases} 16, & \chi = 0, \ M \text{ is non-orientable} \\ 20, & \chi = -1, \ M \text{ is non-orientable} \\ 24, & \chi = -2, \ M \text{ is orientable} \\ 2\left\lceil \frac{7+\sqrt{49-24\chi}}{2} \right\rceil - 2\chi, & \text{otherwise} \end{cases}$$

and every other triangulation of M does not break this bound.

Hence the minimum number of vertices v_χ of a triangulation of a manifold M with Euler characteristic χ is :

$$v_\chi = \begin{cases} 8, & \chi = 0, \ M \text{ is non-orientable} \\ 9, & \chi = -1, \ M \text{ is non-orientable} \\ 10, & \chi = -2, \ M \text{ is orientable} \\ \left\lceil \frac{7+\sqrt{49-24\chi}}{2} \right\rceil, & \text{otherwise} \end{cases}$$

If we define the genus g of a 2-manifold M with Euler characteristic χ as :

$$g = \begin{cases} -\chi + 2, & M \text{ is non-orientable} \\ -\frac{\chi}{2} + 1, & M \text{ is orientable} \end{cases}$$

then we see that the number of vertices in the smallest triangulation of a 2-manifold is in $\Theta\left(\sqrt{g}\right)$.

On the other hand little is known about upper bounds. Barnette and Edelson [BaEd 88] proved the following :

Theorem 4 [BaEd 88] *The orientable 2-manifolds of any genus have finitely many minimal triangulations.*

Sketch of the proof Let T be a minimal triangulation of a manifold M, with genus $g > 0$. Let C be a non-planar 3-cycle in T. Cut M along C. Then we get one (two) manifold(s), M_1, (and M_2). We have two boundary components. Each bounding cycle is filled with a cell. The induced manifold(s) has (have) genus less then g. Let T_1 be the triangulation of M_1. Barnette and Edelson prove that T_1 can be reduced to a minimal triangulation of M_1 with at most $1458g - 648$ edge shrinkings. If M_2 exists this case is similar. Because there is one minimal triangulation of the manifold with genus 0 it follows by induction that the number of minimal triangulations of M_1 and M_2 is bounded. Since we can identify a triangle of T_1 with another triangle of T_1 (T_2) (i. e. gluing to get M back) in finitely many ways, the theorem follows. $\qquad \square$

Corollary 17 *All minimal triangulations of an orientable 2-manifold M with genus g have $O(g2^g)$ vertices.*

Proof Let T be the largest minimal triangulation of M. Let $g > 0$ (otherwise it is obvious). Let C be a non-planar 3-cycle of M. Cut T along C. In the worst case, we get two manifolds M_1 and M_2 with genus $g - 1$. Let T_i be the triangulation of M_i where the boundary is filled with a triangle. By the argument above T_i can be obtained from a minimal triangulation of M_i by performing at most $1458g - 648$ vertex splittings. Let $v(g)$ be the number of vertices of the largest minimal triangulation of an orientable 2-manifold with genus g. Then we have

$$v(g) \leq \begin{cases} 4, & g = 0 \\ 2v(g-1) + 2916g - 1299, & g > 0 \end{cases}$$

It can easily be seen that $v(g) = O(g2^g)$. $\qquad\qquad$ q.e.d.

This makes the constant of lemma 15 $C(M) = O(g16^g)$, if M is a orientable surface with genus g.

Barnette and Edelson [BaEd 89] proved that also non-orientable 2-manifolds have finitely many minimal triangulations by similar arguments.

A number of open problems remain. First of all there is a large gap between the smallest minimal triangulation of a manifold and the proven upper bound for the largest minimal triangulation. For instance the relation in the proof of corollary 17 gives an upper bound of 1625 vertices for a minimal triangulation of the torus. In fact the largest minimal triangulation of the torus counts only 10 vertices ([La 87]). Barnette and Edelson only gave a rough estimate and it is not clear how close these bounds are.

Another problem is the number of minimal triangulations for a certain manifold. Known is that the sphere has one minimal triangulation, the projective plane two, and the torus 22. The number of minimal triangulations of other surfaces or general formulae are unknown.

An even more difficult problem is the problem of finding all minimal triangulations.

Acknowledgement

The author would like to thank Paul Lucassen and Gert Vegter for reading previous drafts of this paper.

References

[Ba 82]   Barnette, D. W. : "Generating the triangulations of the projective plane." *Journal of Combinatorial Theory, Series B, Vol. 33*, 1982, pp. 222 - 230.

[BaEd 88]   Barnette, D. W. and Edelson, A. L. : "All orientable 2-manifolds have finitely many minimal triangulations." *Israel Journal of Mathematics, Vol. 62, Nr. 1*, 1988, pp. 90 - 98.

[BaEd 89]   Barnette, D. W. and Edelson, A. L. : "All 2-manifolds have finitely many minimal triangulations." *Israel Journal of Mathematics, Vol. 67, Nr. 1*, 1989, pp. 123 - 128.

[DuGr]   Duke, R. A. and Grünbaum, B. : *Private communication.*

[Ed 87]   Edelsbrunner, H. : *Algorithms in combinatorial geometry.* Berlin, Heidelberg, New York, London, Paris, Tokyo : Springer-Verlag, 1987.

[JuRi 80]   Jungerman, M. and Ringel, G. : "Minimal triangulations on orientable surfaces." *Acta Mathematica, Vol. 145*, 1980. pp. 121 - 154.

[He 79]   Henle, M. : *A combinatorial introduction to topology.* San Francisco : W. H. Freeman and Company, 1979.

[La 87]   Lavrenchenko, S. A. : "Неприводимые триангыляции тора". ("The irreducible triangulations of the torus".) Украинскии Геометрическии Сборник, *Vol. 30*, 1987, pp. 52 - 62.

[Me 84]   Mehlhorn, K. : "Data structures and algorithms 1 : sorting and searching." in Brauer, W., Rozenberg, G. and Salomaa, A. (Ed) : *EATCS monographs on theoretical computer science.* Berlin, Heidelberg, New York, Tokyo : Springer-Verlag, 1984.

[Ri 55]   Ringel, G. : "Wie man die geschlossenen nichtorientierbaren Flächen in möglichst wenig Dreiecke zerlegen kann." *Mathematische Annalen, Vol. 130, Nr. 1*, 1955, pp. 317 - 326.

[Ru]   Rusnak, K. : *Private communication.*

[StRa 34]   Steinitz, E. and Rademacher, H. : *Vorlesungen über die Theorie der Polyeder.* Berlin : Springer, 1934.

The TR*-tree: A new Representation of Polygonal Objects Supporting Spatial Queries and Operations

Ralf Schneider and Hans-Peter Kriegel

Institut für Informatik, Universität München, Leopoldstr. 11B, D-8000 München 40, Germany

Abstract

In application areas such as graphics and image processing, computer aided design (CAD) as well as geography and cartography complex and time consuming spatial queries and operations have to be performed on polygonal objects. In the area of computational geometry different specialized data structures and techniques, such as plane sweep or divide and conquer, are used to design efficient algorithms for the different queries and operations. In this paper, we propose a new representation of polygonal objects, called the TR*-tree that efficiently supports various types of spatial queries and operations. The TR*-tree is a dynamic data structure that represents single objects as well as scenes of objects. The TR*-tree representations of polygonal objects are persistently stored on secondary storage and they are completely loaded into main memory for efficient query processing. In an experimental performance analysis, we demonstrate the fruitfulness of using TR*-trees investigating the 'Point in Polygon'-query and the 'Intersection of two Polygons'-operation.

1 Introduction

In application areas such as graphics and image processing, computer aided design (CAD) as well as geography and cartography complex spatial objects have to be managed. An important class of such objects are two dimensional polygonal objects. The management of polygonal objects imposes stringent new requirements on spatial database systems. One of the most challenging requirements is efficient processing of spatial queries and operations, e.g., 'Point in Object'-test or 'Intersection of Objects'.

In this paper, we present a new representation of polygonal objects, called the TR*-tree that efficiently supports various types of spatial queries and operations. The TR*-tree is designed to represent single polygonal objects as well as sets of objects, called scenes. These representations are persistently stored on secondary storage. In order to perform the query processing they are completely loaded into main memory. Additionally, a TR*-tree is completely dynamic, updates, insertions and deletions within a scene can be intermixed with queries and no periodic global reorganization is required.

The basic idea of our approach is to handle complex polygonal objects more efficiently by decomposing them into a set of simple components. On these simple components the geometric queries and operations mentioned above can be solved relatively simple. The success of such processing depends on the ability to narrow down quickly the set of components that are affected by the queries and operations. In order to design an efficient 'Point in Object'-test the line of argumentation in the area of computational geometry is as follows (see [PS 88]): The fastest known search methods are based on bisection, or binary search. We easily realize, after a little reflection, that from the viewpoint of query time the ability to use binary search is more important than the minimization of the size of the set to be searched, due to the logarithmic dependence of search time upon the latter. Therefore, the fundamental idea is for the theoretical analysis to create new geometric objects in a preprocessing step to permit binary searching.

In our approach, we want to efficiently support the average case of various types of queries and operations in real applications where only one single representation of the objects is used. Therefore, we decompose in a preprocessing step the polygonal objects into a minimum set of disjoint trapezoids using the plane sweep algorithm, proposed by Asano & Asano [AA 83]. We cannot define a complete spatial order on the set of trapezoids that are generated by this decomposition process. Thus binary search on these trapezoids is not possible. Therefore, we propose to use spatial access methods for the spatial search. The R*-tree, an optimized variant of the well known R-tree [Gut 84], exhibits efficient query processing on spatial objects (see [BKSS 90]). Due to its tree structure the R*-tree also permits logarithmic searching in the average case but due to the overlap within its directory the search is not restricted to one path and thus logarithmic search time cannot be guaranteed in the worst case. The R*-tree was designed as a spatial access method for secondary storage. In order to speed up the queries and operations mentioned above we developed the TR*-tree which is a variant of the R*-tree minimizing the main memory operations and explicitly storing the trapezoids of the decomposed objects. We want to emphasize that in a TR*-tree the trapezoids of the decomposed objects are explicitly stored and not approximated by minimum bounding boxes.

The performance of the TR*-tree cannot be analytically proven because the TR*-tree is a data structure that uses heuristic optimization strategies. Therefore, we want to perform an experimental performance analysis investigating synthetically generated data as well as real cartography data in a systematic framework.

In the next chapter we define the spatial queries and operations that we want to support. Chapter 3 explains the decomposition approach of polygonal objects. In chapter 4 the structure as well as the performance of the TR*-tree is presented. Finally, in chapter 5 we describe how the spatial queries and operations defined in chapter 2 are algorithmically performed using the TR*-tree representation of polygonal objects. An experimental performance analysis that indicates the efficiency and performance of our approach is presented in section 6. The paper concludes with a summary that points out the main contributions of our proposal and gives an outlook to future activities.

2 Which set of spatial queries and operations should be supported?

First, we have to define on which types of objects the operations and queries should be performed. Our base-objects are *simple polygons with holes* where simple polygonal holes may be cut out from the enclosure polygon (see figure1). A polygon is called simple if there is no pair of nonconsecutive edges sharing a point. From our experience, the class of simple polygons with holes is adequate for GIS applications (see [Bur 86]) and most 2D CAD/CAM applications.

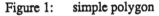

Figure 1:   simple polygon   simple polygon with holes

We define an *object* as a set of disjoint base-objects. Thus, holes may contain further base-objects. Disjointness refers to the regular intersection (see [Til 80]), therefore the edges of base-objects may touch each other. For example, using our definition of object the state Bremen that consists of the disjoint areas 'Bremen-City' and 'Bremerhaven' can be modeled as one single object. Finally, a *scene* is a set of objects where the objects may overlap each other.

From the literature, no standard set of geometric queries and operations fulfilling all requirements of spatial applications is known [SV 89]. Thus it is necessary to provide a small set of basic spatial queries and operations that are efficiently supported by a data structure. Application specific queries and operations, e.g., in [Oos 90], typically using more complex query conditions, can be decomposed into a sequence of such basic spatial queries. In the following we list a set of basic spatial queries and operations that should be supported on the objects and scenes defined above.

- *PointQuery:*   scene × point →   list of objects
  The PointQuery selects all objects of the scene that contain the given query-point.

- *CollisionQuery:*   scene × object →   BOOLEAN
  The 'CollisionQuery' tests whether a given query object is colliding with a scene, i.e., the query-object overlaps one or more objects of the scene.

- *OverlapJoin:*   scene × scene →   list (object × object)
  The 'OverlapJoin' reports for two given scenes T and R all pairs of objects (t,r) where
  t ∈ T and r ∈ R and t overlaps r.

- *CoverQuery:*   object × object →   BOOLEAN
  The 'CoverQuery' tests whether the first object is completely covered by the second object.

- *EnclosureQuery:*   scene × object →   list of objects
  The 'EnclosureQuery' reports all objects of the scene that completely cover the query-object.

- *ContainmentQuery:*   scene × object →   list of objects
  The 'ContainmentQuery' reports all objects of the scene that are completely covered by the query-object.

- *IntersectionJoin:*   scene × scene →   scene
  The 'IntersectionJoin' computes the regular intersection of all objects of one scene with all objects of the other scene.

- *Difference:*   scene × scene →   scene
  The 'Difference' computes the regular difference of the first scene with the second scene.

In this paper, we will present a data structure that is not optimized for one special type of query or operation but a data structure on which all these queries and operations can efficiently be performed. Furthermore, the data strucure supports a dynamic environment, i.e., updates, insertions and deletions of objects of a scene require no global reorganization of the data structure.

3 Decomposition of polygonal objects into a set of trapezoids

The basic idea of our decomposition approach can be summarized as follows: We decompose a complex polygonal object into a set of disjoint simple components where computational geometry concepts are used in order to guarantee efficiency. Due to this preprocessing step, queries and operations can be simplified. For example, a complex 'Point in Object'-test is replaced by a 'Point in Simple Component'-test or 'Intersection of Objects' is replaced by 'Intersections of Simple Components'. A general consequence of the decomposition approach is that complex algorithms are replaced by a set of simple algorithms. As a further advantage, the decomposition approach facilitates the exploitation of the spatial locality of queries and operations (see figure 2). However, the following effect is obvious: If the number of invokations of the simple algorithms is arbitrarily growing, then the performance of the decomposition approach degenerates. This reflection crucially influences the choice of the decomposition.

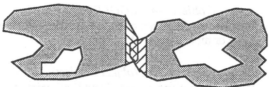

Figure 2: Spatial locality of the operation 'Intersection of Objects'

According to our extensive comparison of decomposition techniques (see [KHS 91]) we decided to decompose a polygonal object into a set of trapezoids introduced by Asano/Asano [AA 83]. The components produced by this algorithm are formed as trapezoids containing two horizontal sides. The algorithm uses the plane sweep technique known from the field of computational geometry. The basic idea is, to send out for each vertex one or two horizontal rays into the interior of the polygon to the first edge encountered. In the following we give a brief description.

As mentioned before the algorithm uses the plane sweep technique, i.e., the vertices will be passed and handled with increasing y-coordinates, for example. Thus, the entire algorithm consists of two steps:

Step 1: Sorting the vertices of the given polygonal object (enclosure polygon and holes) builds up the *event point list*, i.e., a sorted list of all the points that have to be treated by the algorithm.

Step 2: Processing the event point list by switching from one event point (vertex) to the next sending out partition rays (*event point scheduling*). Within this process, each ray and its successor form one trapezoid. In some cases degenerated trapezoids, i.e., triangles with one horizontal side may be produced. Thus, the set of component objects is gradually generated.

Applying step 1 and step 2 to the original polygonal object produces the resulting set of trapezoids. Figure 3 shows an example of the effect of the algorithm.

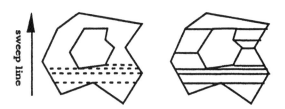

Figure 3: Decomposition of a polygonal object into a set of trapezoids

Summarizing, the main properties of the presented decomposition technique are:
- the trapezoids possess two horizontal edges by construction; thus the possibilities of the decomposition process are restricted to the choice of the sweep line
- the trapezoids are simple components with fixed length
- a polygon with n vertices is decomposed into p trapezoids where $p \leq n-1$
- the decomposition is easy to implement and has a good run time performance of $O(n \log n)$

After decomposing the polygonal objects into trapezoids it is possible to perform the queries and operations on these trapezoids. The success of such processing depends on the ability to narrow down quickly the set of trapezoids that are affected by the queries and operations. In the next sections we will demonstrate how this problem can be efficiently solved using the TR*-tree. By the way, the inverse operation of the decomposition process generating the sorted list of vertices of the polygon is no problem and can efficiently be performed using a plane sweep merge algorithm similar to the map overlay algorithm proposed in [KBS 91].

4 The TR*-tree

We cannot define a complete spatial order on the set of trapezoids that are generated by the decomposition process described in section 3 [PS 88]. Therefore, binary search on these trapezoids is not possible. A new approach is necessary that permits efficient spatial search on the trapezoids that is suitable for all the queries and operations described in section 2 and that is completely dynamic, i.e., updates, insertions and deletions of objects of a scene do not influence the performance of the structure and require no global reorganization. Our TR*-tree realizes these demands. The TR*-tree is derived from the R*-tree proposed in [BKSS 90]. The R*-tree is an optimized variant of the well known R-tree [Gut 84], a data structure for storing multidimensional rectangles.

A TR*-tree is a B\+-tree like structure that stores trapezoids as complete objects without clipping them or transforming them to higher dimensional points. A non-leaf node contains entries of the form (cp, Rectangle) where cp is the address of a child node in the TR*-tree and Rectangle is the minimum bounding rectangle of all rectangles that are entries in that child node. A leaf node contains entries of the form (Oid, Trapezoid) where Trapezoid is a component of the decomposed object with object identifier Oid. A TR*-tree is completely dynamic, insertions and deletions can be intermixed with queries and no periodic global reorganization is required. Obviously, the structure must allow overlapping directory rectangles. Thus it cannot guarantee that only one search path is required for retrieval queries. For further information we refer to [Gut 84]. Here and in the following, we use the term directory rectangle that is geometrically the minimum bounding rectangle of the underlying rectangles or trapezoids.

As mentioned above the TR*-tree is a variant of the R*-tree for efficiently handling sets of trapezoids in main memory. The following tradeoff determines the efficiency of the TR*-tree. On one hand, the number of entries in a node has to be big enough in order to dynamically cluster rectangles and trapezoids. On the other hand the overlap of directory rectangles should be minimized such that the search time in the TR*-tree approaches logarithmic search. Tests demonstrated that for each node the maximum number of 4 entries and a minimum number of one entry are best choices. Summarizing the TR*-tree satisfies the following properties:
- The root has at least two children unless it is a leaf.
- Every non-leaf node has between 1 and 4 children unless it is the root.
- Every leaf node contains between 1 and 4 entries unless it is the root.
- All leaves appear on the same level.

Using our standardized testbed [KSSS 89], we performed an exhaustive performance comparison in a systematic framework. It turned out that we can use the algorithms developed in [BKSS 90] to store the trapezoids of decomposed polygonal objects in an R*-tree. Only little modifications were necessary handling trapezoids. The interested reader will find the algorithms in detail in [BKSS 90].

In order to demonstrate the performance of our TR*-tree we investigated the real cartography data depicted in figure 4. We thankfully acknowledge receiving real data representing national administrative divisions of the European Community by the Statistical Office of the European Communities. The african lakes and islands are taken from World Data Bank II (see [GC 87]).

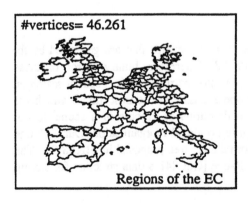

Regions of the EC

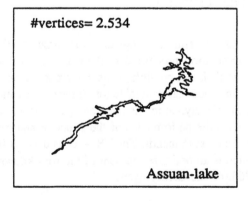

Assuan-lake

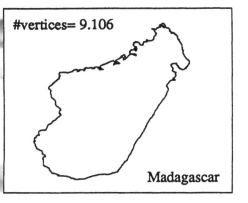

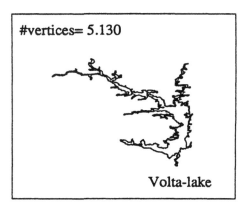

Figure 4: Examples of real cartography data

In table1 we present the results of the tests that we ran with Modula-2 implementations on SUN 3/60 workstations under UNIX. For CPU-time measurements we used the 'time'-command of UNIX. We investigated the following performance parameters:

- CPU-time of the decomposition algorithm (*decompose [sec]*)
- Insertion-time of the set of trapezoids into the TR*-tree (*insert [sec]*)
- *storage-index* =
  (# bytes of the TR*-tree) / (# bytes of the list of the vertices of the polygons)
- *Height* of the TR*-tree
- *approximation-index* =
  (area of the directory-rectangles on the lowest non-leaf level that are clustering the stored trapezoids) / (area of the polygon)

Polygons	#vertices	decompose [sec]	insert [sec]	storage index	height	approx. index
South Uist	85	0.8	1.2	6.07	3	1.11
Mainland	257	1.6	3.6	6.11	4	1.09
Donegal	422	2.2	6.6	6.24	5	1.04
Strathclyde	588	3.1	9.9	6.17	5	1.04
Highlands	869	4.1	14.7	6.28	6	1.02
Assuan Lake	2534	10.6	44.6	5.89	6	1.03
Volta Lake	5130	24.0	108.0	6.01	7	1.02
Contour of EC	7655	35.5	163.9	6.38	8	1.01
Madagascar	9106	38.8	186.6	6.01	8	1.00
scene: Europe	29274	115.3	642.4	6.09	9	1.26
scene: EC-Regions	46261	168.7	1040.7	6.02	9	1.14

Table1: Performance parameters of the TR*-tree

5 Algorithmic realization of queries and operations using the TR*-tree

This chapter will demonstrate that different spatial queries and operations on polygonal objects can be performed in a rather simple way if TR*-trees are used to represent the polygonal objects.

5.1 PointQuery: scene × point →   list of objects

Given a TR*-tree whose root node is N, find all objects of the scene that contain the query-point P.

P1 [PQ subtrees] : IF N is not a leaf, check each entry E to determine whether E.Rectangle contains P. For all entries that are found, invoke 'PointQuery' on the tree whose root node is pointed by E.cp .

P2 [PQ leaf node] : If N is a leaf, check each entry E to determine whether E.Trapezoid contains P. If this is the case, E.Trapezoid is a qualifying trapezoid and E.Oid specifies the object that fulfills the 'PointQuery'.

5.2 CollisionQuery: scene × object →   BOOLEAN

Given a TR*-tree A that stores the scene and a TR*-tree B that stores the query-object, the 'CollisionQuery' reports 'TRUE' if any trapezoid of A overlaps a trapezoid of B, 'FALSE' otherwise.

Invoke 'SpatialJoin' with M:=A and S:=B. If the 'SpatialJoin' reports the first overlapping pair of trapezoids (EM,ES), terminate the 'SpatialJoin' and answer the 'CollisionQuery' with 'TRUE'. IF no overlapping pair of trapezoids is found, the'CollisionQuery' is answered with 'FALSE'.

SpatialJoin:
Given two TR*-trees, the master TR*-tree (=M) and the slave TR*-tree (=S), whose root nodes are N_M and N_S, respectively.

SJ1: If N_M and N_S are non-leaves, check each entry E_M of N_M to determine whether E_M.Rectangle overlaps an E_S.Rectangle of N_S. For all overlapping pairs (E_M,E_S) invoke 'SpatialJoin' on the trees whose root nodes are pointed to by E_M.cp and E_S.cp .

SJ2 : If N_M is a leaf and N_S is a non-leaf, check each entry E_M of N_M to determine whether E_M.Trapezoid overlaps an E_S.Rectangle of N_S. For all overlapping pairs (E_M,E_S) invoke 'OverlapQuery' with the query-object E_M.Trapezoid on the subtree of S whose root node is pointed to by E_S.cp .

SJ3 : If N_M is a non-leaf and N_S is a leaf, check each entry E_S of N_S to determine whether E_S.Trapezoid overlaps an E_M.Rectangle of N_M. For all overlapping pairs (E_M, E_S) invoke 'OverlapQuery' with the query-object E_S.Trapezoid on the subtree of M whose root node is pointed to by E_M.cp.

SJ4 : If N_M and N_S are leafs, check each entry E_M of N_M to determine whether E_M.Trapezoid overlaps an E_S.Trapezoid of N_S. In this case, (E_M, E_S) is a qualifying pair of overlapping trapezoids. Report (E_M, E_S).

OverlapQuery:

Given a TR*-tree whose root node is N and a query-object O=(Oid,Object) where Object ∈ {Trapezoid, Rectangle}, find all trapezoids of the TR*-tree that overlap the query-object O.Object.

O1 : If N is not a leaf, check each entry E to determine whether E.Rectangle overlaps O.Object. For all overlapping entries, invoke OverlapQuery on the tree whose root is pointed to by E.cp.

O2 : If N is a leaf, check each entry E to determine whether E.Trapezoid overlaps O.Object. In this case, (O,E) is a qualifying pair of overlapping objects. Report the pair (O,E)

5.3 OverlapJoin: scene × scene   →   list (object × object)

Given two TR*-trees A and B: A stores the first scene T and B stores the second scene R. The 'OverlapJoin' reports all pairs of objects(t,r) where t∈ T and r ∈ R and t overlaps r.

Invoke 'SpatialJoin' with M:=A and S:=B. Each qualifying pair (E_M, E_S) of the 'SpatialJoin' indicates an overlap of the object E_M.Oid with the object E_S.Oid. These pairs $(E_M$.Oid,E_S.Oid) are inserted into a hash table that can be used to quickly determine if the same pair has already been reported. (It is likely, that the same pair will be encountered multiple times, because each object is decomposed into a set of trapezoids).

5.4 CoverQuery: object × object   →   BOOLEAN

Given two TR*-trees: M (=master) stores the first object that is decomposed into the set of trapezoids $\{E_1,..,E_n\}$. S (=slave) stores the second object. The 'CoverQuery' tests whether the first object is completely covered by the second object.

Invoke 'OverlapQuery' with E_i on the tree S, starting with i=1. Collect all qualifying trapezoids of S. If E_i is covered by these trapezoids, invoke 'OverlapQuery' with E_{i+1}, else terminate the 'CoverQuery' and report 'FALSE'. If E_n is tested with a positive result, the

'CoverQuery' can be answered with 'TRUE'.

5.5 EnclosureQuery: scene × object → list of objects

Given two TR*-trees A and B: A stores the scene and B stores the query-object that is decomposed into the set of trapezoids $\{E_1,..,E_n\}$. The 'EnclosureQuery' reports all objects of the scene that completely cover the query-object.

Invoke 'OverlapQuery' with E_1 on the tree B. Collect all intersecting trapezoids of A with respect to their Oid's. If E_1 is covered by trapezoids with the same Oid, the objects with these Oid's are candidates fulfilling the 'EnclosureQuery'. The Oid's of the candidates are stored in a hash table. Invoke 'OverlapQuery' with E_{i+1}, i=1,..,n-1, on the tree A. Using the hash table for each candidate we test whether E_{i+1} is covered by trapezoids of this candidate. In this case, this candidate remains candidate for the next iteration, else the candidate is deleted from the hash table.

5.6 ContainmentQuery: scene × object → list of objects

Given two TR*-trees A and B: A stores the scene and B stores the query-object. The 'Containment-Query' reports all objects of the scene that are completely covered by the query-object.

Invoke 'OverlapQuery' with rectangle = minimum bounding box of the query object on the tree A. Collect all qualifying trapezoids of A with respect to their Oid's. For each Oid, where the number of the qualifying trapezoids $\{E_1,..,E_m\}$ of an Oid is the same as the number of components of this Oid, invoke 'CoverQuery' where $M := \{E_1,..,E_m\}$ and $S := B$.

5.7 IntersectionJoin: scene × scene → scene

Given two TR*-trees A and B: A stores the first scene and B stores the second scene. The 'Intersection-Join' computes the regular intersection of all objects of the one scene with all objects of the other scene.

Invoke 'SpatialJoin' where $M:=A$ and $S:=B$. For each qualifying pair (E_M,E_S) compute the regular intersection E_M.Trapezoid $\cap$ E_S.Trapezoid. Decompose the result of the intersection into a set of trapezoids $\{I_1.\text{Trapezoid},..,I_m.\text{Trapezoid}\}$, m ≤ 3, and insert $\{I_1,..,I_m\}$ into the result tree where $I_j.\text{Oid}=E_M.\text{Oid}$ concatenated with $E_S.\text{Oid}$ for all j=1,..,m.

5.8 Difference: scene × scene → scene

Given two TR*-trees A and B: A stores the first scene (= set of trapezoids $\{E_1,..,E_n\}$) and B stores the second scene. The 'Difference' computes the regular difference of the first scene with the second scene.

For i=1,..,n DO: Invoke 'OverlapQuery' with query-object E_i on the tree B. For the set Σ of qualifying trapezoids of B compute the regular difference of E_i and Σ. Decompose the result of the difference into a set of trapezoids $\{D_1.\text{Trapezoid},..,D_m.\text{Trapezoid}\}$ and insert $\{D_1,..,D_m\}$ into the result tree where $D_j.\text{Oid}=E_i.\text{Oid}$ for all $j=1,..,m$.

6 Experimental performance analysis

After proposing to represent polygons using TR*-trees and after presenting the algorithmic realization of different queries and operations using the TR*-tree, it is the main goal of this chapter to demonstrate the performance of this approach. The performance of the TR*-tree cannot be investigated using analytical methods because the TR*-tree use heuristics. Therefore, we want to perform an experimental performance analysis.

One basic problem of each experimental performance analysis is the selection of an appropriate standardized set of test data. The best choice is to examine data files used daily in real applications. Thus, we tried to provide complex spatial data, e.g., digitized maps used in existing geographic information systems. Additionally, synthetic polygons were generated by a tool in order to emphasize the characteristics of our approach. We will present series of tests that we ran with Modula-2 implementations on SUN 3/60 workstations under UNIX. The analysis consists of two parts: Part one investigates the 'PointQuery' on polygons and scenes varying the number of vertices. In part two of the test we consider the operation 'Intersection of two Polygons' assuming the worst case with respect to spatial locality, i.e., the two polygons completely overlap each other.

6.1 The PointQuery

For the 'PointQuery' we used real cartography data as depicted in figure 4. The tests are standardized as follows:

We performed 1000 'PointQueries' where the query-points are randomly distributed over the minimum bounding box of the object or scene in order to design realistic unsuccessful 'PointQueries'. The TR*-trees are residing in main memory. CPU-time is measured using the UNIX 'time'-command.

Table 2 presents the average CPU-time of the successful 'PointQuery' and the unsuccessful 'PointQuery'. Because CPU-time measurements are generally not comparable since different processors or different implementations lead to different CPU-times, we calculated the number of procedure calls of the two most time consuming procedures of a 'PointQuery', namely 'Point in Rectangle'-test (passing through the directory of the TR*-tree) and 'Point in Trapezoid'-test (searching in the leaves of the TR*-tree). The time which is spent to follow pointers is negligible. Additionaly, for a successful 'PointQuery' we compute the value c=(# 'Point in Rectangle'-tests) / (ln n) where n is the number of vertices of the polygon or scene. The normalization with respect to ln n is done in order to demonstrate that the number of 'Point in Rectangle'-tests is growing logarithmically with the number of vertices (see table 2). This results in a logarithmic growth of the CPU-time of a successful 'PointQuery' because the number of 'Point in Trapezoid'-tests is nearly constant to 3 (see table 2). The results of the unsuccessful 'PointQuery' are extremely good. This is based on the fact that the TR*-tree approximates the polygons more accurate than the minimum bounding box and thus an unsuccessful 'PointQuery' is often detected on a high level of the tree, sometimes even on the root level.

Polygons	n =number of vertices	successful				unsuccessful		
		#PiR	c∗ln n	#PiT	CPU[ms]	#PiR	#PiT	CPU[ms]
South Uist	85	10.2	2.30	2.68	3.1	7.35	0.42	1.1
Mainland	257	14.9	2.69	3.19	3.8	8.17	0.12	1.5
Donegal	422	15.7	2.60	3.00	4.2	6.70	0.07	1.5
Strathclyde	588	17.8	2.80	2.98	4.7	8.53	0.08	1.8
Highlands	869	17.5	2.59	2.96	4.7	7.71	0.11	1.9
Assuan Lake	2534	20.4	2.60	3.02	6.8	5.86	0.03	1.1
Volta Lake	5130	23.3	2.73	2.65	6.8	8.00	0.06	1.6
Contour of EC	7655	25.8	2.85	2.81	6.7	6.20	0.26	1.3
Madagascar	9106	23.2	2.54	2.88	6.3	6.50	0.04	1.1
scene: Europe	29274	30.1	2.93	2.96	7.4	11.3	0.02	2.3
scene: EC-Regions	46261	41.3	3.84	3.13	8.5	12.2	0.03	2.9

Table 2: Performance parameters of the 'PointQuery'

Considering the results presented in table 2 we can postulate that the TR*-tree representation is very efficient for point queries. For example, on the scene 'Regions of the EC' that contains 860 objects digitized with 46261 vertices, a 'PointQuery' is answered in 8.5 msec in the successful case and in 2.9 msec in the unsuccessful case.

6.2 The intersection of polygonal objects

Using TR*-trees, the performance of the operation 'Intersection of two Polygonal Objects' stringently depends on the spatial relation of the two objects to each other. Since we do not want to design tests that are advantageous for the TR*-tree representation, we assume the worst case with respect to spatial locality, i.e., the two polygons completely overlap each other. For the operation 'Intersection of two polygons' we randomly generated synthetic polygons of varying numbers of vertices (n ∈ {100, 200, ..., 1000}). The shape and the extension of the objects are nearly the same independently of the number of vertices. Figure 5 depicts an example of a synthetic polygon (n = 100) and the number of intersection-points of two overlapping polygons with the same number of vertices.

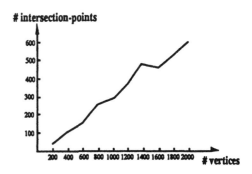

intersection-points

Figure 5: Example of a synthetic polygon.   #Intersection points of two overlapping polygons

The results of the operation 'Intersection of two Polygons' presented in figure 6 can be summarized as follows: The number of intersections of trapezoids is linearly growing if the #vertices of one polygon is fixed whereas the #vertices of the other polygon is varying from 100 to 1000 (see figure 6a). Using the TR*-tree representation about two tests whether trapezoids are overlapping each other have to be performed in order to detect one intersection of trapezoids (compare figure 6a and 6b). The number of directory tests whether two directory rectangles are overlapping are almost linearly growing (see figure 6c). These facts result in an almost linear growing of the CPU-time of the operation 'Intersection of two Polygons' (see figure 6d). The results demonstrate that the operation 'Intersection of two Polygons' is efficiently performed if the TR*-tree representation is used although we assumed the worst case with respect to spatial locality. The following extracted example of the test series confirms this statement: If we intersect the two polygons with 1000 vertices respectively and 596 intersection-points using the TR*-tree representation, 3349 trapezoids intersect each other that are detected by performing 11844 directory tests and 6311 trapezoid tests within the TR*-tree structure. This results in an overall CPU-time of 2.9 sec.

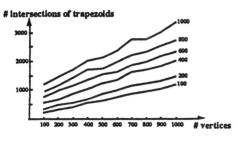

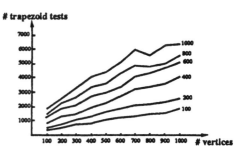

Figure 6a: Number of intersecting trapezoids   Figure 6b: Number of trapezoid tests

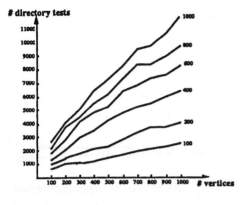

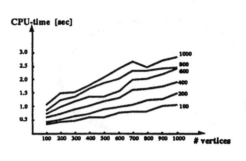

Figure 6c: Number of directoy tests   Figure 6d: CPU-time of the intersection operation

7. Conclusions

In our approach polygonal objects are represented as TR*-trees. In a preprocessing step the polygonal objects are decomposed into a set of trapezoids. These trapezoids are inserted into the TR*-tree that is persistently stored on secondary storage. For query processing the TR*-trees are completely loaded into main memory. Using the TR*-tree representation of polygonal objects, various types of queries and operations on polygonal objects can efficiently be performed. This is based on the fact that the TR*-tree is a dynamic structure that supports efficient spatial search on trapezoids. An experimental performance analysis of the point query and the intersection operation that is based on real cartography and synthetically generated data indicates the good run time performance of our approach in the average case. In the future, we want to compare the performance of the queries and operations when using the TR*-tree with other structures and algorithms, e.g., the optimum point location structure 'the trapezoid tree' described in [PS 88], or the plane sweep algorithm for the operation intersection of polygons, proposed by [NP 82]. Additionally, we want to integrate the TR*-tree representation of polygonal objects for efficient query processing in a spatial database system (see [KHHSS 91]), particularly in a geographic information system.

References

[AA 83]   Asano, Ta. & Te. Asano, 'Minimum Partition of Polygonal Regions into Trapezoids', in Proc. 24th IEEE Annual Symposium on Foundations of Computer Science, 233-241, 1983.

[BKSS 90]   N. Beckmann, H.-P. Kriegel, R. Schneider, B. Seeger: 'The R*-tree: An Efficient and Robust Access Method for Points and Rectangles´, Proc. ACM SIGMOD Int. Conf. on Management of Data, 322-331, 1990.

[Bur 86]   P.A. Burrough: 'Principles of Geographical Information Systems for Land Resources Assessment´, Oxford University Press, 1986

[GC 87] D.G. Gorny & Russ Carter: World Data Bank II, General users Guide. Technical Report, U.S. Central Intelligence Agency, 1987.

[Gut 84] Guttman A., 'R-trees: a dynamic index structure for spatial searching', in Proc. ACM SIGMOD Int. Conf. on Management of Data, 47-57, June 1984.

[KBS 91] H.-P. Kriegel, T. Brinkhoff, R. Schneider: 'An Efficient Map Overlay Algorithm based on Spatial Access Methods and Computational Geometry', Proc. Int. Workshop on DBMS's for geographical applications, Capri, May 16-17, 1991

[KHHSS 91] H.-P. Kriegel, P. Heep, S. Heep, M. Schiwietz, R. Schneider: 'An Access Method Based Query Processor for Spatial Database Systems', Proc. Int. Workshop on DBMS's for geographical applications, Capri, May 16-17, 1991

[KHS 91] Kriegel, H.P., H. Horn & M. Schiwietz: 'The Performance of Object Decomposition Techniques for Spatial Query Processing', Proc. 2nd Symposium on Large Spatial Databases, Zurich, August 28-30, 1991

[KSSS 89] H.-P. Kriegel, M. Schiwietz, R. Schneider, B. Seeger: 'Performance Comparison of Point and Spatial Access Methods', Proc. 1st Symp. on the Design of Large Spatial Databases, 1989 (Lecture Notes in Computer Science 409, Springer, 89-114, 1990)

[NP 82] J. Nievergelt, F.P. Preparata: 'Plane-Sweep Algorithms for Intersecting Geometric Figures', Comm. of the ACM, Vol. 25, No. 10, 739-747, 1982

[Oos 90] P.J.M. Oosterom: 'Reactive Data Structures for Geographic Information Systems', PhD-thesis, Department of Computer Science at Leiden University, 1990

[PS 88] F.P. Preparata, M.I. Shamos: 'Computational Geometry', Springer, 1988

[SV 89] Scholl, M. & A. Voisard, 'Thematic Map Modelling', in Proc. "Symposium on the Design and Implementation of Large Spatial Databases", 167-190, Santa Barbara, USA, July 1989.

[Til 80] R.B. Tilove: 'Set Membership Classification: A Unified Approach to Geometric Intersection Problems', IEEE Trans. on Computers, Vol. C-29, No. 10, 874-883, 1980

A Voronoi Diagram Based Adaptive K-Means-Type Clustering Algorithm for Multidimensional Weighted Data

Thomas Schreiber

Universität Kaiserslautern
Fachbereich Informatik
Postfach 3049
D–6750 Kaiserslautern

Abstract

This paper describes a solution to the following problem: Given a set of weighted data points, find the cluster center points, which minimize the least squared errors. The k-means-type methods produce good results, but usually the quality of the representation depends on an initial cluster configuration. Also this does not allow a variable number of clusters for a given error tolerance.

The proposed method removes these disadvantages by an adaptive sequential insertion of new clusters in those areas, where the largest errors occur. This can be done more efficiently by using multidimensional Voronoi diagrams and local procedures. The data points can be weighted and arbitrarily distributed in the Euclidean space. The weight of each point may be chosen by the user depending on the importance or correctness of that point. At the same time the method produces a hierarchical multidimensional triangulation of the data at different levels of accuracy.

Keywords: Clustering algorithm, Delaunay triangulation, hierarchical representation, k-means algorithm, Voronoi diagram.

1 Introduction

Clustering means grouping of similar objects by optimizing a certain criterion function or other object dependent properties. Clustering techniques are very common and useful in many applications, like data analysis, data reduction, digital image processing and pattern recognition. In the past many algorithms have been developed and extensively illustrated in [3]. In 1982 Heckbert [4] introduced the median-cut algorithm for color reduction in computer graphics which subdivided a hyperbox recursively by a hyperplane through the median and perpendicular to the coordinate axis with the greatest expansion. The mean-split algorithm from Wu and Witten [13] chooses the partition point to be the mean rather than the median, and the hyperplane perpendicular through the coordinate axis with the largest spread. Wan, Wong and Prusinkiewicz [11] evaluate an optimal cut-point for each axis, and subdivide the data set by a hyperplane perpendicular to that axis, on which the overall reduction of expected variance is the largest among all projected distributions. A comparison of these three algorithms is given in [11]. The knot selection algorithm from Franke and McMahon [6][7] uses the k-means method which is reviewed for weighted data in Section 3. They determine the clusters with the most and the least numbers of data points, and start moving these cluster center points towards each other and executing the k-means method, until all clusters contain approximately the same number of data points. Dierks [2] extended this method to three dimensions.

In this paper an algorithm is presented to solve the following problem: Given a set $\mathcal{P}$ of arbitrarily distributed data points $P_1, P_2, \ldots, P_N$ with weights $W_1, W_2, \ldots, W_N$ ($P_i \in \mathbb{E}^d$, $0 < W_i \in \mathbb{R}$), find a smaller set $\mathcal{Q}$ of cluster center points $q_1, q_2, \ldots, q_m$ ($q_j \in \mathbb{E}^d$, $m \ll N$), which optimizes a certain cost function. We want to restrict ourselves to the most widely used function, the minimizing of the least squared errors.

Let n ($m \ll n \leq N$) denote the number of distinct input points $p_1, p_2, \ldots, p_n$. The new weights $w_1, w_2, \ldots, w_n$ of $p_1, p_2, \ldots, p_n$ are the sum of the weights W_i of equal points P_i ($i = 1, 2, \ldots, N$). Thus, the task of this work is to determine the cluster center points q_j which minimize the following error S:

$$S = \sum_{j=1}^{m} s_j \rightarrow \min \tag{1}$$

with

$$s_j = \sum_{i \in I_j} w_i \|p_i - q_j\|^2 \Big/ \sum_{i \in I_j} w_i$$

and the index sets

$$I_j = \left\{ i : p_i \text{ is closer to } q_j \text{ than to any other } q_k, \forall i \neq k \right\}. \tag{2}$$

$\|\cdot\|$ denotes the Euclidean norm. A necessary condition for minimization is that the partial derivations $\frac{\partial s_j}{\partial q_j} = 0$ and $\frac{\partial^2 s_j}{\partial q_j^2} \geq 0$. The global minimum in each cluster is achieved, if the cluster point q_j lies in the weighted mean of all data points p_i ($i \in I_j$).

$$\frac{\partial s_j}{\partial q_j} = -2 \sum_{i \in I_j} w_i (p_i - q_j) = 0$$

$$\Longleftrightarrow \quad q_j = \sum_{i \in I_j} w_i p_i \Big/ \sum_{i \in I_j} w_i \tag{3}$$

$$\frac{\partial^2 s_j}{\partial q_j^2} = 2 \sum_{i \in I_j} w_i > 0 \qquad (j = 1, 2, \ldots, m).$$

The computation of the global minimum of S is a NP-complete problem [5]. There exist $\frac{m^n}{m!}$ possibilities of grouping n data points to m clusters. The boundaries of the clusters form a multidimensional Voronoi diagram (see Section 2), which is very useful for the efficient calculation of the index sets. To compare clustering algorithms quantities for the experimentally determination of their performance were given in Section 4. A theoretical analysis seems to be infeasible because the result depends heavily on the distribution of the input data.

2 Multidimensional Voronoi Diagrams

A multidimensional Voronoi diagram is a partitioning of the space $\mathbb{E}^d$ in regions R_j with the following properties: Each point q_j lies in exactly one region R_j, which consists of all points x of $\mathbb{E}^d$ that are closer to q_j than to any other point q_k $(j \neq k)$. The points q_j are also called Voronoi points.

$$R_j = \left\{ x \in \mathbb{E}^d : \|x - q_j\| < \|x - q_k\|, \forall j \neq k \right\}. \tag{4}$$

With this definition the index set I_j can be defined as:

$$I_j = \{i : p_i \text{ lies in region } R_j\}. \tag{5}$$

Bowyer [1], Watson [12] and Palacios-Velez and Renaud [8] describe algorithms to compute a multidimensional Voronoi diagram by sequential insertion of new points. The insertion and also the deletion or movement of a point are local procedures. That means that the change of one point affects only a small region of the diagram and can be computed independent of the number of Voronoi points, if the region is known, and all points are scattered over the entire space. The Delaunay triangulation results from the Voronoi diagram by joining all Voronoi points, which have a common border, with straight lines. For more details on Voronoi diagrams and triangulations see [9].

3 The Method of K-Means

The method of k-means is an iterative procedure of moving cluster points into the centroid of its related data points. This ensures the global minimum of s_j for each cluster (see Section 1), but only a local minimum of S. If a cluster point has been moved, some data points near the boundary may change the cluster. This causes a new iteration step because the pertaining cluster points do not lie in the centroids any more, and therefore they have to be relocated. Selim and Ismail [10] show that this process converges after a finite number of iterations towards a local minimum of S, if a quadratic metric is used.

Algorithm 1: Weighted K-Means

Step 1: Initialize a good starting cluster configuration $q_1, q_2, \ldots, q_m$ with at least one data point for each cluster.

Step 2: For each cluster $(j = 1, 2, \ldots, m)$ do:

Move the cluster point q_j in the weighted mean of the data points p_i $(i \in I_j)$ and update the index sets.

Step 3: Repeat Step 2 until no cluster is changed.

This method has the following main disadvantages:

1) The error depends heavily on the initial cluster configuration, i. e. other configurations cause other end configurations with different errors.

2) The number of clusters is fixed. Thus, it is not possible to use as many clusters as necessary to yield a given error tolerance.

3) The update of the index sets requires a great computational effort. For each iteration, the least squared errors from all data points to all cluster centers have to be determined.

4) A later change of cluster and/or data points could not efficiently be done.

Now, we introduce a new method which avoids all these disadvantages.

4 The Proposed Method

In this section a Voronoi diagram based multidimensional clustering algorithm is described to find a local minimum for formula (1). The idea is based on an adaptive sequential insertion of a new cluster point in the region with the largest error s_j of the Voronoi diagram of all points which have already been inserted.

Algorithm 2: Adaptive Clustering

Step 1: Initialize the first cluster point with the weighted mean of all data points. The corresponding region of the Voronoi diagram is the entire space.

Step 2: Divide the set of data points of the region R_e with the largest error into two sets and determine the new index sets and their center points.

First, compute the coordinate axis k with the largest projected variance:

$$k = \max_{l=1,2,\ldots,d} \left\{ \sum_{i \in I_e} w_i (p_i^l - q_e^l) \right\} \tag{6}$$

where x^l denotes the l-th component of vector x and q_e the cluster point of region R_e.

Then divide all points p_i $(i \in I_e)$ by a hyperplane through q_e perpendicular to the k-th coordinate axis. The new index sets I_{e1}, I_{e2} and their cluster centers m_1, m_2 are determined as follows:

$$I_{e1} = \{i : p_i^k \le q_e^k, i \in I_e\} \quad \text{and} \quad I_{e2} = \{i : p_i^k > q_e^k, i \in I_e\} \qquad (7)$$

$$m_1 = \sum_{i \in I_{e1}} w_i p_i \bigg/ \sum_{i \in I_{e1}} w_i \quad \text{and} \quad m_2 = \sum_{i \in I_{e2}} w_i p_i \bigg/ \sum_{i \in I_{e2}} w_i \qquad (8)$$

Step 3: Update the Voronoi diagram:

Move the cluster point q_e to the center m_1, insert a new cluster point at m_2 and update the index sets of the affected regions.

Step 4: For all modified regions:

Move the cluster point into the weighted centroid of its related data points. Update the Voronoi diagram, the index sets and the set of changed regions.

Step 5: Repeat steps 2–4 until the clustering requirement is met.

This requirement could be met, if a given number of cluster points are inserted, and/or the mean or maximum error is lower than a given value and/or each cluster contains no more than a given number of data points.

Notice, that a different cost function can be used to determine the region that should be subdivided (step 2). Thus, it is possible to optimize a second function, like the demand of a nearly equal number of data points for each cluster.

After each iteration the cluster points and their Delaunay triangulation may be stored for future use. This leads to a triangular based hierarchical surface representation of the data points at different levels of accuracy, which may be used in computer graphic applications for the rendering of the same object at different resolutions.

To point out the performance of our algorithm, we choose 9 data points with equal weights and compute their global minimum. For many distributions we achieved this minimum or were very close to it. The following example was chosen because it shows two things: The global minimum is not always achieved if the region of the largest error is subdivided (see Figure 5), and the initial subdivision of step 2 does not lead (as expected) to an optimal clustering (see Figure 6). The author proposes this subdivision method because it yields good results and does not require too much computational effort. But the user could employ any other method, especially, if he knows more about the distribution of the data. If rotational invariance is desired, a subdivision perpendicular to the best fit hyperplane could be adapted.

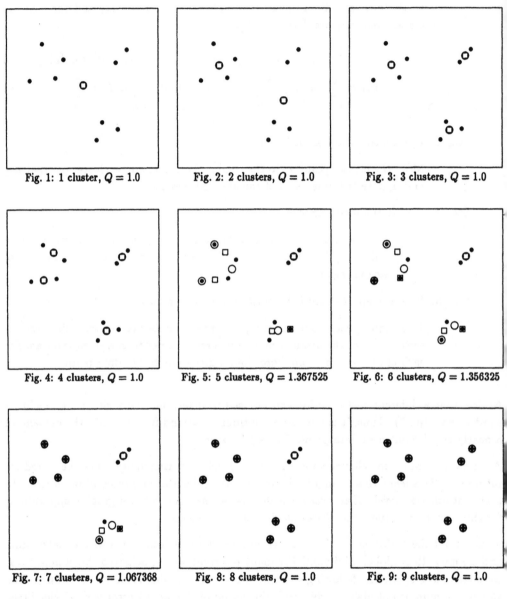

Fig. 1: 1 cluster, $Q = 1.0$

Fig. 2: 2 clusters, $Q = 1.0$

Fig. 3: 3 clusters, $Q = 1.0$

Fig. 4: 4 clusters, $Q = 1.0$

Fig. 5: 5 clusters, $Q = 1.367525$

Fig. 6: 6 clusters, $Q = 1.356325$

Fig. 7: 7 clusters, $Q = 1.067368$

Fig. 8: 8 clusters, $Q = 1.0$

Fig. 9: 9 clusters, $Q = 1.0$

with :   •   data points,
         ○   cluster center points, which reaches the global minimum of S,
         □   cluster center points, which were computed by our method.

The following formula computes the quality Q $(1 \leq Q)$ of a cluster representation:

$$Q = \frac{\text{error of the clustering algorithm}}{\text{global minimum error}} \tag{9}$$

The quantities Q_{max}, Q_{mean}, σ_Q^2 are the maximum, mean and expected variance of the clustering quality:

$$Q_{max} = \max_{i=1,2,\ldots,t} \{Q_i\} \tag{10}$$

$$Q_{mean} = \frac{1}{t} \sum_{i=1}^{t} Q_i \tag{11}$$

$$\sigma_Q^2 = \sum_{i=1}^{t} (Q_i - Q_{mean})^2 . \tag{12}$$

The following table shows the results of our algorithm after $t = 100$ tests with $n = 9$ randomly generated data points in 2 dimensions.

m	Q_{max}	Q_{mean}	σ_Q^2
1	1.000000	1.000000	0.000000
2	1.813297	1.064167	1.692273
3	1.696984	1.097048	2.961433
4	1.881154	1.097230	4.335665
5	2.135782	1.144528	4.621574
6	2.073583	1.145597	5.396631
7	2.230519	1.105344	6.486076
8	2.618419	1.027692	3.168752
9	1.000000	1.000000	0.000000

Table 1: Clustering quality ($t = 100, n = 9, d = 2$)

The next example shows 192 data points in 12 and 33 clusters with their Voronoi diagrams and Delaunay triangulations. The data points in Figures 10–13 have equal weights, in Figures 14–15 the weights correspond to the radius of the points.

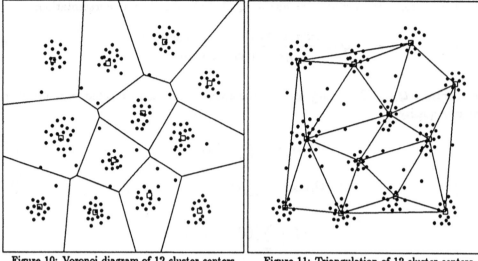

Figure 10: Voronoi diagram of 12 cluster centers   Figure 11: Triangulation of 12 cluster centers

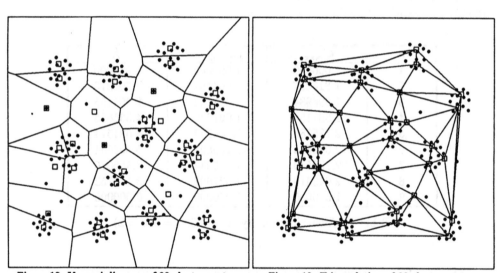

Figure 12: Voronoi diagram of 33 cluster centers   Figure 13: Triangulation of 33 cluster centers

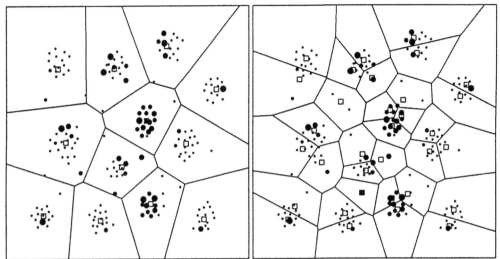

Figure 14: Voronoi diagram of 12 cluster centers   Figure 15: Voronoi diagram of 33 cluster centers

5 Conclusion

An adaptive multidimensional clustering algorithm was introduced which has the following properties:

The algorithm
1) allows weighted data.
2) needs no initial cluster configuration.
3) yields different levels of accuracy.
4) produces a triangular based hierarchical surface representation.
5) allows the optimization of a second cost function.

The following table shows a comparison of the computational effort for various cluster algorithms:

algorithm	computational effort
median-cut (Heckbert)	$O(d \cdot n \cdot \log n \cdot \log m)$
mean-split (Wu et. al.)	$O(d \cdot n \cdot \log m)$
optimal-cut (Wan et. al.)	$O(d \cdot n \cdot m)$
k-means	$O(d \cdot n \cdot m \cdot t)$
knot selection (McMahon et. al.)	$O(d \cdot n \cdot m \cdot t \cdot s)$
our method	$O(d \cdot n \cdot \log m \cdot t)$

Table 2: Computational effort

with: d dimension,
          n number of data points,
          m number of cluster points,
          t number of iterations,
          s number of movements.

The number t of iterations of our algorithm is larger than in other methods since it computes m instead of 1 cluster representations. But t decreases with the number of inserted cluster points. Table 3 shows the CPU time necessary for the computation of various cluster representations for 1000 randomly generated data points on a HP835 workstation. The update of the Voronoi diagram gets more influence in higher dimensions, because the calculations become more difficult.

d	m	t	memory management [%]	error calculation [%]	Voronoi diagram update [%]	total CPU time [sec]
2	5	70	19.0	43.5	37.5	2.12
	25	806	23.5	37.4	39.1	13.93
	50	1165	28.8	35.7	35.5	19.01
	75	1400	31.7	31.9	36.4	22.47
	100	1551	31.8	31.2	37.0	24.74
	200	1770	35.4	29.9	34.7	30.41
	300	1860	40.6	26.7	32.7	35.76
	500	1907	46.9	23.7	29.4	47.28
	1000	1923	61.8	17.9	20.3	92.74
3	5	131	16.1	37.8	46.1	6.37
	25	1183	18.3	19.2	62.5	126.50
	50	1662	16.6	15.7	67.7	245.37
	75	1852	17.9	14.9	67.2	327.85
	100	2036	17.1	13.8	69.1	397.10
	200	2313	19.6	13.5	66.9	592.55
	300	2370	22.5	11.8	65.7	721.39
	500	2394	29.0	10.6	60.4	970.12
	1000	2406	42.8	8.6	48.6	1795.01
4	5	63	17.0	29.1	53.9	7.50
	10	348	13.9	13.9	72.2	104.11
	25	866	8.2	8.7	83.1	1351.25
	50	1338	5.9	7.8	86.3	6546.68
	75	1611	5.3	7.7	87.0	14692.05

Table 3: Time necessary for $n = 1000$ random data points

The current implementation uses dynamic memory structures. The use of an array implementation, a median-cut or mean-split algorithm for a first part of cluster points and a tolerance for moving a cluster point or not, may speed up the algorithm (with a loss of quality and flexibility).

References

[1] A. Bowyer: *Computing Dirichlet tesselations*, Comp. Journal, Vol. 24, No. 2, 1981, 162–166.

[2] T. Dierks: *The Modelling and Visualization of Scattered Volumetric Data*, Masters Thesis, Arizona State University, USA, Dec. 1990.

[3] J. A. Hartigan: *Clustering Algorithms*, Wiley, New York, 1975.

[4] P. Heckbert: *Color image quantization for frame buffer display*, ACM Trans. Computer Graphics 16, 3 (July 1982), 297–304.

[5] I. Hyafil, R. L. Rivest: *Construction optimal binary decision trees is NP-complete*, Inf. Process. Lett. 5, May 1976, 15–17.

[6] J. R. McMahon: *Knot Selection for Least Squares Approximation using Thin Plate Splines*, Masters Thesis, Naval Postgraduate School, Monterey, USA, June 1986.

[7] J. R. McMahon, R. Franke: *An Enhanced Knot Selection Algorithm for Least Squares Approximation using Thin Plate Splines*, ARO Report 90–1, Trans. of the Seventh Army, Conf. on Applied Math. and Computing.

[8] O. Palacios-Velez, B. C. Renaud: *A Dynamic Hierarchical Subdivision Algorithm for Computing Delaunay Triangulations and Other Closest-Point Problems*, ACM Trans. on Math. Soft., Vol. 16, No. 3, Sept. 1990, 275–292.

[9] F. P. Preparata, M. I. Shamos: *Computational Geometry*, Springer, 1985.

[10] S. Z. Selim, M. A. Ismail: *K-means-type algorithms: A generalized convergence theorem and characterization of local optimality*, IEEE Trans. Pattern Anal. Mach. Intell. PAMI-6, 1 (1986), 81–87.

[11] S. J. Wan, S. K. M. Wong, P. Prusinkiewicz: *An Algorithm for Multidimensional Data Clustering*, ACM Trans. on Math. Soft., Vol. 14, No. 2, June 1988, 153–162.

[12] D. F. Watson: *Computing the n-dimensional Delaunay tesselation with application to Voronoi polytops*, Comp. Journal, Vol. 24, No. 2, 1981, 162–166.

[13] X. Wu, I. H. Witten: *A fast k-means type clustering algorithm*, Dept. Computer Science, Univ. of Calgary, Canada, May 1985.

A Generalization of Staircase Visibility *

Sven Schuierer[†]   Gregory J. E. Rawlins [‡]   Derick Wood [§]

Abstract

Let $\mathcal{O}$ be some set of orientations, i.e., $\mathcal{O} \subseteq [0°, 360°)$. In this paper we look at the consequences of defining visibility based on curves that are monotone w.r.t. to the orientations in $\mathcal{O}$. We call such curves $\mathcal{O}$-staircases. Two points p and q in a polygon $\mathbf{P}$ are said to $\mathcal{O}$-see each other if there exists an $\mathcal{O}$-staircase from p to q that is completely contained in $\mathbf{P}$. The $\mathcal{O}$-kernel of a polygon $\mathbf{P}$ is then the set of all points which $\mathcal{O}$-see all other points. We show that the $\mathcal{O}$-kernel of a simple polygon can be obtained as the intersection of all $\{\theta\}$-kernels, with $\theta \in \mathcal{O}$. With the help of this observation we are able to develop an $O(n \log |\mathcal{O}|)$ algorithm to compute the $\mathcal{O}$-kernel in a simple polygon, for finite $\mathcal{O}$.

1 Introduction

Visibility problems play an important role in computational geometry. Apart from the usual line segment visibility several other notions of visibility have been investigated in the past years: *staircase visibility* [2,6], *rectangular visibility* [4,8,10] and *periscope visibility* [3]. In this paper we introduce a new definition of sight called $\mathcal{O}$-*visibility*. It is based on the framework of *restricted orientation convexity* which was first considered by G. Rawlins in [12].

Restricted orientation convexity tries to bridge the gap between Euclidean convexity and $\{0°, 90°\}$-convexity. Recall that a set $\mathbf{S}$ is called $\{0°, 90°\}$-*convex* or *orthogonally convex* if the intersection of $\mathbf{S}$ with any axis-parallel line is connected; $\{0°, 90°\}$-convexity is a well-studied area; see [2,4,7,9,16,17]. In [12,13,14,15] Rawlins and Wood take the idea of $\{0°, 90°\}$-convexity one step further by developing the theory of restricted orientation convexity or $\mathcal{O}$-*convexity*. Instead of considering only axis-parallel lines we allow lines with orientation in some fixed set $\mathcal{O}$. So a set $\mathbf{S}$ is called $\mathcal{O}$-*convex* if the intersection of $\mathbf{S}$ with any line whose orientation is in $\mathcal{O}$ is connected. Note that restricted orientation convexity encompasses both $\{0°, 90°\}$-convexity—if $\mathcal{O} = \{0°, 90°\}$—and Euclidean convexity—if $\mathcal{O}$ is the set of all orientations—as special cases.

The framework of $\{0°, 90°\}$-convexity spawns a new definition of visibility called *staircase visibility* or $\{0°, 90°\}$-*visibility* which is based on $\{0°, 90°\}$-convex paths. Two points p and q in a set $\mathbf{S}$ are *staircase visible* from each other if there exists a $\{0°, 90°\}$-convex path from p to q that is completely contained in $\mathbf{S}$. Staircase visibility has been considered by Reckhow and Culberson [2] and Motwani et al. [6]. Both papers deal with covering polygons with the minimum number of $\{0°, 90°\}$-starshaped sets, i.e., sets that contain one point p such that all other points are staircase visible from p. In the same way $\mathcal{O}$-convexity gives rise to a new definition of visibility we call $\mathcal{O}$-*visibility*. Two points p and q in a set $\mathbf{S}$ are $\mathcal{O}$-*visible* from each other or $\mathcal{O}$-*see* each other if there exists an $\mathcal{O}$-convex path from p to q that is completely contained in $\mathbf{S}$. Note that $\mathcal{O}$-visibility again encompasses $\{0°, 90°\}$-visibility and Euclidean visibility as special cases.

---

*This work was supported by the Deutsche Forschungsgemeinschaft under Grant No. Ot 64/5–4 and by Natural Sciences and Engineering Research Council Grant No. A-5692.

[†]Institut für Informatik, Universität Freiburg, Rheinstr. 10–12, D-7800 Freiburg, Fed. Rep. of Germany; email: sven.schuierer@informatik.uni-freiburg.dbp.de

[‡]Department of Computer Science, Indiana University, 101 Lindley Hall, Bloomington, Indiana 47405, USA. email: rawlins@iuvax.cs.indiana.edu

[§]Department of Computer Science, University of Waterloo, Waterloo, Ontario, N2L3G1, Canada. email: dwood%watdaisy@waterloo.csnet

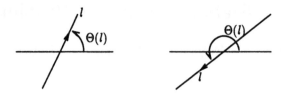

Figure 1: *The definition of the orientation of a line.*

One of the central visibility problems is the computation of the *kernel* of a polygon. Recall that the kernel of a set **S** is the set of points that see all other points in **S**. While for Euclidean visibility this is a well-studied problem where several optimal algorithms have been developed [1,5], the corresponding problem for $\{0°, 90°\}$-visibility has not been considered. $\mathcal{O}$-visibility offers the opportunity to develop an algorithm that computes the Euclidean kernel as well as the $\{0°, 90°\}$-kernel depending on the input parameter $\mathcal{O}$ and that is competitive in both cases. The algorithm we present here is a first step in this direction.

Visibility questions can also be seen as *reachability* questions. Consider the problem of "guarding" a polygon with one robot whose motion is restricted in such a way that the robot's path must be *monotone* in some set of orientations. In order to decide whether it is possible to guard a polygon and, if so, where to place the guard, we consider the $\mathcal{O}$-*kernel* of a polygon, i.e., the set of points that $\mathcal{O}$-see all other points.

An important requirement in the geometry of fixed orientations is to try to find algorithms that *do not* solve the problem separately, for each given orientation, and then put the obtained solutions together; instead one is interested in algorithms that exploit the "coherence" of the orientations in the solution of the problem. Our algorithm shows that this can be done very efficiently in the computation of the $\mathcal{O}$-kernel.

The rest of this paper is organized as follows. We start off with a precise definition of $\mathcal{O}$-convexity and $\mathcal{O}$-visibility in the next section. In Section 3 we turn to the computation of the $\mathcal{O}$-kernel of a simple polygon. We show that that the $\mathcal{O}$-kernel is the intersection of the $\{\theta\}$-kernels, where $\theta \in \mathcal{O}$. This observation can be exploited to derive a simple algorithm to compute the $\mathcal{O}$-kernel in time $O(n \log |\mathcal{O}|)$, for finite $\mathcal{O}$.

2 Basic Definitions for Restricted Orientation Convexity

If we are given an oriented line l in the plane, we define its *orientation* to be the angle it forms with the x-axis and denote it by $\Theta(l)$. Of course, we can speak in the same way of the orientation of a line segment or a ray (see Figure 1).

As already stated in the introduction the restricted oriented convex sets can now be defined as follows.

Definition 2.1 *Let $\mathcal{O}$ be a subset of $[0°, 360°)$. A set $C \subseteq I\!E^2$ is $\mathcal{O}$-convex if $l \cap C$ is connected, for all lines l with $\Theta(l) \in \mathcal{O}$.*

An example is shown in Figure 2. In order not to have to deal with orientations that are greater than $360°$, we assume from now on that the addition and subtraction of two orientations is done modulo $360°$. The first thing to note about the definition of $\mathcal{O}$-convexity is that we can assume that $\mathcal{O}$ is symmetric w.r.t. $180°$, that is, if the orientation θ is in $\mathcal{O}$, then a set is $\mathcal{O}$-convex if and only if it is $\mathcal{O} \cup \{\theta + 180°\}$-convex. We denote the orientation $\theta + 180°$ by θ^{-1}. So there is no loss in generality if we assume that, for all $\theta \in \mathcal{O}$, we also have $\theta^{-1} \in \mathcal{O}$. We say orientation θ_1 is *counterclockwise before* θ_2 if $\theta_2 - \theta_1 \leq 180$. We write $\theta_1 \leq \theta_2$ in this case. Since $\mathcal{O}$ always contains either both orientations θ and θ^{-1} or none of them, we use the notation $|\mathcal{O}|$ to denote *half* the cardinality of $\mathcal{O}$. Furthermore,

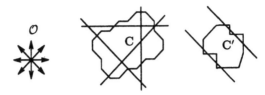

Figure 2: C *is* $\mathcal{O}$-convex *while* C' *is not.*

we will only specify the orientations in $[0°, 180°)$ to define a specific $\mathcal{O}$ though it should be kept in mind that $\mathcal{O}$ also always contains the opposite orientations in $[180°, 360°)$.

We say a range (θ_1, θ_2) is $\mathcal{O}$-*free* if $(\theta_1, \theta_2) \cap \mathcal{O} = \emptyset$. A range (θ_1, θ_2) is called a *maximal $\mathcal{O}$-free range* if (θ_1, θ_2) is $\mathcal{O}$-free and there is no other range (θ_1', θ_2') that is also $\mathcal{O}$-free and contains (θ_1, θ_2). If θ is some orientation in $[0°, 360°)$, the *maximal $\mathcal{O}$-free range of θ* is the maximal $\mathcal{O}$-free range that contains θ or if such a range does not exist, the emptyset. In the same way we will speak of the maximal $\mathcal{O}$-free range of a line, ray, line segment, or point which is, of course, meant to denote the maximal $\mathcal{O}$-free range of the orientation of the geometric object.

If p and q are two points in the plane, we denote the line segment between p and q by $\overline{pq}$. It can be shown that if $\Theta(\overline{pq}) \notin \mathcal{O}$ and (θ_1, θ_2) is the maximal $\mathcal{O}$-free range of $\overline{pq}$, then a curve S from p to q is $\mathcal{O}$-convex if and only if S is $\{\theta_1, \theta_2\}$-convex. Note that if $\Theta(\overline{pq}) \in \mathcal{O}$, then the only $\mathcal{O}$-staircase from p to q is $\overline{pq}$.

Since $\mathcal{O}$-free ranges play a significant role in the theory of $\mathcal{O}$-convex sets, it is important to note that we can map any two linearly independent unit vectors $\vec{u_1}$ and $\vec{u_2}$ that correspond to two orientations θ_1 and θ_2 with a bijective linear mapping f to the unit vectors $\vec{e_1}$ and $\vec{e_2}$ that have orientation $0°$ and $90°$, respectively. Since f is linear and bijective, all incidences with lines are preserved and none are introduced. Hence, we will assume in the following that whenever we consider a particular non-empty $\mathcal{O}$-free range (θ_1, θ_2), then $\theta_1 = 0°$ and $\theta_2 = 90°$.

It is easy to see that $\mathcal{O}$-convex sets are closed under intersection. Thus, the notion of a *hull* is well-defined and we obtain the following definition of the $\mathcal{O}$-hull of a set **S**.

$$\mathcal{O}\text{-}hull(\mathbf{S}) := \bigcap \{ \mathbf{C} \subseteq I\!\!E^2 \mid \mathbf{C} \text{ contains } \mathbf{S} \text{ and } \mathbf{C} \text{ is } \mathcal{O}\text{-convex} \}.$$

2.1 $\mathcal{O}$-Visibility

As we already mentioned it is the aim of this paper to investigate some properties of visibility in the context of $\mathcal{O}$-convexity. So the next thing we need is a "natural" generalization of the usual line segment visibility which is given in the following definition (see Figure 3).

Definition 2.2 *Let* **S** *be a subset of the plane and* p, q *be two points in* **S**; *we say* p *and* q $\mathcal{O}$-*see each other or are* $\mathcal{O}$-*visible from each other if there is an* $\mathcal{O}$-*convex path from* p *to* q *that is completely contained in* **S**.

From now on we will call an $\mathcal{O}$-convex path an $\mathcal{O}$-*stairsegment*. An $\mathcal{O}$-stairsegment that consists of a finite number of edges is called an $\mathcal{O}$-*staircase*.

There is an interesting and rather obvious connection with monotone paths. Let $\mathcal{O}^{\perp}$ denote the set of orientations which are orthogonal to those in $\mathcal{O}$. Then, a path $\mathcal{P}$ is $\mathcal{O}$-convex if and only if $\mathcal{P}$ is monotone with respect to all orientations in $\mathcal{O}^{\perp}$ (for an exact definition of monotonicity refer to [11, Definition 2.2]).

As we already mentioned we are interested in the $\mathcal{O}$-kernel of a polygon. The definition of it is completely analogous to the usual definition of the kernel of a set.

Definition 2.3 *Let* $\mathcal{O}$ *be some set of orientations and* **S** *a set in the plane. The* $\mathcal{O}$-*kernel is defined as the set of all points in* **S** *that* $\mathcal{O}$-*see all other points in* **S**. *We denote the* $\mathcal{O}$-*kernel of* **S** *by* $\mathcal{O}$-*kernel*(**S**).

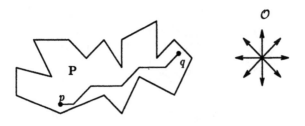

Figure 3: *The definition of $\mathcal{O}$-visibility.*

Though the theory of $\mathcal{O}$-convexity can be developed for arbitrary sets we are mainly concerned with *simple polygons*. If $\mathcal{P}$ is a simple closed curve consisting of (a finite number of) line segments (called edges) such that no two consecutive edges are collinear, we define a *simple polygon* **P** to be the area that is interior to $\mathcal{P}$ together with the curve $\mathcal{P}$. If we want to refer to the curve that surrounds a simple polygon **P**, we speak of the *boundary of* **P** which is denoted by $\partial\mathbf{P}$, i.e., $\partial\mathbf{P} = \mathcal{P}$. The *exterior* of **P** is defined as the points of $I\!\!E^2$ that do not belong to **P** and is denoted by $ext(\mathbf{P})$.

3 Computing the $\mathcal{O}$-Kernel of a Polygon

In the following we turn to the computational aspects of $\mathcal{O}$-visibility. We show in the first part of this section that the $\mathcal{O}$-kernel of a simple polygon can be obtained as the intersection of the $\{\theta\}$-kernels, where $\theta \in \mathcal{O}$. Thus, the computation of the $\mathcal{O}$-kernel is reduced to the task of computing the $\{\theta\}$-kernel, for all $\theta \in \mathcal{O}$, and of intersecting the resulting kernels. A naive application of this observation leads to an $\Omega(n \cdot |\mathcal{O}|)$ algorithm. With a slight refinement of this approach we show how to obtain an algorithm that runs in time $O(n \log |\mathcal{O}|)$, for finite $\mathcal{O}$.

3.1 A Theorem about Monotone Curves in Simple Polygons

The above mentioned intersection property of the $\mathcal{O}$-kernel is based on a by far more general theorem about $\mathcal{O}$-convex curves in simple polygons. Recall that $\mathcal{O}$-convex curves are curves that are monotone w.r.t. to the orientations in $\mathcal{O}^\perp$. In the following theorem we claim that if we are given a simple polygon **P** and two points p and q contained in **P**, there exists an $\mathcal{O}$-staircase from p to q if we only have that, for all $\theta \in \mathcal{O}$, there is a θ-staircase from p to q.

Theorem 3.1 *Let* **P** *be a simple polygon, p and q two points in* **P**, *and $\mathcal{O}$ a set of orientations. We have that p $\mathcal{O}$-sees q if and only if p $\{\theta\}$-sees q, for all $\theta \in \mathcal{O}$.*

Proof: Since the proof of this theorem requires some results about $\mathcal{O}$-staircases which are not developed here, we give only a brief sketch of the proof. The "only if" direction of the claim is obvious since any $\mathcal{O}$-staircase from p to q is $\{\theta\}$-convex, for any $\theta \in \mathcal{O}$. So we turn immediately to the if direction.

Recall that we denote the line segment from p to q by $\overline{pq}$. If $\Theta(\overline{pq}) \in \mathcal{O}$, then there is a $\theta \in \mathcal{O}$, with $\Theta(\overline{pq}) = \theta$. Since p $\{\theta\}$-sees q, we have that $\overline{pq} \subseteq \mathbf{P}$ and, hence, p $\mathcal{O}$-sees q.

So suppose that $\Theta(\overline{pq}) \notin \mathcal{O}$. W.l.o.g. we assume that the maximal $\mathcal{O}$-free range of $\Theta(\overline{pq})$ is $(0°, 90°)$ and that p is to the left and below q. As we mentioned before it can be shown that a staircase from p to q is $\mathcal{O}$-convex if and only if it is $\{0°, 90°\}$-convex.

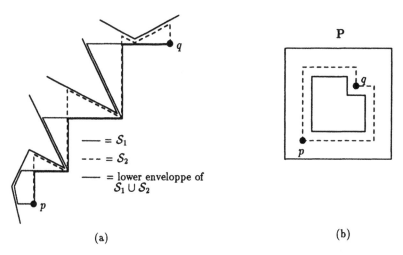

(a)   (b)

Figure 4: *An illustration of the proof of Theorem 3.1.*

By assumption there exist a $\{0°\}$- and a $\{90°\}$-staircase from p to q that are contained in **P**. Let S_1 be the "leftmost" of the $\{0°\}$-staircases and S_2 the "uppermost" of the $\{90°\}$-staircases from p to q. Consider the lower envelope S of S_1 and S_2 as shown in Figure 4a. It can be proven that S is $\{0°, 90°\}$-convex and, thus, satisfies the conditions of the claim. □

Note that the claim is false for polygons with holes. A counterexample is depicted in Figure 4b. q can be reached from p with a $\{0°\}$- and a $\{90°\}$-staircase but there exists no $\{0°, 90°\}$-staircase from p to q that does not intersect the hole of **P**.

The main implication of Theorem 3.1 is the following corollary about the $\mathcal{O}$-kernel.

Corollary 3.2 *If $\mathcal{O}$ is some set of orientations and $\mathbf{P} \subseteq I\!\!E^2$ a simple polygon, then*

$$\mathcal{O}\text{-}kernel(\mathbf{P}) = \bigcap_{\theta \in \mathcal{O}} \{\theta\}\text{-}kernel(\mathbf{P}).$$

Proof: Let q be a point in **P**. If p is in $\mathcal{O}\text{-}kernel(\mathbf{P})$, then p $\mathcal{O}$-sees q; since every $\mathcal{O}$-stairsegment is also an θ-stairsegment, for all $\theta \in \mathcal{O}$, we have $p \in \bigcap\{\theta\}\text{-}kernel(\mathbf{P})$. On the other hand, if p is in $\bigcap\{\theta\}\text{-}kernel(\mathbf{P})$, then p $\{\theta\}$-sees q, for all $\theta \in \mathcal{O}$ and Theorem 3.1 implies that p $\mathcal{O}$-sees q. Hence, $p \in \mathcal{O}\text{-}kernel(\mathbf{P})$. □

As we noted earlier Corollary 3.2 leaves us with the computation of the $\mathcal{O}$-kernel in the case $|\mathcal{O}| = 1$. Of course, we can assume that $\mathcal{O} = \{0°\}$. But before we continue, we need to introduce some notation. Let **P** be a polygon, p a point on the boundary of **P**, and h_p the horizontal line through p. We say that p is a *maximum of* **P** if there exists a neighbourhood N of the component c_p of $h_p \cap \mathbf{P}$ that contains p such that there is no point in $\partial \mathbf{P} \cap N$ that is higher than p. p is called a *reflex maximum* if p is a maximum and there exists a neighbourhood N of p such that $ext(\mathbf{P}) \cap N$ is below h_p otherwise p is called a *convex maximum*. The definition of minima is analogous to the definition of maxima. Note that a maximum or minimum may consist of a vertex or an edge of **P** (see Figure 5). In the following we give a characterization the $\{0°\}$-kernel of a polygon. In order to prove it, we need the following very basic lemma about Jordan-curves, i.e., simple closed curves, which relates reflex and convex minima on the curve (for illustration refer to Figure 6a).

Lemma 3.3 *Let $\mathcal{J}$ be a Jordan-curve. If p and q are two (different) convex minima (maxima) of $\mathcal{J}$, then there is at least one reflex maximum (minimum) of $\mathcal{J}$ above (below) them.*

Proof: Omitted. □

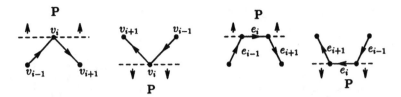

Figure 5: *Possible maxima and minima of a polygon.*

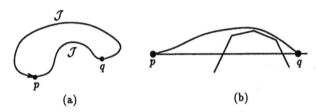

(a)   (b)

Figure 6: *There is a reflex maximum above any two convex minima of a Jordan-curve.*

We now turn to the characterization of the $\{0°\}$-kernel of a polygon. The conditions stated in the following theorem can be easily tested which results in a simple algorithm to compute the $\{0°\}$-kernel of a simple polygon.

Theorem 3.4 *If* **P** *is a simple polygon, then* $p \in \{0°\}$-*kernel*(**P**) *if and only if*

1. *p is above every reflex maximum of* **P** *and*

2. *p is below every reflex minimum of* **P**.

Proof: The only if part of the claim is very easy to see. W.l.o.g. assume that p is below a reflex maximum and let q be some point of the same height as p on the other side of the maximum. Any curve that connects p and q cannot be $\{0°\}$-convex (see Figure 6b).

Now let p be above every reflex maximum and below every reflex minimum and q some point in **P**. W.l.o.g. assume that q is above p. We construct a $\{0°\}$-convex curve $\mathcal{C}$ that connects p and q (for illustration refer to Figure 7). In order to do so let r_q be the horizontal ray towards the left starting in q. We follow r_q until we reach the exterior of **P**. Let the cross point with the boundary of **P** be p_1. Consider a sufficiently small neighbourhood N of p_1 (see Figure 7). Since we cross ∂**P**, there is some part of ∂**P** $\cap N$ below r_q. We follow the part of ∂**P** below r_q until we encounter a reflex minimum or intersect the maximal horizontal line segment h_p through p in p_2. If the latter is the case we just follow h_p to p which is always possible since h_p is a chord in **P**. Otherwise, we continue from p_2 to the left until we again cross ∂**P** and continue the process. We claim that the resulting curve $\mathcal{C}$ finally reaches p and is $\{0°\}$-convex.

To see this note that h_p splits the polygon **P** into a number of components $\mathbf{P}_1,\ldots,\mathbf{P}_k$ that are above and below h_p. We denote the component that contains q by $\mathbf{P}_1$. By construction $\mathcal{C}$ does not cross h_p and, hence, $\mathcal{C}$ is contained in $\mathbf{P}_1 \cup h_p$. We first show that $\mathcal{C}$ is $\{0°\}$-convex. For assume the contrary. Then, there is a horizontal line h that intersects $\mathcal{C}$ in two disconnected components c_1 and c_2. Clearly, both components belong to $\mathbf{P}_1$ since the part of $\mathcal{C}$ that is contained in h_p is connected by construction.

We claim that there is a convex minimum m in the part of $\mathcal{C}$ between c_1 and c_2 that belongs to ∂**P**. To see this just note that there is no minimum of $\mathcal{C}$ between c_1 and c_2 only if both c_1 and c_2 are

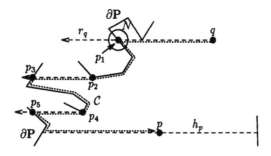

Figure 7: *The construction of a {0°}-convex path from p to q.*

reflex minima. Since the reflex minima of $\partial \mathbf{P}$ are avoided by introducing horizontal edges that form no longer minima, this cannot happen. Hence, there is a minimum m in the part of $\mathcal{C}$ from c_1 to c_2 and, by the construction, of $\mathcal{C}$ m is a convex minimum that belongs to $\partial \mathbf{P}$. If p equals m, then $\mathcal{C}$ does not continue after m which contradicts the fact that h intersects $\mathcal{C}$ in c_1 and c_2. Thus, suppose that p differs from m.

Note that h_p is a convex minimum of $\partial \mathbf{P}_1$. In the following we show that there is no other convex minimum in $\partial \mathbf{P}_1$. This immediately yields the desired contradiction since we already noted above that $\mathcal{C}$ is contained in $\mathbf{P}_1 \cup h_p$. So assume there is a convex minimum m in $\mathbf{P}_1$ that differs from h_p. By Lemma 3.3 we have that there is a reflex maximum r of $\partial \mathbf{P}_1$ between h_p and m that is above both h_p and m. This contradicts Conditions 1 of the claim. Hence, there can be no line that intersects $\mathcal{C}$ in two components and is {0°}-convex.

Since $\mathcal{C}$ is a {0°}-staircase, i.e., monotone, and we follow each edge to its lower end point, we encounter each edge of $\mathbf{P}$ at most once. Hence, the procedure has to stop after a finite number of steps. Since we stop the construction only if we reach p, $\mathcal{C}$ is a {0°}-convex path from q to p as claimed. □

The above theorem immediately yields an algorithm to compute {0°}-*kernel*($\mathbf{P}$). We just have to find the lowest reflex minimum v_l and highest reflex maximum v_h and then output the left and right parts of the boundary of $\mathbf{P}$ with a y-range in the interval $[v_h, v_l]$. All of this can be done in linear time. We summarize our results in the following theorem.

Theorem 3.5 *If* $\mathbf{P}$ *is a simple polygon, then* {0°}-*kernel*($\mathbf{P}$) *can be computed in time linear in the number of edges of* $\mathbf{P}$.

3.2 A Fast Algorithm for the $\mathcal{O}$-Kernel of a Polygon

As we mentioned before the algorithm for the {0°}-kernel enables us to compute $\mathcal{O}$-*kernel*($\mathbf{P}$) by first computing the {θ}-kernel of a polygon $\mathbf{P}$, for each $\theta \in \mathcal{O}$, and then intersecting the kernels obtained to yield $\mathcal{O}$-*kernel*($\mathbf{P}$). Of course, we have to restrict ourselves to finite $\mathcal{O}$. Since we have to apply the {θ}-kernel subroutine $|\mathcal{O}|$-times, the algorithm needs at least time $\Omega(n \cdot |\mathcal{O}|)$ to compute $\mathcal{O}$-*kernel*($\mathbf{P}$) even if we do not take the time to intersect the {θ}-kernels of $\mathbf{P}$ into account.

In the following we present an algorithm to compute $\mathcal{O}$-*kernel*($\mathbf{P}$) that only takes $O(n \log |\mathcal{O}|)$ time if $\mathcal{O}$ is given as a sorted set. In order to develop an algorithm that obtains the above time bound, we have to view the {0°}-kernel a little differently. Let v_1 be the lowest reflex minimum and v_2 the highest reflex maximum of $\mathbf{P}$. If we denote the horizontal line through v_i by h_i, for $i = 1, 2$, then Theorem 3.4 implies that {0°}-*kernel*($\mathbf{P}$) is the intersection of the slab $S(0°)$ between h_1 and h_2 with $\mathbf{P}$ (see Figure 8). Hence, if we are given the slabs $S(\theta)$ that lie between the θ-oriented lines through the lowest θ-reflex minimum and the highest reflex θ-maximum, for all $\theta \in \mathcal{O}$, then we have

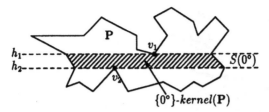

Figure 8: $\mathcal{O}^s$-kernel(P) is the intersection of $S(0°)$ and P.

that

$$\mathcal{O}\text{-}kernel(\mathbf{P}) = \bigcap_{\theta \in \mathcal{O}} (S(\theta) \cap \mathbf{P})$$

$$= \left(\bigcap_{\theta \in \mathcal{O}} S(\theta) \right) \cap \mathbf{P}.$$

In the following we denote $\bigcap_{\theta \in \mathcal{O}} S(\theta)$ by $S_{\mathcal{O}}(\mathbf{P})$. Since $S(\theta)$ is the intersection of two θ-oriented halfplanes, $S_{\mathcal{O}}(\mathbf{P})$ is convex and consists of at most $2|\mathcal{O}|$ edges. Hence, $S_{\mathcal{O}}(\mathbf{P}) \cap \mathbf{P}$ can be computed in time $O(n \log |\mathcal{O}|)$ by testing, for each edge e of $\mathbf{P}$, if e intersects an edge of $\partial S_{\mathcal{O}}(\mathbf{P})$. This yields the intersection points of $\partial S_{\mathcal{O}}(\mathbf{P})$ with $\partial \mathbf{P}$ and enables us to compute the vertices of $S_{\mathcal{O}}(\mathbf{P}) \cap \mathbf{P}$ in additional linear time.

So we only have to show how to compute $S_{\mathcal{O}}(\mathbf{P})$ in time $O(n \log |\mathcal{O}|)$. In order to do so we introduce the following notation. Given a point p in the plane we denote the (closed) halfplane to the left of the θ-oriented line through p by $h^+(p, \theta)$. For a given orientation θ, we denote the highest reflex θ-maximum of $\mathbf{P}$ by v_θ.

The first observation we make is that $S(\theta)$ is the is the intersection of $h^+(v_\theta, \theta)$ with $h^+(v_{\theta-1}, \theta^{-1})$ since the lowest reflex θ-minimum is, of course, the highest reflex θ^{-1}-maximum. Since we always have that θ^{-1} is in $\mathcal{O}$ if $\theta \in \mathcal{O}$, we only have to find the intersection of the θ-oriented halfplanes above the reflex θ-maxima, taken over all $\theta \in \mathcal{O}$. Therefore, if v_θ is the highest reflex θ-maximum of $\mathbf{P}$, for each $\theta \in \mathcal{O}$, we want to compute $\bigcap_{\theta \in \mathcal{O}} h^+(v_\theta, \theta)$.

The idea of the algorithm is to scan the boundary of $\mathbf{P}$ counterclockwise and, thus, successively process the vertices of $\mathbf{P}$. During the scan we keep a list $\mathcal{L}$ of the highest reflex θ-maxima v_θ we have encountered so far, for each $\theta \in \mathcal{O}$. When we process the next reflex vertex v, we check for which orientations $\theta \in \mathcal{O}$ v is a reflex a θ-maximum. We replace those θ-maxima v_θ for which v is higher than v_θ (in the θ-coordinate system) by v. The problem is that there may be $\Omega(|\mathcal{O}|)$ of orientations θ which have v_θ replaced by v. Hence, the algorithm may still take time $\Omega(n|\mathcal{O}|)$ if there are $\Omega(n)$ reflex vertices, each being a reflex θ-maximum for $\Omega(|\mathcal{O}|)$ orientations θ.

The important observation to reduce the running time of the algorithm is that we are only interested in the intersection of the halfplanes $h^+(v_\theta, \theta)$, with $\theta \in \mathcal{O}$. If we have that the so far encountered highest reflex θ-maxima v_θ are lower than v, for a subset $\mathcal{O}'$ of orientations θ in $\mathcal{O}$, the following lemma shows that there exist two orientations θ_1 and θ_2 such that $h^+(v, \theta_1) \cap h^+(v, \theta_2)$ already equals $\bigcap_{\theta \in \mathcal{O}'} h^+(v, \theta)$. If we can find θ_1 and θ_2 in logarithmic time, we only have to replace v_{θ_1} and v_{θ_2} in $\mathcal{L}$ since only these may influence the final intersection. If $\mathcal{L}$ is a dictionary, then replacing v_{θ_1} and v_{θ_2} can be achieved in time $O(\log |\mathcal{O}|)$. Hence, it only has to be shown that θ_1 and θ_2 exist and can be found in time $O(\log |\mathcal{O}|)$.

Lemma 3.6 *If $\mathbf{P}$ is a polygon, $\mathcal{O}$ a finite set of orientations, and v a reflex vertex of $\mathbf{P}$, then there are two orientations θ_1 and θ_2 in $\mathcal{O}$ such that*

$$h^+(v, \theta_1) \cap h^+(v, \theta_2) = \bigcap_{\theta} h^+(v, \theta)$$

if the intersection on the right hand side is taken over all $\theta \in \mathcal{O}$ for which v is a reflex θ-maximum. Furthermore, θ_1 and θ_2 can be found in time $O(\log |\mathcal{O}|)$.

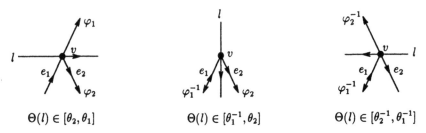

$\Theta(l) \in [\theta_2, \theta_1]$ $\Theta(l) \in [\theta_1^{-1}, \theta_2]$ $\Theta(l) \in [\theta_2^{-1}, \theta_1^{-1}]$

Figure 9: v is a reflex θ-maximum only for the orientations $\theta \in [\varphi_2, \varphi_1]$.

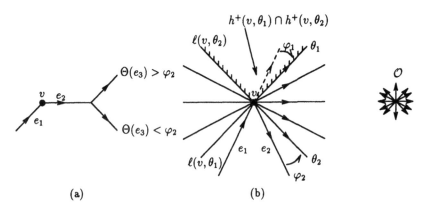

(a) (b)

Figure 10: *How to find θ_1 and θ_2.*

Proof: Let e_1 be the edge that is (counterclockwise) before v and e_2 the edge after v. To see the claim we denote the orientation of e_i by φ_i. We claim that v is a reflex θ-maximum only for the orientations θ that are in the interval $[\varphi_2, \varphi_1]$.[1] This can be shown as follows (see Figure 9). We denote the line through a point p with orientation θ by $\ell(p, \theta)$. If $\theta \in [\varphi_1, \varphi_2^{-1}]$ or $\theta \in [\varphi_1^{-1}, \varphi_2]$, then the start point of e_1 is on the other side of $\ell(v, \theta)$ than the end point of e_2. Hence, v is no θ-extremum. If θ is in $[\varphi_2^{-1}, \varphi_1^{-1}]$, then e_1 and e_2 are above $\ell(v, \theta)$, and v is no θ-maximum. Only if $\theta \in (\varphi_2, \varphi_1)$, then both of e_1 and e_2 are completely below $\ell(v, \theta)$. There is a little problem if φ_2 is in $\mathcal{O}$, since then it depends on the edge after e_2 if v is a φ_2-maximum or not (refer to Figure 10). To see this suppose that $\varphi_2 \in \mathcal{O}$ and let e_3 be the edge after e_2. If $\Theta(e_3) < \varphi_2$, then e_2 is a φ_2-maximum since $\varphi_2 < \varphi_1$ and if $\Theta(e_3) > \varphi_2$, then e_2 merely contains some φ_2-saddlepoints. Since we only have to spend a constant amount of time to find the edge after e_2, we will not consider this special case any more. Clearly, it can be checked in logarithmic time if φ_2 is in $\mathcal{O}$ if $\mathcal{O}$ is sorted.

So let θ_1 be the last orientation in $\mathcal{O}$ that is (counterclockwise) before φ_1 and θ_2 be the first orientation in $\mathcal{O}$ after φ_2 if they exist (see Figure 10). We denote the interval (φ_2, φ_1) by I. If there is one θ in $\mathcal{O}$ with $\theta \in I$, then, obviously, we have that θ_1 and θ_2 both exist. Hence, if they are not defined, then there is no θ in $\mathcal{O}$ for which v is a θ-maximum and we are done. So assume that $\theta_1, \theta_2 \in I$ and that they are different (if $\theta_1 = \theta_2$, then there is only one element in $\mathcal{O} \cap I$ and the lemma is trivially true.)

Consider the wedge w that is formed by $h^+(v, \theta_1) \cap h^+(v, \theta_2)$. It can easily be seen that w is contained in any halfplane $h^+(v, \theta)$ if $\theta \in [\theta_2, \theta_1]$. Since we have that $\mathcal{O} \cap I = \mathcal{O} \cap [\theta_2, \theta_1]$ by our choice of θ_1 and θ_2 and the orientations in $\mathcal{O} \cap I$ are the only ones for which v is a θ-maximum, we

---
[1] Since v is reflex, we have that $\Theta(e_1) > \Theta(e_2)$.

have that

$$h^+(v, \theta_1) \cap h^+(v, \theta_2) = \bigcap_{\theta \in \mathcal{O} \cap [\theta_1, \theta_2]} h^+(v, \theta) = \bigcap_{\theta \in \mathcal{O} \cap I} h^+(v, \theta).$$

Hence, it suffices to just look at v_{θ_1} and v_{θ_2} and possibly replace them with v. Note that if v_{θ_1} is higher than v in the θ_1-coordinate system, i.e., if we *do not* replace v_{θ_1} with v, then $h^+(v_{\theta_1}, \theta_1) \subseteq h^+(v, \theta_1)$, and, hence, $h^+(v_{\theta_1}, \theta_1) \cap h^+(v, \theta_2)$ is still contained in each $h^+(v, \theta)$, with $\theta \in [\theta_2, \theta_1]$.

If $\mathcal{O}$ is given as a sorted array, it is easy to see that θ_1 and θ_2 can be found in time $O(\log |\mathcal{O}|)$ given φ_1 and φ_2. □

So the scanning algorithm computes a number of vertices v_θ such that

$$\bigcap_{\theta \in \mathcal{O}} h^+(v_\theta, \theta) \cap \mathbf{P}$$

yields the $\mathcal{O}$-kernel of $\mathbf{P}$. Above we have shown that we can compute the vertices v_θ in time $O(n \log |\mathcal{O}|)$. Observe that there are at most $O(|\mathcal{O}|)$ such vertices, one for each orientation $\theta \in \mathcal{O}$. Furthermore, note that the above computation yields the vertices v_θ sorted according to θ. It is easy to see that the intersection of halfplanes that are sorted according to slope can be computed in time linear in the number of halfspaces. Hence, the intersection $\bigcap_{\theta \in \mathcal{O}} h^+(v_\theta, \theta)$ can be computed in time $O(|\mathcal{O}|)$. This proves the following theorem.

Theorem 3.7 *The $\mathcal{O}$-kernel of a simple polygon with n vertices can be computed in time $O(n \log |\mathcal{O}| + |\mathcal{O}|)$, for finite $\mathcal{O}$, given $O(|\mathcal{O}| \log |\mathcal{O}|)$ preprocessing time to sort $\mathcal{O}$.*

References

[1] Richard Cole and Michael T. Goodrich. Optimal parallel algorithms for polygon and point-set problems. In *Proceedings of the Fourth Annual Symposium on Computational Geometry*, pages 201–210, ACM, ACM Press, Champaign, Illinois, June 1988.

[2] Joseph Culberson and Robert Reckhow. *A Unified Approach to Orthogonal Polygon Covering Problems via Dent Diagrams.* Technical Report TR 89-6, Department of Computing Science, University of Alberta, Edmonton, Alberta, Canada, February 1989.

[3] Laxmi P. Gewali and Simeon Ntafos. Minimum covers for grids and orthogonal polygons by periscope guards. In Jorge Urrutia, editor, *Proceedings of the Second Candian Conference in Computational Geometry*, pages 358–361, University of Ottawa, 1990.

[4] J. Mark Keil. Minimally covering a horizontally convex polygon. In *Proceedings of the Second Annual Symposium on Computational Geometry*, pages 43–51, ACM, Yorktown Heights, New York, June 1986.

[5] D. T. Lee and F. P. Preparata. An optimal algorithm for finding the kernel of a polygon. *Journal of the ACM*, 26(3):415–421, July 1979.

[6] Rajeev Motwani, Arvind Raghunathan, and Huzur Saran. Covering orthogonal polygons with star polygons: the perfect graph approach. In *Proceedings of the Fourth Annual Symposium on Computational Geometry*, pages 211–223, ACM, Urbana, Illinois, June 1988.

[7] Rajeev Motwani, Arvind Raghunathan, and Huzur Saran. Perfect graphs and orthogonally convex covers. In *Fourth Annual SIAM Conference on Discrete Mothematics*, page , 1988.

[8] J. Ian Munro, Mark Overmars, and Derick Wood. Variations on visibility. In *Proceedings of the Third Annual Computational Geometry*, ACM, Waterloo, Ontario, Canada, June 1987.

[9] T. M. Nicholl, D. T. Lee, Y. Z. Liao, and C. K. Wong. Constructing the x-y convex hull of a set of x-y convex polygons. *BIT*, 23:456–471, 1983.

[10] Mark Overmars and Derick Wood. On rectangular visibility. *Journal of Algorithms*, 9:372–390, 1988.

[11] F.P. Preparata and M.I. Shamos. *Computational Geometry — an Introduction.* Springer Verlag, 1985.

[12] Gregory J. E. Rawlins. *Explorations in Restricted-Orientation Geometry*. PhD thesis, University of Waterloo, 1987.

[13] Gregory J. E. Rawlins and Derick Wood. Computational geometry with restricted orientations. In *Proceedings of the 13th IFIP Conference on System Modelling and Optimization*, Springer Verlag, 1988. Lecture Notes in Computer Science.

[14] Gregory J. E. Rawlins and Derick Wood. On the optimal computation of finitely-oriented convex hulls. *Information and Computation*, 72:150–166, 1987.

[15] Gregory J. E. Rawlins and Derick Wood. Ortho-convexity and its generalizations. In Godfried T. Toussaint, editor, *Computational Morphology*, pages 137–152, Elsevier Science Publishers B. V., (North-Holland), 1988.

[16] Robert A. Reckhow and Joseph Culberson. Covering a simple orthogonal polygon with a minimum number of orthogonally convex polygons. In *Proceedings of the Third Annual Symposium on Computational Geometry*, pages 268–277, ACM, Waterloo, Ontario, Canada, June 1987.

[17] Derick Wood and Chee K. Yap. The orthogonal convex skull problem. *Discrete and Computational Geometry*, 3(4):349–365, 1988.

A New Simple Linear Algorithm to Recognize Interval Graphs

Klaus Simon
Institut für Theoretische Informatik
ETH-Zentrum
CH-8092 Zürich

Abstract

The first linear algorithm for recognizing interval graphs was presented by
BOOTH and LEUKER[4] in 1976. The first phase of this algorithm finds a
perfect elimination scheme and determines all maximal cliques $A_1, \ldots, A_s$ of a
given graph using the fact that interval graphs are a proper subclass of chordal
graphs. This part is based on a lexicographic breadth first search (for short
lexBFS) which is also used in other areas such as scheduling problems [18][5].
In the second phase the PQ-tree data structure is used to get a representation
of all possible consecutive arrangements of all maximal cliques $A_1, \ldots, A_s$.
KORTE and MÖHRING[13], 1989, improved this algorithm by a more adaptive
version of PQ-trees, so called MPQ-trees, for the second phase. In this paper
we show a new solution of the second phase by repeated use of lexBFS, which
produces a linear arrangement of the maximal cliques $A_1, \ldots, A_s$, if there is
one.

1 Introduction

An undirected graph $G = (V, E)$, $V = \{1, \ldots, n\}$, $E \subseteq V \times V$, is called an *interval
graph* for a family $F = \{I_1, \ldots, I_n\}$ of intervals on a line if and only if

(1) $$\forall v, w \in V \qquad (v, w) \in E \iff I_v \cap I_w \neq \emptyset,$$

i.e. two vertices are adjacent if and only if the corresponding intervals intersect.
Such a family[1] is called an *interval representation* or an *intersection model* of G.
Interval graphs were first introduced by HAJÓS [12]. Interest in interval graphs
arises out of many applications in biology [1], psychology [16], from traffic light

---

[1] In general we call a graph G which fulfills (1) the intersection graph of the family F without
any condition on F.

sequencing [8], VLSI-layout problems [6] and many others[2]. The main reason to study interval graphs is that some classical combinatorial problems like maximum clique, coloring [9], stable set, total domination [2] or Hamiltonian cicruit [3] can be solved efficiently for interval graphs.

Recognition algorithms existing so far are based on one of the three characterizations in theorem 1. In order to characterize interval graphs we need some further notation. An undirected graph is called *triangulated* or *chordal* if every cycle of length strictly greater than 3 possesses a *chord*, that is an edge joining two nonconsecutive vertices of the cycle. A *comparability graph* is an undirected graph $G = (V, E)$ to every edge of which can be assigned a direction such that the resulting oriented graph $T = (V, F)$ represents a partial order of the nodes, i.e. T is acyclic and

$$\forall x, y, z \in V \quad (x, y) \in F \land (y, z) \in F \Rightarrow (x, z) \in F.$$

A subset $A \subseteq V$ is a *clique* if the subgraph of G induced by A is complete. A clique A is *maximal* if there is no clique of G which properly contains A as a subset. A graph G is called *asteroidal* if it contains three distinct nodes a_0, a_1, a_2 such that, for $i = 0, 1, 2$, there is a path from a_{i-1} to a_{i+1} in the graph $G - (\{a_i\} + adj(a_i))$, where the arithmetic will be done mod 2. The three vertices a_0, a_1, a_2 fulfilling the preceding property will form an *asteroidal triple*. Now we are able to come to the characterizations of interval graphs due to GILMORE and HOFFMAN [10], LEKKERKERKER and BOLAND [14], FULKERSON and GROSS [8].

Theorem 1 *Let $G = (V, E)$ be an undirected graph. Then the following statements are equivalent.*

1. *G is an interval graph.*

2. *G is triangulated and its complement $\bar{G}$ is a comparability graph.*

3. *The maximal cliques of G can be linearly ordered such that, for every vertex x of G, the maximal cliques containing x occur consecutively.*

4. *G is triangulated and G is not asteroidal.*

An ordering of the maximal cliques of G which fulfills point 3 of theorem 1 is called a *linear* or *consecutive arrangement*. The first polynomial recognition algorithms based on these characterizations all have at least $O(n^3)$ worst case complexity. The first linear algorithm was developed by BOOTH and LEUKER [4]. They improved the FULKERSON-GROSS test with a special data structure called a *PQ*-tree. To obtain the input of their test, namely the maximal cliques of G, they used a labeling method [17], known as lexicographic breadth first search — lexBFS for short — which leads to a special permutation of the nodes called lexBFS-ordering. From a lexBFS-ordering the maximal cliques can be constructed in linear time using the fact that an interval graph is always chordal. A useful property of chordal graphs is the fact that the number of maximal cliques is bounded by the number of nodes n, see [11].

---

[2]See the survey on interval graphs used in ROBERTS [16], Chap. IV.

A drawback of this method is the complexity of PQ-tree algorithms. Since many cases must be differentiated, this is true for its implementation and intuitive under-standing, although PQ-trees perform quite well in practice [15]. This complexity was reduced by KORTE and MÖHRING using a more adaptive version of PQ-trees noted as modified PQ-trees (MPQ-trees). This new data structure uses the lexBFS-ordering not only to compute maximal cliques but also to simplify the updates of MPQ-trees.

Now in this paper we show that a consecutive arrangement of the maximal cliques of an interval graph can be computed without using PQ-trees or related data structures but only by iterating lexBFS. Experiments with an implementation of our simple algorithm indicate that in practice it is competitive with the PQ-tree-method. In our analysis we extend the concept of asteroidal triple to maximal cliques. This leads to a clear and direct correspondence between geometric and combinatorial properties which can be implemented in a straight forward manner.

The rest of the paper is organized as follows: Section 2 contains notations and a short description of lexicographic breadth first search. Section 3 describes some properties of maximal cliques in an interval graph and the relationship to the lexBFS-ordering. Section 4 outlines the algorithm.

2 Notations, Chordal Graphs and lexBFS

A undirected graph $G = (V, E)$ consists of a finite set $V = \{1, \ldots, n\}$ of elements called vertices or nodes and a symmetric, irreflexive adjacency relation E whose elements are called edges or arcs. The set of vertices adjacent to v is given by $adj(v)$,

$$adj(v) = \{w \in V \mid (v, w) \in E\}.$$

As a shorter form of $\{v\} \cup adj(v)$ we will use $adj^*(v)$. We say that the edge (v, w) leaves v, enters w and joins v with w. A node b is adjacent to a vertex set $A \subseteq V$ if there is a vertex $a \in A$ with $(a, b) \in E$. The complement $\bar{G} = (V, \bar{E})$ of G is given by

$$\bar{E} = \{(v, w) \mid v \neq w \wedge (v, w) \notin E\}.$$

For a set $A \subseteq V$ the graph $G_A = (A, E_A)$ is the subgraph induced by the vertex set A, where

$$E_A = \{(v, w) \in E \mid v, w \in A\}.$$

A path P in G from vertex v_0 to vertex v_s is a sequence of vertices $v_0, v_1, \ldots, v_s$ such that (v_{i-1}, v_i) is an edge for $i \in \{1, \ldots, s\}$; s is the length of the path. P is called simple if the nodes $v_0, \ldots, v_s$ are pairwise distinct. A path $v_0, \ldots, v_s$, $s \geq 2$, is chordless if $(v_{i-1}, v_{i+1}) \notin E$ for $1 \leq i \leq s - 1$. A path $v_0, \ldots, v_s$ is a cycle if $s > 1$ and $v_0 = v_s$ and a simple cycle if in addition $v_0, \ldots, v_{s-1}$ are pairwise distinct. Two nodes in graph G are connected if there exists a path in G joining them; further, G is connected if any two nodes are connected; finally, two sets $A, B \subseteq V$ are called connected if there are nodes $a \in A$ and $b \in B$ which are connected. Unless we specify otherwise, in the following we will assume that the graph G is connected. A connected component or shorter a component is a maximal[3] connected subgraph of G. A subset $A \subseteq V$ is a (vertex) separator of a connected graph $G = (V, E)$ if $G - A$ is not connected. The set A is called a a, b-separator, $a, b \in V - A$, $(a, b) \notin E$, if in $G - A$ the vertices a and b are contained in distinct connected components. If no proper subset of A is an a, b-separator, then A is a minimal separator for a and b.

As noted above a graph G is called triangulated if for every single cycle $K = v_0, \ldots, v_k$ of length $k > 3$, there is an edge of G joining two nonconsecutive vertices of K. The following characterization was given[4] by DIRAC[7].

Theorem 2 *An undirected graph $G = (V, E)$ is triangulated if and only if every minimal vertex separator is a clique.*

A vertex v of G is noted as simplicial if its adjacency set $adj(x)$ induces a complete subgraph, i.e. $adj(x)$ is a clique and therefore $adj^*(v)$ is a clique, too. Let $\varphi = [\varphi(v), \varphi(2), \ldots, \varphi(n)]$ be a permutation of the nodes $V = \{1, \ldots, n\}$. We say that φ is a *perfect elimination scheme* if each $\varphi(i)$, $1 \leq i \leq n$ is a simplicial node in the induced subgraph

$$G_i = (V_i, E_i) \stackrel{\text{def}}{=} G - \{\varphi(1), \ldots, \varphi(i-1)\}.$$

FULKERSON and GROSS [8] have shown the following characterization of chordal graphs.

Theorem 3 *An undirected graph $G = (V, E)$ is triangulated if and only if there is a perfect elimination scheme φ for G.*

This characterization was used by ROSE, TARJAN and LEUKER [17] to introduce their linear recognation algorithm for chordal graphs. In order to find a perfect elimination scheme they developed a method like the well-known breadth first search (BFS) procedure with the additional rule that vertices with earlier visited neighbors are preferred. In the lexicographic breadth-first search version the usual queue of vertices is replaced by a queue of sets noted as $label(v)$, $v \in V$. Thereby a label is a subset of $\{1, \ldots, n\}$ and two labels $label(v)$ and $label(w)$, $v, w \in V$, are ordered as follows:

$$label(v) \stackrel{lex}{<} label(w) \stackrel{\text{def}}{\Longleftrightarrow} \max(label(v) - label(w)) < \max(label(w) - label(v)).$$

If the members in the labels are listed in decreasing order then the lexicographic order on the corresponding strings is just our order[5] on the labels. The method is given in figure 1. The following theorem about lexBFS was proved in ROSE, TARJAN and LUECKER [17].

---

[3]In this paper maximal always refers to set inclusion.
[4]See [11] for all proofs not in this paper.
[5]But sets are easier to handle than strings.

Theorem 4

1. *A lexBFS-ordering φ can be computed in linear time $O(n + m)$.*

2. *For every chordal graph a lexBFS-ordering is a perfect elimination scheme.*

3. *Let φ be a perfect elimination scheme for a chordal graph $G = (V, E)$ then the maximal clique $A_1, \ldots, A_s$, $1 \leq s \leq n$, can be determined from φ in time complexity $O(|V| + |E|)$.*

4. *For each value i, $1 \leq i \leq n$, let $L_i(v)$, $v \in V$, denote the label of x immediately before line (4) is executed, i.e. right before the node $\varphi(i)$ is numbered. Then for all $x, y \in V$:*

   (L1) $L_i(x) \leq L_j(x)$ for $1 \leq j \leq i \leq n$.
   (L2) $L_i(x) < L_i(y) \Rightarrow L_j(x) < L_j(y)$ for $1 \leq j < i \leq n$.
   (L3) If $\varphi^{-1}(x) < \varphi^{-1}(y) < \varphi^{-1}(z)$ and $z \in adj(x) - adj(y)$ then there is a node $u \in adj(y) - adj(x)$ with $\varphi^{-1}(z) < \varphi^{-1}(u)$.

5. *Let A be a maximal clique of a chordal graph $G = (V, E)$ and φ a perfect elimination scheme for G. Then for a node v, $v \in A$, with*

$$\varphi^{-1}(v) = \min \{ \varphi^{-1}(w) \mid w \in A \}$$

   it is valid

$$A = \{ x \} \cup \{ w \in adj(v) \mid \varphi^{-1}(v) < \varphi^{-1}(w) \}.$$

--- lexBFS ---

```
(1)      forall v ∈ V do label(v) ← ∅ od;
(2)      for i ← n downto 1 do
(3)         let v be an unnumbered vertex with largest label;
(4)         φ(i) ← v;
            (* update *)
(5)         forall unnumbered vertices w ∈ adj(v) do
(6)            label(w) ← label(w) + { i };
(7)         od;
(8)      od;
```

--- lexBFS ---

Figure 1: Algorithm A

Corollary 5 *Let φ be a lexBFS-ordering of an interval graph $G = (V, E)$. For any chordless path $v_0, \ldots, v_s$, $s \geq 2$ in G there is an index l, $0 \leq l \leq s$ such that*

$$\varphi^{-1}(v_0) < \cdots < \varphi^{-1}(v_l) > \cdots > \varphi^{-1}(v_s).$$

Especially, if $\varphi^{-1}(v_0) < \varphi^{-1}(v_s)$ then

$$\forall h, 1 \leq h \leq s \qquad \varphi^{-1}(v_1) < \varphi^{-1}(v_h).$$

Proof: The corollary follows immediately from the observation

$$\not\exists\, h, 1 < h < s - 1 \qquad \varphi^{-1}(v_h) < \min(\varphi^{-1}(v_{l-1}), \varphi^{-1}(v_{l+1})).$$

Assume this is not true, then in the graph

$$G_i = G - \{\, \varphi(1), \ldots, \varphi(i-1)\,\},$$

where $\varphi(i) = v_h$, the node v_h is simplicial and both vertices v_{h-1} and v_{l+1} are adjacent to v_l. This would imply $(v_{l-1}, v_{l+1}) \in E$, against the definition of a chordless path.

□

Corollary 6 *Let x be a vertex of an undirected graph $G = (V, E)$ with lexBFS-ordering φ. If x is adjacent to all other nodes in G then there exists a lexBFS-ordering φ_1 for $G' = G - x$ with*

$$\forall y \in V, y \neq x \qquad \varphi_1^{-1}(y) \le \varphi^{-1}(y).$$

Proof: If we apply lexBFS to G' with the same initialisation as for G then we obtain an ordering[6] φ_1 given by

$$\forall y \in V, y \neq x \qquad \varphi_1^{-1}(y) = \begin{cases} \varphi^{-1}(y) - 1 & \text{for } \varphi^{-1}(x) < \varphi^{-1}(y) \\ \varphi^{-1}(y) & \text{otherwise.} \end{cases}$$

□

If φ_1 is a lexBFS-ordering for G_X, $X \subseteq V$, and φ_1 arises from φ by inductive use of corollary 6 then we call φ_1 a reduced ordering with respect to X and φ.

In relation to theorem 4 there exists a linear ordering $\overset{\varphi}{<}$ of the maximal cliques $A_1, \ldots, A_s$ of a chordal graph $G = (V, E)$ induced by a perfect elimination scheme φ

$$\forall 1 \le i, j \le s \qquad A_i \overset{\varphi}{<} A_j \overset{\text{def}}{\Longleftrightarrow} \min_{x \in A_i} \varphi^{-1}(x) < \min_{y \in A_j} \varphi^{-1}(y).$$

3 Interval Graphs and lexBFS

In this section we want to analyse the relations between an interval graph $G = (V, E)$, its intersection model $F = \{\, I_1, \ldots, I_n \,\}$ and its maximal cliques $A_1, \ldots, A_s$. We consider without loss of generality[7] only closed intervals on the real line. The boundaries of an interval I_v are given by $a(v)$ and $b(v)$, $a(v) \le b(v)$, hence

$$\forall v \in V \qquad I_v = [a(v), b(v)].$$

A family of intervals is connected if their union forms an unique interval. Note that, a set of intervals is connected if and only if the corresponding subgraph of G is

---

[6]This fact is also true if $G - x$ is no longer connected.

[7]w.l.o.g. for short

connected. Clearly, a maximal clique A_i of G corresponds to a maximal subset of F with no empty intersection. The intersection of the intervals I_v, $v \in A_i$, is noted as

$$D(A_i) = \bigcap_{v \in A_i} I_v \neq \emptyset.$$

Since

$$D(A_i) \cap D(A_j) = \emptyset$$

for $1 \leq i \neq j \leq n$ the maximal cliques $A_1, \ldots, A_s$ can be ordered as follows

$$\forall i, j, \ 1 \leq i \neq j \leq n \qquad A_i \overset{F}{<} A_j \overset{\text{def}}{\Longleftrightarrow} \max D(A_i) < \min D(A_j).$$

Due to theorem 1 for each interval graph $G = (V, E)$ there is a *standard interval representation* such that

$$\{\, a(v), b(v) \mid v \in V \wedge I_v = [a(v), b(v)]\,\} \subseteq \{\, 1, \ldots, s \,\}$$

and the numbering of the cliques is consistent with the intersection of their intervals, i.e.

$$\forall 1 \leq i \leq s \qquad D(A_i) = [i, i] = i.$$

Two intervals overlap if they intersect but neither properly contains the other. Evidently, for every two overlapping intervals $I' = [a', b']$ and $I'' = [a'', b'']$

$$\min \{\, a', a'' \,\} < \max \{\, b', b'' \,\}.$$

Let $[l, h]$ be an interval such that

$$\forall v \in V \qquad (I_v \cap [l, h] \neq \emptyset) \Rightarrow [l, h] \subseteq I_v;$$

then we gain a new interval representation $F' = \{\, I'_1, \ldots, I'_n \,\}$ for G by *reflecting* the intervals contained in $[l, h]$, i.e.

$$I'_v = \begin{cases} I_v & \text{if } I_v \not\subseteq [l, h] \\ [l + h - b(v), l + h - a(v)] & \text{otherwise.} \end{cases}$$

A clique A_i, $1 \leq i \leq s$, is called an *inner clique* if and only if for every intersection model $F = \{\, I_v = [a(v), b(v)] \mid v \in V \,\}$ of G

$$\min_{v \in V} b(v) < \min_{v \in A_i} a(v)$$

and

$$\max_{v \in A_i} b(v) < \max_{v \in V} a(v)$$

hold. A maximal clique of G which is not an inner clique is called an *outer clique*.

Corollary 7 *If A_i is an inner clique of G then the subgraph $G - A_i$ is not connected.*

Proof: Immediately from the definition.

$\square$

Our first aim is to derive a combinatorial characterization for an inner clique of G. The following notation will be used to reach this. When no confusion is possible, we will sometimes identify a vertex set A and the corresponding set of intervals.

Definition 8 *Let B', B, B'' be maximal cliques of an interval graph $G = (V, E)$. Then we call (B', B, B'') a linked triple with inner clique B if and only if the following hold:*

1. *$B' - B$ and $B'' - B$ are not connected in $G - B$.*

2. *$(B' \cap B) - B'' \neq \emptyset \neq (B'' \cap B) - B'$.*

3. *For all vertices $x_1 \in B' - B$, $x_2 \in (B' \cap B) - B''$, $x_3 \in (B \cap B'') - B'$, $x_4 \in B'' - B$ the vertex sequence x_1, x_2, x_3, x_4 is a chordless path in G.*

Note that, since B', B, B'' are maximal, we get

$$(B' - B'') \neq \emptyset \neq (B - B'') \text{ and } (B'' - B') \neq \emptyset \neq (B - B'')$$

and therefore the nodes $x_1, \ldots, x_4$ always exist. The first lemma demonstrates the correlation between a linked triple and an intersection model of G.

Lemma 9 *For all linked triples (B', B, B'') of an interval graph $G = (V, E)$ and for every interval representation $F = \{ I_1, \ldots, I_n \}$ of them it is true that*

$$B' \overset{F}{<} B \overset{F}{<} B'' \text{ or } B'' \overset{F}{<} B \overset{F}{<} B'.$$

Proof: Suppose the lemma is wrong. Then there is a linked triple (B', B, B'') and an interval representation F with B as the smallest or the largest clique. Let w.l.o.g. B be the leftmost clique and further let B' be less than B'', then

$$B \overset{F}{<} B' \overset{F}{<} B'' \iff \max D(B) < \min D(B') \leq \max D(B') \leq \min D(B'').$$

Now consider a path $v_1, \ldots, v_l$ from $v_1 \in B$ to $v_l \in B''$. The corresponding sequence of intervals must intersect $D(B')$. Therefore B' is a separator for every couple of vertices $v_1 \in B$ and $v_l \in B''$ in opposition to our definition of a linked triple (B', B, B''), which requires that $v_1 \in (B \cap B'') - B' \neq \emptyset$ and $v_l \in B'' - B \neq \emptyset$ are connected by an edge in $G - B'$.

$\square$

Corollary 10 *Let (B', B, B'') be a linked triple in an interval graph $G = (V, E)$ and let x, $x \notin B''$, be a node adjacent to B'. Then the set $(B' \cap B) - B''$ is a x, y-separator in $G - (B' \cap B'')$ for any $y \in B - B''$.*

The concept of a linked triple also gives us a combinatorial characterization of an inner clique, which is used in the next section to prove the correctness of our algorithm.

Lemma 11 *Let B be an inner clique of an interval graph $G = (V, E)$; then there exists a linked triple (B', B, B'') of G or there are linked triples (B_1, B_2, B) and (B, B_3, B_4) such that $B - (B_1 \cup B_4)$ is a x, y-separator in $G - (B_1 \cap B_4)$ for all $x \in B_1 - B$ and $y \in B_4 - B$.*

Proof: Let G be an interval graph with maximal cliques $A_1, \ldots, A_s$ and standard intersection model $F = \{ I_1, \ldots, I_n \}$. Then the intersection of intervals corresponding to A_i, $1 \leq i \leq s$, is given by

$$D(A_i) = i$$

and there is an index l, $1 < l < s$, with $A_l = B$. W.l.o.g. let F be selected in such a way that l is as large as possible.

First of all we observe that with

(1) $\exists t, 1 \leq t < l$ $A_t - A_s$ is connected to $A_l - A_s$ in $G - A_s$.

Supposing the contrary, no interval of F overlaps with the interval $[l, s]$. For this reason $[l, s]$ can be reflected producing a new legal interval representation F' for G with A_l as an outer clique against our precondition. This contradiction shows our proposition.

Now let us choose the index t as small as possible; then we find out that

(2) $A_l - A_t$ is connected to $A_s - A_t$ in $G - A_t$.

Assume the statement is wrong. Let h be the largest index, $l \leq h < s$ such that $A_l - A_t$ is connected to $A_h - A_t$ in $G - A_t$. Then no interval $I = [a, b]$ with $h < b \leq s$ overlaps the interval $[t, h]$, because such an interval would violate the maximality of h. On the other hand, according to the choice of t every interval $I = [a, b]$, $1 \leq a < t$, which intersects $[t, l]$ is an element of A_s and therefore does not overlap $[t, s]$. Hence the intervals $[t, s]$ and $[t, h]$ can be reflected. Applying both reflections we observe a new intersection model of G,

$$A_1, \ldots, A_{t-1}, A_s, A_{s-1}, \ldots, A_{h+1}, A_t, A_{t+1}, \ldots, A_l, \ldots, A_h,$$

with larger index for $A_l = B$ against our choice of l. This contradiction proves the validity of fact (2).

Combining the facts (1) and (2) we obtain that

(3) $\exists t, 1 \leq t < l \leq s$ $A_t - A_s$ is connected to $A_s - A_t$ in $G - (A_t \cap A_s)$.

Let now $P = x_1, \ldots, x_p$ be a shortest path in $G - (A_t \cap A_s)$ between $x_1 \in A_t - A_s$ and $x_p \in A_s - A_t$. With (3) such a path exists and by definition its length p is greater than 2. From the minimality of P we infer that $x_1, \ldots, x_p$ corresponds to a sequence of overlapping intervals $I_{x_1}, \ldots, I_{x_p}$. Next we extend this sequence by $I_{x_0} = [a(x_0), b(x_0)]$ and $I_{x_{p+1}} = [a(x_{p+1}), b(x_{p+1})]$, where I_{x_0} and $I_{x_{p+1}}$ are selected such that

$$b(x_0) = t \quad \text{and} \quad a(x_{p+1}) = h.$$

The existence of such intervals follows from our definition of F as a standard intersection model of G. Now it is easy to complete the proof. We distinguish two cases.

Case 1. The number l is contained in $I_{x_q} \cap I_{x_{q+1}}$ for $1 \leq q < p$. Then the choice $B' = A_{a(x_q)}$, $B = A_l$ and $B'' = A_{b(x_{q+1})}$ satisfies our conditions.

Case 2. There is an interval I_{x_q}, $1 \leq q \leq p$ with

$$l \in I_{x_q} \text{ and } l \notin (I_{x_{q_1}} \cup I_{x_{q+1}}).$$

Then $B_1 = A_{a(x_{q-1})}$, $B_2 = A_{a(x_q)}$, $B = A_l$, $B_3 = A_{b(x_q)}$, and $B_4 = A_{b(x_{q+1})}$ fulfill our proposition.

$\square$

The next statement shows the central motivation for our definition of a linked triple.

Theorem 12 *Let φ be a* lexBFS-*ordering of an interval graph $G = (V, E)$ such that*

$$B' \overset{\varphi}{\lesssim} B \overset{\varphi}{\lesssim} B'' \text{ or } B'' \overset{\varphi}{\lesssim} B \overset{\varphi}{\lesssim} B'$$

for all linked triples (B', B, B'') then $\overset{\varphi}{\lesssim}$ induces a consecutive arrangement of G.

Proof: In the contrary, suppose the lexBFS-odering $A_1 \overset{\varphi}{\lesssim} A_2 \overset{\varphi}{\lesssim} \cdots \overset{\varphi}{\lesssim} A_s$ is not a linear arrangement of G. Now let l, $l \leq s - 2$, be the greatest index such that $A_l, A_{l+1}, \ldots, A_s$ is not a linear arrangement of G_A, the induced subgraph by A, where A stands for $A_l \cup A_{l+1} \cup \cdots \cup A_s$. Let x be the least numbered node in A_l, i.e.

(1) $$\varphi^{-1}(x) = \min \{ \varphi^{-1}(v) \mid v \in A_l \}.$$

Since $A_l, \ldots, A_s$ is not a linear arrangement of G_A, there are maximal cliques A_k and A_h such that

(2) $$A_l \overset{\varphi}{\lesssim} A_k \overset{\varphi}{\lesssim} A_h$$

and

(3) $$\exists z \in V \quad z \in (A_l \cap A_h) \wedge z \notin A_k.$$

Because $A_{l+1}, \ldots, A_s$ occur consecutively we can choose k for any h such that

(4) $$k = h - 1.$$

Further we select the index h as great as possible. Now let y be the node in A_k with smallest number; then it follows by our definitions that

(5) $$\varphi^{-1}(x) < \varphi^{-1}(y) < \varphi^{-1}(z)$$

and

(6) $$(x, y) \notin E \text{ and } (y, z) \notin E.$$

But statements (4) and (5) satisfy the preconditions of theorem 4.4, (L3). Therefore there exists a node u with

(7) $$u \in (adj(y) - adj(x)) \text{ and } \varphi^{-1}(z) < \varphi^{-1}(u).$$

Select the node u as large as possible with respect to φ^{-1}. Now, for the same reason as above there exists a vertex u' with

(8) $$u' \in (adj(z) - adj(y)) \text{ and } \varphi^{-1}(u) < \varphi^{-1}(u').$$

Next we see that

(9) $$u' = u \Rightarrow (u, z) \in E.$$

Since the vertices u and u' are connected in $G_{\varphi^{-1}(u)}$ by a path $v_1, \ldots, v_a$ with $v_1 = u$ and $v_a = u'$. Let $A_{k'}$ be a maximal clique containing the edge (u, v_2). From the maximality of u and $(y, v_2) \notin E$ we infer $A_k \neq A'_k$ and moreover $A_k \overset{\varphi}{<} A_{k'}$, since $\varphi^{-1}(u) < \varphi^{-1}(v_2)$. Further, with $(z, u) \in E$ we get $A_h \neq A_k$ and by (4), $k = h - 1$,

$$A_k \overset{\varphi}{<} A_h \overset{\varphi}{<} A_{k'}.$$

But this is a contradiction to the maximality of the choice of l, since $u \in A_k \cap A_{k'}$ and $l < k$. Therefore the statement (9) must be valid, hence $u' = u$ and $(u, z) \in E$. The final step in our construction is the observation that

$$(u, z) \text{ is an edge of } A_h$$

which follows from the maximality of the index h. Altogether we have found that A_l, A_k, A_h contains the path x, z, u, y as an induced subgraph of G and for this reason they build the linked triple (A_l, A_k, A_h). Therefore statement (2)

$$A_l \overset{\varphi}{<} A_k \overset{\varphi}{<} A_h$$

is a contradiction to the precondition of the theorem.

□

Finally, we give two technical statements used in the next section.

Lemma 13 *Let y be a vertex of an interval graph $G = (V, E)$ with lexBFS-ordering φ. Then for all vertices x, $x \neq y$, $x \notin adj(y)$ such that y is connected to $\varphi(n)$ in $G' = G - adj(x)$ we obtain*

$$\varphi^{-1}(x) < \varphi^{-1}(y).$$

Proof: Note firstly that due to our precondition the vertex set $adj(y)$ is a $x, \varphi(1)$-separator. Further, in every intersection model of G all intervals I_x are all left of I_y or all right of I_y. Now let x be as large as possible with respect to φ^{-1}. Therefore by induction on the numberings of the vertices we find

(1) $$L_i(x) \leq L_i(y)$$

for all $i \geq \max\{\varphi^{-1}(x), \varphi^{-1}(y)\}$. Next, let $P = z_s, \ldots, z_0$ be a shortest path from $y = z_s$ to $\varphi(n) = z_0$. Now we use induction on s to show $\varphi^{-1}(x) < \varphi^{-1}(z_s)$. For $s = 0$ it is nothing to prove. Let now $s \geq 1$. Then we get

$$\varphi^{-1}(x) < \varphi^{-1}(z_{s-1})$$

as the induction hypothesis. If now $\varphi^{-1}(y) > \varphi^{-1}(x_{s-1})$ then we have finished. On the other hand, with $j = \varphi^{-1}(z_{s-1}) > i$ we find

$$L_{j-1}(x) = L_j(x) \leq L_j(y) < L_j(y) \cup \{j\} = L_{j-1}(y)$$

and the lemma follows by theorem 4.

$\square$

Corollary 14 *Let $G = (V, E)$ be an interval graph with lexBFS-ordering φ. Then for all vertices x, y, z, $y \neq x \neq z$, such that $adj(y)$ is a $x, \varphi(n)$-separator and in $G - adj(x)$ the vertex y is connected to z, where z is $\varphi^{-1}(z)$ is as large as possible, we obtain*

$$\varphi^{-1}(x) < \varphi^{-1}(y).$$

Proof: Since y and $\varphi(n)$ are not connected in $G - adj(x)$, there exists a corresponding minimal separator $S \subseteq (x)$. Let X be the nodes which are not connected to $\varphi(n)$ in $G - S$ and let Z be the subset of X with

$$Z = \{v \in X \mid S \subseteq adj(v)\}.$$

By our precondition we have $x, y, z \in Z$. Then by corollary 5

$$\forall x', s \in S \qquad \varphi(x) < \varphi(s)$$

and further for $i = \varphi^{-1}(z)$

$$\forall x' \in X - S, s \in S \qquad L_i(x') < L_i(s)$$

and

$$\forall s', s'' \in S \qquad L_i(s') = L_i(s'').$$

For that reason and by corollary 6 it is consitent with a numbering φ' of $G'' = G_Z - S$. Now our proposition follows from the preceding lemma with $n' = |Z - S|$, $\varphi'(n') = z$.

$\square$

4 The Algorithm

In this section we want to construct a linear arrangement by finding a lexBFS-ordering which fulfills theorem 12. In general a lexBFS-ordering does not satisfy the preconditions of this theorem but we observe the following.

Lemma 15 *Let φ be a lexBFS-ordering of an interval graph G containing a linked triple (B', B, B''). Further, let b, b', b'', b''' be the vertices with*

$$
\begin{aligned}
\varphi^{-1}(b) &= \max\{\varphi^{-1}(v) \mid v \in B - (B' \cap B'')\} \\
\varphi^{-1}(b') &= \max\{\varphi^{-1}(v) \mid v \in B' - B\} \\
\varphi^{-1}(b'') &= \max\{\varphi^{-1}(v) \mid v \in B'' - B\} \\
\varphi^{-1}(b''') &= \max\{\varphi^{-1}(v) \mid v \in (B' \cup B \cup B'') - (B' \cap B'')\}.
\end{aligned}
$$

1. *If $b''' \in (B \cup B'') - B'$ then*

$$
\varphi^{-1}(b') < \min\{\varphi^{-1}(v) \mid v \in B\}.
$$

   In particular $B' \overset{\varphi}{<} B$.

2. *If $b''' \in B'' - B$ then*

$$
\varphi^{-1}(b) < \min\{\varphi^{-1}(v) \mid v \in B''\}.
$$

Proof: 1) First we show that on condition $b''' \in (B \cup B'') - B'$

(1) $\qquad\qquad b \in B - B'$ and $\varphi^{-1}(b') < \varphi^{-1}(b)$.

This is clear for $b''' \in B$. On the other hand, if

$$
b''' \in B'' - B \iff b''' = b''
$$

then by definition of a linked triple the sequence $b' = x_1, x_2, x_3, x_4 = b''$ is a chordless path in G for all $x_2 \in B - B'$ and $x_3 \in B - B''$. By corollary 5 we get

$$
\varphi^{-1}(x_1) < \cdots < \varphi^{-1}(x_h) > \cdots > \varphi^{-1}(x_4).
$$

Since $h \neq 4$ would imply a contradiction to the maximality of $\varphi^{-1}(b''')$ we find that

$$
\varphi^{-1}(b') < \varphi^{-1}(x_2) < \varphi^{-1}(x_3) < \varphi^{-1}(b'''),
$$

which shows (1). Let now $i = \varphi^{-1}(b)$ then we claim

(2) $\qquad\qquad\qquad\qquad L_i(b') \subseteq B.$

Conversely, suppose the label $L_i(b')$ contains a node $x \notin B' \cup B$ with $\varphi^{-1}(b) = i < \varphi^{-1}(x)$. Since $x \notin B' \cup B$ there is a path $P = y_0, \ldots, y_s$, $s \geq 2$, in graph $G - (B' \cap B'')$ from $b' = y_0$ to $x = y_s$. Because a shortest path is chordless and a chordless path in G_A, $A \subseteq V$, is also chordless in G, we obtain from corollary 5

$$
\varphi^{-1}(b') = \varphi^{-1}(x_0) < \cdots < \varphi^{-1}(x_s) = \varphi^{-1}(b).
$$

By corollary 14 at least one of the nodes $x_1, \ldots, x_s$ is an element of $(B \cap B') - B''$. Let x_j be this node; then we have

$$
\varphi^{-1}(b) < \varphi^{-1}(x_j) \text{ and } x_j \in (B' \cap B) - B'' \subseteq B - (B' \cap B'')
$$

against the maximality of b. By this contradiction the statement (2) must be right. But from $L_i(b') \subseteq B$ we infer that

$$(3) \qquad L_i(b') \subseteq L_i(z) \Rightarrow L_i(b') \leq L_(z)$$

for all unnumbered nodes $z \in B$ at moment $i = \varphi^{-1}(b)$. Next note $(b, b') \notin B$ which implies

$$L_{i-1}(b') = L_i(b') \leq L_i(z) < L_i(z) \cup \{\varphi^{-1}(b)\} = L_{i-1}(z).$$

For that reason the proposition follows immediately by theorem 4.

2) Now we have $b''' \in B'' - B$. With the same argumentation as for 1) we find out

$$L_i(b) \subseteq B'',$$

where $i = \varphi^{-1}(b''')$, since $L_i(v) - B'' \neq \emptyset$ for $v \in B$ would produce a contradiction to the maximality of b''. Therefore we get

$$\forall v \in B'' \qquad L_i(b) < L_i(v).$$

Form this we infer that

$$\forall v \in B'', v \neq b''' \qquad L_{i-1}(b) = L_i(b) \leq L_i(v) < L_i(v) \cup \varphi^{-1}(b''') = L_{i-1}(v)$$

and theorem 4 completes 2).

□

The preceding lemma demonstrates that any lexBFS-ordering does half of the work to our aim. To complete the work we have to guarantee that an inner clique B of a linked triple (B', B, B'') will never be numbered before B' or B'' is numbered, thereby a maximal clique is numbered when all of its vertices are numbered. An evident step on this way is to start lexBFS with a node not element of B. The next statement shows us the way to find such a vertex.

Lemma 16 *Let φ be a lexBFS-ordering of an interval graph $G = (V, E)$, then the set $adj^*(\varphi(1))$ is an outer clique of G.*

Proof: Suppose the contrary. Then

$$B \stackrel{def}{=} \{\varphi(1)\} \cup adj(\varphi(1))$$

is an inner clique of G. According to lemma 9, B is an inner clique of a linked triple (B', B, B'') or there are linked triples (B_1, B_2, B) and (B, B_3, B_4) such that $B - (B_1 \cup B_4)$ is a x, y-separator in $G - (B_1 \cap B_4)$ for all $x \in B_1 - B$ and $y \in B_4 - B$. Since $\varphi(1)$ is the least numbered node we infer in the first case that

$$B \stackrel{\varphi}{\lesssim} B' \text{ and } B \stackrel{\varphi}{\lesssim} B''$$

in opposition to lemma 15 which requires

$$B' \stackrel{\varphi}{\lesssim} B \text{ or } B'' \stackrel{\varphi}{\lesssim} B.$$

In the second case, let b'' be the largest numbered node in

$$(B_1 \cup B_2 \cup B \cup B_3 \cup B_4) - (B_1 \cap B_4)$$

and let further w.l.o.g. $b'' \in B \cup B_3 \cup B_4$. Then we infer that the largest numbered node in $(B_1 \cup B_2 \cup B) - (B_1 \cap B_4)$ is an element of $(B_2 \cup B) - B_1$. In the opposite way, suppose this vertex b' is element of $(B_2 \cup B) - B_1$. Then there exists a shortest path in $G - (B_1 \cap B_4)$ from b' to b''. This path is chordless in G and contains a node $b \in B - (B_1 \cap B_4)$ by our precondition. With corollary 5 we have $\varphi^{-1}(b') < \varphi^{-1}(b)$, against the minimality of b'. Therefore $b' \in (B_2 \cup B) - B_1$ which leads us together with lemma 15.1 to

$$\min_{v \in B_1} \varphi^{-1}(v) < \min_{v \in B} \varphi^{-1}(v).$$

But this contradicts the precondition $\varphi(1) \in B$. For this reason the assumption "B is an inner clique" is wrong and therefore the lemma is true.

$\square$

Corollary 17 *Let φ be a lexBFS-ordering of an interval graph $G = (V, E)$ and $\varphi(n)$ be a simplicial node of an outer clique A. Then there is a standard interval representation for G with $B = adj(\varphi(n)) \cup \varphi(n)$ as leftmost clique and A as rightmost clique.*

Proof: We use induction on n. For $n = 1$ there is nothing to show. Let now $n \geq 2$. First we consider the case where $G - B$ is connected. Then B is the leftmost clique or the rightmost clique in every intersection model of G. In particular, B is the leftmost clique in an intersection model F which contains A as rightmost clique, and by definition of A there is such F. If $G - B$ is not connected then there is a minimal separator S, $S \subseteq adj(\varphi(1))$ such that $G - S$ is not connected. Furthermore, by the preceding lemma there is an interval representation $F = \{I_1, \ldots, I_n\}$ containing B as leftmost clique. Let Z be the set nodes which are connected to $\varphi(1)$ in $G - S$ and X be set vertices which are not connected to $\varphi(1)$ in $G - S$. By our induction hypothesis the graph G_i, $i = \min\{\varphi^{-1}(w) \mid w \in V - X\}$, has a standard interval model with A as rightmost and

$$C = \{w \in adj^*(\varphi(i)) \mid \varphi^{-1}(w) \geq i\}$$

as leftmost clique. By corollary 5 the set C contains S. Therefore the intersection model

$$F' = \{I_x \mid x \in X\}$$

of G can be placed to the left of C. Then by extending the intervals

$$I_s = [a(s), b(s)] \text{ to } I_s = [a(\varphi(1)), b(s)]$$

for all $s \in S$, we get an interval representation of G which fulfills our proposition.

$\square$

However, a lexBFS-ordering started in a simplicial point of an outer clique does not always get the desired result. The problem arises from the fact that for some linked triples (B', B, B'') the first numbered node in $B' \cup B \cup B''$ is an element of $B' \cap B$ or $B \cap B''$. To avoid having B as the first completely numbered clique we need a method to choose the right vertex if the lexBFS algorithm has more than one choices in line 3. This method is described in the following.

Definition 18 Let $G = (V, E)$ be an interval graph and let φ' be a lexBFS-ordering Let further φ'' be a second lexBFS-ordering resulting from the lexBFS-algorithm with the additional rule for the line 3:

- If the largest label does not get a unique vertex then select the smallest numbered vertex according to φ'.

In particular, we get $\varphi''(n) = \varphi'(1)$. The map φ'' is noted as the lexBFS-ordering induced by φ'.

Lemma 19 Let $G = (V, E)$ be an interval graph and let φ be a lexBFS-ordering produced by a run started in a simplicial point of an outer clique of G. If φ_1 be a lexBFS-ordering induced by φ then $\varphi(n) = \varphi_1(1)$.

Proof: (Induction on n.) For $n = 1$ there is nothing to show. For the induction hypothesis we assume that the lemma is true for all interval graphs with fewer vertices. Let $x = \varphi(n)$, $z = \varphi_1(1)$ and $V' = V - z$. Now, we have to show

(1) $$\forall v \in V, v \neq x \qquad \varphi_1^{-1}(x) < \varphi_1^{-1}(v).$$

First we consider the case $(z, x) \in E$. Then G is complete and the lemma follows from our definition of φ_1. In the opposite way, let now $z \notin adj(x)$. Then we can divide the vertices $V' = V - z$ into 3 groups Z, X, S. First we define S as minimal subset of the x, y-separator $adj(x)$ such that x and z are not connected in $G - S$. Next, the the group Z contains all vertices which are connected to z in the graph $G - S$ and X is given by $V' - (Z \cup S)$, in particular $x \in X$. Since G is an interval graph and x is an simplicial node of an outer clique of G, by corollary 17 there is a standard intersection model F of G such that $I_x = [1, 1]$ and all intervals $I_{x'}$, $x' \in X$, are left of all intervals $I_{z'}$, $z' \in Z \cup \{z\}$. Because every interval I_s, $s \in S$, intersects I_x and at least one $I_{z'}$, $z' \in Z$, it intersectes all $I_{x'}$, $x' \in X$, i.e.

(2) $$\forall x' \in X \qquad S \subseteq adj(x').$$

Now by lemma 13 we get

$$\forall x' \in X, z' \in Z \qquad \varphi_1^{-1}(x') < \varphi_1^{-1}(z')$$

and by corollary 5 we get

(3) $$\forall x' \in X, s \in S \qquad \varphi_1^{-1}(x') < \varphi_1^{-1}(s),$$

especially

$$\forall v \in Z \cup S \qquad \varphi_1^{-1}(x) < \varphi_1^{-1}(v).$$

—————————————— is_interval_graph ——————————————

Input: A Graph $G = (V, E)$.
Output: A consecutive arrangement of the maximal cliques of G if there is one.

1. Produce with lexBFS a first numbering φ_1.

2. Produce a lexBFS-ordering φ_2 induced by φ_1.

3. Produce a lexBFS-ordering φ_3 induced by φ_2.

4. Produce a lexBFS-ordering φ_4 induced by φ_3.

5. With φ_4 compute the maximal cliques $A_1 \overset{\varphi_4}{\lesssim} \cdots \overset{\varphi_4}{\lesssim} A_s$ of G and test if $A_1, \ldots, A_s$ are consecutive.

—————————————— is_interval_graph ——————————————

Figure 2: Algorithm B.

Now we consider the graph G_X induced by X. Since S is a x', z-separator, $x' \in X$, we infer from (2) and (3)

$$\forall x' \in X \qquad L_i(x') = \{\varphi^{-1}(s) \mid s \in S\},$$

where $i = \min\{\varphi^{-1}(s) \mid s \in S\}$, and therefore

$$\forall x', x'' \in X \qquad L_i(x') = L_i(x''),$$

holds, i.e. all nodes in G_X has the same label before any node in G_X is numbered. Hence for G_X the map φ_1 is a lexBFS-ordering in respect to X and φ given by corolary 6. For this reason the graph G_X fulfills the precondition of our induction and hence we get by the induction hypothesis

$$\forall x' \in X, x' \neq x \qquad \varphi_1^{-1}(x) < \varphi_1^{-1}(x').$$

This completes the lemma.

□

Now our preparation is complete and our alogorithm is outlined in figure 2. The final theorem sums up our analysis.

Theorem 20 *A linear arrangement of the maximal cliques of an interval graph $G = (V, E)$ can be computed in linear time $O(n + m)$ by four runs of* lexBFS.

Proof: First we prove the correctness of the algorithm. By theorem 12 and lemma 15 it is sufficient to show

$(*)$ $\qquad\qquad\qquad B \overset{\varphi_4}{\lesssim} B'$ or $B \overset{\varphi_4}{\lesssim} B''$

for every linked triple (B', B, B''). By lemma 17 we observe that $\varphi_2(n)$ and $\varphi_3(n)$ are simplicial points of outer cliques and further by lemma 19 we have

$$\varphi_2(n) = \varphi_3(1) = \varphi_4(n) \text{ and } \varphi_2(1) = \varphi_3(n) = \varphi_4(1).$$

To prove $(*)$ we use induction on n. The case $n = 1$ is clear and for $n \geq 2$ we assume that the lemma is true for all interval graphs with fewer vertices. By lemma 17 there is a standard intersection model F of G which contains $C' = adj^*(\varphi_2(1))$ as leftmost clique and $C'' = adj^*(\varphi_2(n))$ as rigthmost clique. W.l.o.g. we assume

$$C' \overset{F}{<} \cdots \overset{F}{<} B' \overset{F}{<} B \overset{F}{<} B'' \overset{F}{<} \cdots \overset{F}{<} C''.$$

By our definition of a linked triple we deduce that B is $B'' - B, \varphi_2^{-1}(1)$-separator. Therefore B contains a subset S' which is a minimal $B'' - B, \varphi_2(1)$-separator. Similarly we define a minimal $B' - B, \varphi_2(n)$-separator S''. Then we have 3 cases.

Case 1. $B'' - S'''$ is connected to an maximal clique $X - S'''$ in $G - S'''$ with

$$B \overset{F}{<} X \text{ and } S' \not\subseteq X.$$

On this condition we find $X \overset{\varphi_3}{<} B''$. For this reason and the fact that X is a $B'' - S''', \varphi_4(n)$-separator we find for the smallest numbered node x of X

$$\forall v \in B - S''' \qquad \varphi_4^{-1}(v) < \varphi_4^{-1}(x).$$

From this we infer that

$$B' \overset{\varphi_4}{<} B \overset{\varphi_4}{<} B''$$

by lemma 13, corollary 14, lemma 15 and the definition of φ_4.

Case 2. $B' - S'$ is connected to a maximal clique $Y - S'$ of $G - S'$ with

$$Y \overset{F}{<} B \text{ and } S'' \not\subseteq Y.$$

With case 1 we get

$$B'' \overset{\varphi_3}{<} B \overset{\varphi_3}{<} B'$$

By definition B'' is a $(B' \cup B) - B'', \varphi_4(n)$-separator. Therefore

$$\forall b \in (B' \cup B) - B'', \ b'' \in B'' \qquad L_i(b) \leq L_i(b'')$$

for all $i \geq \max \{ \varphi_4^{-1}(v) \mid v \in (B' \cup B \cup B'') - (B' \cap B'') \}$. Hence the largest numbered node with respect to φ_4^{-1} is in $B'' - B$ and $(*)$ follows from lemma 15.

Case 3. Neither case 1 nor case 2 is true. Then let $G' = (V', E')$ be the component which contains $(B' \cup B \cup B'') - (S' \cup S''')$ in the graph $G - (S' \cup S''')$. By our precondition all nodes in G' are adjacent to all nodes in $S' \cup S'''$. By corollary 6 we infer that the orderings $\varphi_1, \varphi_2, \varphi_3, \varphi_4$ of G are all consistent with corresponding numberings of G'. Therefore our proposition follows from the induction hypothesis.

307

The running time of the steps 1, 2, 3, 4 and 5 is clearly linear with theorem 4 until one critical point. This point is to realize the extented rule for line 3 of lexBFS . Let us now have a short look inside the implementation of lexBFS. The algorithm does not actually calculate the labels, but rather keeps the unnumbered vertices $V - \{\varphi(n),\ldots,\varphi(i+1)\}$ in lexicographic order. By means of that, nodes with equal label l are summarized in a set S_l represented by a doubly linked list. The lists S_l on their part are also organized as doubly linked list Q reflecting the current linear order on the label l. In lines (5)–(7) of lexBFS updates are made by splitting the set S_l into a set S_l' connected with the node w and $S_l'' = S - S_l' = S_l - adj(v)$. Then the S_l' is added as a new element of Q immediately before the old set S_l, which now contains the new set S_l''. Initially there is only one set $S_\emptyset = V$. See [17][11] for details. Note that this method is very similar to sorting strings in lexicographic order and — very important for us — like sorting strings it is stable. In this context stability means that if a vertex x appears before a vertex y in the list $S_\emptyset = V$ and the ordering of the list $S_\emptyset$ is consistent with the ordering in every adjacency list $adj(v)$, $v \in V$, then the vertex x always appears before vertex y in a list S_l which contains both vertices x and y. Therefore, to implement the additional rule for step 3, it is only necessary to sort the adjacency lists $adj(v)$ in order of φ_2^{-1} and to initialize the set $S_\emptyset$ with $\{\varphi_2(1),\ldots,\varphi_2(n)\}$. This can be acheived in time $O(n+m)$, see [11], theorem 2.3. Moreover the preceding run of lexBFS can do this sorting as a byproduct. For this reason the steps 2,3,4 can also be implemented in linear time and the proof is complete.

$\square$

References

[1] S. Benzer. On the topology of the genetic fine structure. *Proce. Nat. Acad. Sci. U.S.A. 45*, pages 1607–1620, 1959.

[2] A. A. Bertossi. Total domination in interval graphs. *Information Processing Letters*, 23:131–134, 1986.

[3] A. A. Bertossi and M. A. Bonuccelli. Hamiltonian circuits in interval graph generalizations. *Information Processing Letters*, 23:195–200, 1986.

[4] K. S. Booth and G. Leuker. Testing for the consecutive ones property, interval graphs and graph planarity using pq-tree algorithms. *Journal of Computer and System Science*, 13:335–379, 1976.

[5] E. G. Coffman and R. L. Graham. Optimal scheduling for two-processor systems. *Acta Informatica*, 1:200–213, 1972.

[6] N. Deo, M. S. Krishamoorty, and M. A. Langston. Exact and approximate solutions for the gate matrix layout problem. In *IEEE Trans. Computer Aided Design*, pages 79–84, 1987.

[7] G. A. Dirac. On rigid circuit graphs. *Abh. Mathe. Sem. Univ. Hamburg*, 25:71–76, 1961.

[8] D. R. Fulkerson and O. A. Gross. Incidence matrices and interval graphs. *Pacific J. Math.*, 15:835–855, 1965.

[9] F. Gavril. Algorithms for minimum coloring, maximum clique, minimum covering by cliques and maximum independent set of chordal graph. *SIAM J. Comput.*, 1:180–187, 1972.

[10] P. C. Gilmore and A. J. Hoffman. A characterization of comparability graphs and of interval graphs. *Canad. J. Math.*, 16:539–548, 1964.

[11] M. C. Golumbic. *Algorithmic Graph Theory and Perfect Graphs*. Academic Press, London, 1980.

[12] G. Hajós. Über eine Art von Graphen. *Internationale Mathematische Nachrichten*, 11, 1957. Problem 65.

[13] N. Korte and R. H. Möhring. An incremental linear-time algorithm for recognizing interval graphs. *SIAM J. Comput.*, 18:68–81, 1989.

[14] C. G. Lekkerkerker and J. C. Boland. Representation of a finite graph by a line of intervals on the real line. *Fundamenta Mathemticae*, 51:45–64, 1962.

[15] M. Mende. Implementierung von PQ-Bäumen. *Diplomarbeit am Institut für Theoretische Informatik (ETH-Zürich)*, 1991.

[16] F. S. Roberts. *Graph Theory and Its Application to Problems of Society*. Society for Industrial and Applied Mathematics, Philadelphia, 1978.

[17] D. J. Rose, R. E. Tarjan, and G. S. Leuker. Algoritmic aspects of vertex elimination on graphs. *SIAM J. Comput.*, 5:266–283, 1976.

[18] R. Sethi. Scheduling graphs on two processors. *SIAM J. Comput.*, 5:73–82, 1976.

Predictions about Collision Free Paths from Intersection Tests

Sabine Stifter

RISC-Linz

(Research Institute for Symbolic Computation)

Johannes Kepler University

A-4040 Linz, Austria

Testing whether an object in motion collides with another one is, intuitively, much harder than testing whether two stationary objects intersect. Usually, different techniques are used for testing for intersections and for collisions. We show, for the Roider Method, that intersection tests can be converted to collision tests. There is some additional information available from the intersection test, namely a cone that contains one object but no part of the other one or a separating hyperplane. The conversion is based on theorems involving the disjointness of (and some additional information about) the objects in two "consecutive" positions along the path, and some information about the path. In this paper we study piecewise linear motions. We show that the Roider Method *decides* whether a piecewise linear motion is collision free by applying intersection tests to the objects at the endpoints of the line segments.

1 Introduction

In [Roider, Stifter 1987], [Stifter 1988] we introduced the Roider Method as an intersection test, see also [Stifter 1988a, 1989]. The Roider Method is an iterative algorithm that tests whether two objects (that are characterized by the availability of certain "basic operations" on them) intersect. By practical experiences it turned out that the Roider Method is quite efficient and applicable, see [Stifter 1991].

Although the method is originally an intersection test, in certain situations one can also ensure that a whole motion is collision free. These situations can be characterized by the relative positions of two objects in consecutive snapshots along the path and the direction, the length, or the course of the path, [Stifter 1989]. Since the statement for a whole path is based on statements about two snapshots it is indispensable that there is some information available about the relative positions of the objects in the two snapshots. The knowledge that the two objects do not intersect would be not sufficient. In our case the additional information is a wedge that contains one of the objects but nothing of the other, a "witness to disjointness". Also other intersection tests compute such witnesses to disjointness in case the objects do not intersect. For example, [Chazelle, Dobkin 1980, 1987], [Dobkin, Kirkpatrik 1983, 1985], [Mayr 1990].

However, the conversion from an intersection test to a collision test as described in [Stifter 1989] works only for certain situations. By making many snapshots one can hope that the criteria hold for any two consecutive snapshots and that hereby the whole motion can be shown to be collision free. However, there is no guarantee that this splitting of the path into smaller and smaller pieces is always successful, not even for linear motions.

In this paper, we show that by applying the Roider Method in the start and end position of a linear motion one can always decide whether the path is collision free. We also study the complexity of the method.

The paper is organized as follows: In the next section we briefly sketch the Roider Method and the Theorem on Translations (that is used for obtaining the results of this paper). Section 3 states some notations. Section 4 contains the new results.

2   The Roider Method

The Roider Method is an iterative procedure for testing whether two convex compact sets, say A and B, (that satisfy some additional conditions) intersect. In case the objects are disjoint, a *witness to disjointness* is constructed. A witness to disjointness is a wedge formed by two touching lines from some point P on A to B that contains B but nothing of A except P. In case the objects intersect, a *witness to intersection*, a point in common to both objects, is constructed.

Starting from some point P (the *starting point*) on the boundary of A, the two touching lines t_1, t_2 to B are constructed. Let T_1, T_2 be the touching points of t_1 and t_2, respectively. (If there is more than one touching point, the one closest to P is taken.) Let S_1, S_2 be the intersection points of t_1, t_2 with the boundary of A (that are different from P). Figure 1a illustrates the notation. The five points P, S_1, S_2, T_1, T_2 give all the necessary information:

- If T_i is between P and S_i (or equal to S_i), $i \in \{1, 2\}$, then the objects intersect and T_i is a witness to intersection; see Figure 1b.

- If P is between S_1 and T_1 (or equal to S_1) and between S_2 and T_2 (or equal to S_2) then the objects are disjoint and the wedge formed by the two tangents meeting in P is a witness to disjointness; see Figure 1c. (We also say, P *forms a witness to disjointness* and (P, S_1, S_2, T_1, T_2) *specifies the witness to disjointness*.)

- If one of S_i is between P and T_i then one cannot yet decide whether the objects intersect. So S_i is taken as new P and the procedure is repeated.

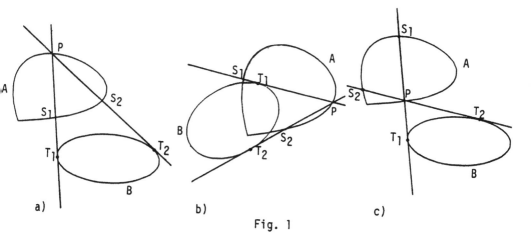

Fig. 1

The algorithm in this form may fail if A has some vertex or B has some straight line segment in its boundary, or the objects touch. These situations are handled by first testing whether one of the vertices of A forms a witness to disjointess or some of the straight line segments of B intersects A, and stopping the algorithm with "close" as soon as the distance between P and some T_i is less than a predefined tolerance ε, respectively.

So the Roider Method decides whether two objects intersect in a relaxed way because there is also the possibility of the answer "close". This answer is output only if the objects intersect or are less than ε apart. So one will treat "close" the same way as "intersect" for practical applications.

Next we state the procedures that are required for the objects. (We assume that we are working in a coordinate system.) For a more formal description of the Roider Method in 2D we refer to [Stifter 1988].

Procedures Required for the Roider Method:

- A procedure $(T_1, T_2) = \text{touch}(B, P)$ that takes an object B and a point P and computes the touching points T_1, T_2 on B of the tangents from P to B; if the touching points are not unique, the ones closest to P are returned.

- A procedure $(S_1, S_2) = \text{intersect-2}(A, l)$ that takes an object A and a straight line l and returns the endpoints S_1, S_2 of the segment $l \cap A$. (Note that $l \cap A$ is a segment because A is convex and compact.)

- A procedure $P = \text{point-on}(A)$ that takes an object A and returns a point on the boundary of A.

- A procedure $v = \text{inside}(P, A)$ that takes a point P and an object A and returns $v = \text{true}$ if $P \in A$, $v = \text{false}$ otherwise.

These procedures are available, for example, for circles, ellipses, and also superellipses, [Stifter 1988].

Because the proof in Section 4 is based on the proof of the Roider Method, we sketch the idea of the proof of the Roider Method here. This proof works in three steps; compare Fig. 2.

- Let $\bar{A}$ be a triangle inscribed to A, $\bar{B}$ a rectangle circumscribed to B, as shown in Fig. 2a. We assume a coordinate system as in Fig. 2a.

One can show by a careful case analysis that the Roider Method applied to $\bar{A}$ and $\bar{B}$ needs at least as many iteration steps as the Roider Method applied to A and B because the points P constructed for $\bar{A}$ and $\bar{B}$ are always "above" the P constructed for A and B in the same number of iteration steps.

- The points P constructed by the Roider Method for $\bar{A}$ and $\bar{B}$ converge to the vertex of the triangle with minimal y (if the objects are disjoint) and to $y = 0$ (if the objects intersect).

- The Roider Method stops if the last P constructed is between the common touching lines to A and B (in case the objects are disjoint) or below the line b (in case the objects intersect); see Fig. 2b.

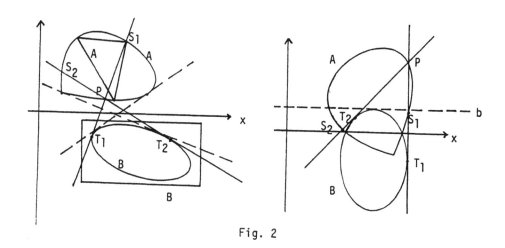

Fig. 2

We close this section with the Theorem on Translations.

Theorem on Translations:
Let P be a point on the boundary of A such that there exists a unique touching-line, say t to A in P. Let P form a witness to disjointness for A and B positioned at p_0 and let P also form a witness to disjointness for A and B positioned at p_1. Assume that t is not a touching line to B. Then, for all p on the segment between p_0 and p_1, P generates a witness to disjointness

for A and B positioned at p. (Compare Fig. 3.)

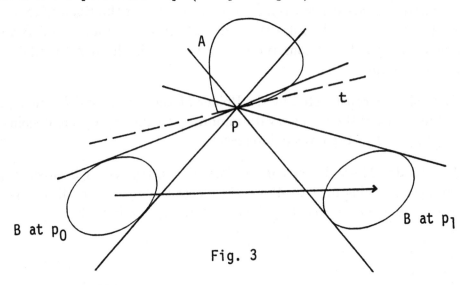

Fig. 3

3   Notations

In the sequel, we use the following notations:

An *object* is a convex, compact set in 2-dimensional space, having inner points.

An *admissible object* is an object for which the procedures required for the Roider Method are available and which has a smooth boundary.

A *separating line* for two objects A and B is a straight line, such that A and B are on different sides of the line.

We assume that for each object a reference point in the interior of the object and a local coordinate system are specified. An *object is positioned at p*, where p is a point, means that the reference point of the object is located at p and its local coordinate system is parallel to the world coordinate system. This is often abbreviated as "the object at p". Together with each object a position is prespecified. If no explicit position is mentioned for an object, the object is positioned at the prespecified position.

A *piecewise linear motion* is a motion along straight line segments without any rotations.

The swept volume of an object B with respect to a path p, $S_p(B)$, is defined as $S_p(B) = \{P | P \in B \text{ at } q, \text{ where } q \text{ is a point on } p\}$.

4 Collision Test for Linear Motions

In this section we show that the Roider Method can be used to decide whether piecewise linear motions are collision free by applying intersection tests to the objects at the endpoints of the line segments. Before stating the respective algorithm and theorem, we give some relatively easy observations that are helpful for the algorithm.

In the sequel, let A and B be admissible objects, p a straight line segment with endpoints p_0 and p_1. Let $\varepsilon > 0$ be the given tolerance. It is clear that by repeated application of the collision test for linear motions one gets a collision test for piecewise linear motions.

Assume that A and B at p_0 are disjoint and that A and B at p_1 are disjoint.

Observation 1:
If A and B do not collide, where B is moving along p, then one of the following is true:
- Each separating line for A and B at p_0 is also a separating line for A and B at p_1.
- Each separating line for A and B at p_1 is also a separating line for A and B at p_0.
- There is a line parallel to p that is a separating line for A and B at p_0 and also for A and B at p_1.

Fig. 4 illustrates the three possibilities.

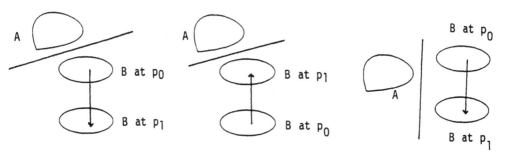

Fig. 4

Observation 2:
If a line t is a separating line for A and B at p_0 and also for A and B at p_1, then the path p is collision free.

Observation 3:
If a line t is a separating line for A and B at p_0 and also for A and B at p_1, then t is also a separating line for A and $S_p(B)$.

The idea for the collision detection algorithm is to construct a witness to disjointness for A and $S_p(B)$ by iteratively constructing witnesses to disjointness for A and B at p_0 and also for A and B at p_1 (especially, without explicitly constructing $S_p(B)$).

Algorithm Collision Test:

1. Apply the Roider Method to A and B at p_0.

   Let $(P^0, S_1^0, S_2^0, T_1^0, T_2^0)$ specify the witness to disjointness found by the Roider Method.

2. If P^0 forms a witness to disjointness for A and B at p_1 then stop with "disjoint".

3. Apply the Roider Method to A and B at p_1 taking P^0 as starting point.

   Let $(P^1, S_1^1, S_2^1, T_1^1, T_2^1)$ specify the witness to disjointness found by the Roider Method.

4. If P^1 forms a witness to disjointness for A and B at p_0 then stop with "disjoint".

5. Let $S_{ij}^1 = P^1 T_k^0 \cap T_i^1 T_j^0$, for $i, j, k \in \{1, 2\}$, $j \neq k$.

   If some S_{ij}^1 is between P^1 and S_k^0 then stop with "intersect".

   If the distance of P^1 and the straight line segment between T_i^1 and T_j^0, $i, j \in \{1, 2\}$, is less than ε, then stop with "close".

6. Apply the Roider Method to A and B at p_0 taking P^1 as starting point.

   Let $(P^0, S_1^0, S_2^0, T_1^0, T_2^0)$ specify the witness to disjointness found by the Roider Method.

7. If P^0 forms a witness to disjointness for A and B at p_1 then stop with "disjoint".

8. Let $S_{ij}^0 = P^0 T_k^1 \cap T_i^0 T_j^1$, for $i, j, k \in \{1, 2\}$, $j \neq k$.

   If some S_{ij}^0 is between P^0 and S_k^1 then stop with "intersect".

   If the distance of P^0 and the straight line segment between T_i^0 and T_j^1, $i, j \in \{1, 2\}$, is less than ε, then stop with "close".

9. Continue with Step 3.

Fig. 5 is an example for a collision free motion, Fig. 6 is an example for a collision.

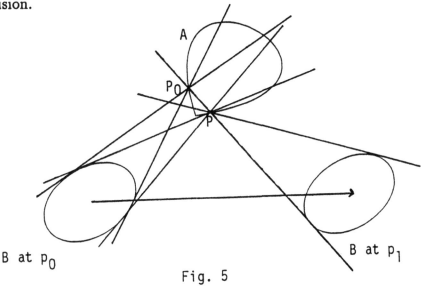

Fig. 5

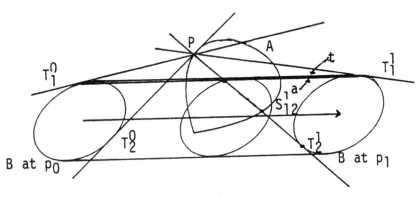

Fig. 6

Comment:

$S_p(B)$ has two straight line segments (parallel to p) in its boundary. We do not have to compute these straight line segments explicitly. The computation of the straight line segments is non-trivial. The procedures required for the Roider Method do not suffice to give an algorithm for its construction. One of these segments intersects A iff the motion along p is not collision free. By the construction of S^1_{ij} and S^0_{ij}, an intersection of A and $S_p(B)$ will be detected, i.e. collisions during the motion will be detected.

Theorem:

The algorithm Collision Test always stops. The answers "disjoint", "intersect", and "close" are subject to the following conditions:

- If the answer is "disjoint" then the motion is collision free.

- If the answer is "intersect" then there is a point q on p such that A and B at q intersect.

- If the answer is "close" then there is a point q on p such that the distance of A and B at q is less than ε.

Proof:

The partial correctness of the algorithm is easy to see: (We argue for the situations in steps 4) and 5). The situations in steps 7) and 8) are symmetric.)

If the algorithm terminates with "disjoint", then P^1 forms a witness to disjointness for A and B at p_0 and for A and B at p_1. Hence, by the Theorem on Translations, the motion is collision free.

If the algorithm terminates with "intersect", then S^1_{ij} is between P^1 and S^0_k. By construction, S^1_{ij} is contained in $S_p(B)$. (Note that B and, hence, $S_p(B)$ are convex.) P^1 and S^0_k belong both to A. So S^1_{ij} is, by convexity, also in A. This means that there is a collision during the motion.

If the algorithm terminates with "close", then the distance between P^1 and a straight line segment inside $S_p(B)$ is less than ε. So especially the distance between A and $S_p(B)$ is less than ε.

It remains to show that the algorithm always terminates. By Observations 1 and 3, there is a line parallel to p that separates A and $S_p(B)$ or the objects intersect.

In case the objects intersect, a common touching line to B at p_0 and to B at p_1 that is parallel to p intersects A. (A segment on this common touching line, say t, is part of the boundary of $S_p(B)$.) Hence, any touching line from some point P on A to $S_p(B)$ is also a touching line from P to B at p_0 or from P to B at p_1.

So in any iteration of the collision test, the P computed is closer to p than the P computed by the Roider Method applied to A and $S_p(B)$ in the same number of iteration steps.

If the motion is collision free, i.e. A and $S_p(B)$ do not intersect, then the Roider Method stops in at least

$$\frac{2\tilde{a}\max\{\tilde{a}, \tilde{b} + l(p)\}}{d_{\min}e}$$

steps, where $\tilde{a}$, $\tilde{b}$ are the maximal extensions of A and B, respectively, $l(p)$ denotes the length of p, $d_{\min}$ is the minimal distance of A and B during the motion, and e is a measure for the relative curvatures of A and B. We refer to [Stifter 1988] for the details of notations.

If the motion is not collision free, then the sequence of P computed by the collision test converges to t. So the distance of P and t gets less than ε and the algorithm stops at least with "close". However, if the objects do not only touch during the motion, there is some $m > 0$ with the following property: If P is less than m apart from t then some touching line from P to B at p_0 intersects A on the opposite side of t (as P is). This follows from the fact that A and B at p_0 are disjoint and the termination of the Roider Method. This is exactly the situation where the algorithm stops with "intersect" in step 8). (The situation in step 5) is symmetric.)

According to the proof of the Roider Method in [Stifter 1988], the number of iteration steps is given by

$$\frac{2\tilde{a}\max\{\tilde{a}, \tilde{b} + l(p)\}}{me.}$$

5 References

Chazelle, B., Dobkin, D.P., 1980:
Detection is easier than computation; 12th ACM Symposium on Theory of Computing, Los Angeles, California, pp. 146-153.

Chazelle, B., Dobkin, D.P., 1987:
Intersection of convex objects in two and three dimensions; Journal of the ACM, vol. 4/3, pp. 1-27.

Dobkin, D.P., Kirkpatrik, D.G., 1983:
Fast detection of polyhedral intersection; Theoretical Computer Science, vol. 27, pp. 241-253.

Dobkin, D.P., Kirkpatrik, D.G., 1985:
A linear algorithm for determining the separation of convex polyhedra; Journal of Algorithms, vol. 6, pp. 381-392.

Mayr, H., 1990:
Highly-efficient collision checking for robots/NC using linear programming techniques; Proc. Operations Research, Vienna, Austria.

Roider, B., Stifter, S., 1987:
Collision of convex objects; Proc. EUROCAL'87, Leipzig, GDR, June 2-5, 1987, J. Davenport (ed.), Springer LNCS 378, 1989, pp. 258-259.

Stifter, S., 1988:
A medley of solutions to the robot collision problem in two and three dimensions; Ph.D. thesis, RISC-Linz, J. Kepler University, A-4040 Linz, Austria. Also: VWGÖ, Dissertations of the University of Linz, vol. 81, 1989.

Stifter, S., 1988a:
A generalization of the Roider Method to solve the robot collision problem in 3D; Proceedings ISSAC'88, Rome, Italy, July 1988, Springer LNCS 358, pp. 332-343.

Stifter, S., 1989:
The Roider Method: a method for static and dynamic collision detection; Issues in Robotics and Non-Linear Geometry, Ch. Hoffmann (ed.), JAI Press, to appear. Also: Technical Report, RISC-Linz series no. 89-40, J. Kepler University, A-4040 Linz, Austria.

Stifter, S., 1991:
Collision detection in the robot simulation system SMART; International Journal of Advanced Manufacturing Technology, to appear.

Lecture Notes in Computer Science

For information about Vols. 1–466
please contact your bookseller or Springer-Verlag

Vol. 509: A. Endres, H. Weber (Eds.), Software Development Environments and CASE Technology. Proceedings, 1991. VIII, 286 pages. 1991.

Vol. 510: J. Leach Albert, B. Monien, M. Rodríguez (Eds.), Automata, Languages and Programming. Proceedings, 1991. XII, 763 pages. 1991.

Vol. 511: A. C. F. Colchester, D.J. Hawkes (Eds.), Information Processing in Medical Imaging. Proceedings, 1991. XI, 512 pages. 1991.

Vol. 512: P. America (Ed.), ECOOP '91. European Conference on Object-Oriented Programming. Proceedings, 1991. X, 396 pages. 1991.

Vol. 513: N. M. Mattos, An Approach to Knowledge Base Management. IX, 247 pages. 1991. (Subseries LNAI).

Vol. 514: G. Cohen, P. Charpin (Eds.), EUROCODE '90. Proceedings, 1990. XI, 392 pages. 1991.

Vol. 515: J. P. Martins, M. Reinfrank (Eds.), Truth Maintenance Systems. Proceedings, 1990. VII, 177 pages. 1991. (Subseries LNAI).

Vol. 516: S. Kaplan, M. Okada (Eds.), Conditional and Typed Rewriting Systems. Proceedings, 1990. IX, 461 pages. 1991.

Vol. 517: K. Nökel, Temporally Distributed Symptoms in Technical Diagnosis. IX, 164 pages. 1991. (Subseries LNAI).

Vol. 518: J. G. Williams, Instantiation Theory. VIII, 133 pages. 1991. (Subseries LNAI).

Vol. 519: F. Dehne, J.-R. Sack, N. Santoro (Eds.), Algorithms and Data Structures. Proceedings, 1991. X, 496 pages. 1991.

Vol. 520: A. Tarlecki (Ed.), Mathematical Foundations of Computer Science 1991. Proceedings, 1991. XI, 435 pages. 1991.

Vol. 521: B. Bouchon-Meunier, R. R. Yager, L. A. Zadek (Eds.), Uncertainty in Knowledge-Bases. Proceedings, 1990. X, 609 pages. 1991.

Vol. 522: J. Hertzberg (Ed.), European Workshop on Planning. Proceedings, 1991. VII, 121 pages. 1991. (Subseries LNAI).

Vol. 523: J. Hughes (Ed.), Functional Programming Languages and Computer Architecture. Proceedings, 1991. VIII, 666 pages. 1991.

Vol. 524: G. Rozenberg (Ed.), Advances in Petri Nets 1991. VIII, 572 pages. 1991.

Vol. 525: O. Günther, H.-J. Schek (Eds.), Advances in Spatial Databases. Proceedings, 1991. XI, 471 pages. 1991.

Vol. 526: T. Ito, A. R. Meyer (Eds.), Theoretical Aspects of Computer Software. Proceedings, 1991. X, 772 pages. 1991.

Vol. 527: J.C.M. Baeten, J. F. Groote (Eds.), CONCUR '91. Proceedings, 1991. VIII, 541 pages. 1991.

Vol. 528: J. Maluszynski, M. Wirsing (Eds.), Programming Language Implementation and Logic Programming. Proceedings, 1991. XI, 433 pages. 1991.

Vol. 529: L. Budach (Ed.), Fundamentals of Computation Theory. Proceedings, 1991. XII, 426 pages. 1991.

Vol. 530: D. H. Pitt, P.-L. Curien, S. Abramsky, A. M. Pitts, A. Poigné, D. E. Rydeheard (Eds.), Category Theory and Computer Science. Proceedings, 1991. VII, 301 pages. 1991.

Vol. 531: E. M. Clarke, R. P. Kurshan (Eds.), Computer-Aided Verification. Proceedings, 1990. XIII, 372 pages. 1991.

Vol. 532: H. Ehrig, H.-J. Kreowski, G. Rozenberg (Eds.), Graph Grammars and Their Application to Computer Science. Proceedings, 1990. X, 703 pages. 1991.

Vol. 533: E. Börger, H. Kleine Büning, M. M. Richter, W. Schönfeld (Eds.), Computer Science Logic. Proceedings, 1990. VIII, 399 pages. 1991.

Vol. 534: H. Ehrig, K. P. Jantke, F. Orejas, H. Reichel (Eds.), Recent Trends in Data Type Specification. Proceedings, 1990. VIII, 379 pages. 1991.

Vol. 535: P. Jorrand, J. Kelemen (Eds.), Fundamentals of Artificial Intelligence Research. Proceedings, 1991. VIII, 255 pages. 1991. (Subseries LNAI).

Vol. 536: J. E. Tomayko, Software Engineering Education. Proceedings, 1991. VIII, 296 pages. 1991.

Vol. 537: A. J. Menezes, S. A. Vanstone (Eds.), Advances in Cryptology – CRYPTO '90. Proceedings. XIII, 644 pages. 1991.

Vol. 538: M. Kojima, N. Megiddo, T. Noma, A. Yoshise, A Unified Approach to Interior Point Algorithms for Linear Complementarity Problems. VIII, 108 pages. 1991.

Vol. 539: H. F. Mattson, T. Mora, T. R. N. Rao (Eds.), Applied Algebra, Algebraic Algorithms and Error-Correcting Codes. Proceedings, 1991. XI, 489 pages. 1991.

Vol. 540: A. Prieto (Ed.), Artificial Neural Networks. Proceedings, 1991. XIII, 476 pages. 1991.

Vol. 541: P. Barahona, L. Moniz Pereira, A. Porto (Eds.), EPIA '91. Proceedings, 1991. VIII, 292 pages. 1991. (Subseries LNAI).

Vol. 543: J. Dix, K. P. Jantke, P. H. Schmitt (Eds.), Nonmonotonic and Inductive Logic. Proceedings, 1990. X, 243 pages. 1991. (Subseries LNAI).

Vol. 544: M. Broy, M. Wirsing (Eds.), Methods of Programming. XII, 268 pages. 1991.

Vol. 545: H. Alblas, B. Melichar (Eds.), Attribute Grammars, Applications and Systems. Proceedings, 1991. IX, 513 pages. 1991.

Vol. 547: D. W. Davies (Ed.), Advances in Cryptology – EUROCRYPT '91. Proceedings, 1991. XII, 556 pages. 1991.

Vol. 548: R. Kruse, P. Siegel (Eds.), Symbolic and Quantitative Approaches to Uncertainty. Proceedings, 1991. XI, 362 pages. 1991.

Vol. 550: A. van Lamsweerde, A. Fugetta (Eds.), ESEC '91. Proceedings, 1991. XII, 515 pages. 1991.

Vol. 551:S. Prehn, W. J. Toetenel (Eds.), VDM '91. Formal Software Development Methods. Volume 1. Proceedings, 1991. XIII, 699 pages. 1991.

Vol. 552: S. Prehn, W. J. Toetenel (Eds.), VDM '91. Formal Software Development Methods. Volume 2. Proceedings, 1991. XIV, 430 pages. 1991.

Vol. 553: H. Bieri, H. Noltemeier (Eds.), Computational Geometry - Methods, Algorithms and Applications. Proceedings, 1991. VIII, 320 pages. 1991.